Thomas Banchoff
John Wermer

Linear Algebra Through Geometry

With 81 Illustrations

Springer-Verlag
New York Heidelberg Berlin

Thomas Banchoff
John Wermer
Department of Mathematics
Brown University
Providence, RI 02912
U.S.A.

Editors

F. W. Gehring
Department of Mathematics
University of Michigan
Ann Arbor, MI 48109
U.S.A.

P. R. Halmos
Department of Mathematics
University of Indiana
Bloomington, IN 47405
U.S.A.

AMS Subject Classification: 15-01

Cover: Computer-generated projection of a four-dimensional cube (see Figure 4.5, p. 223), by Thomas Banchoff with the assistance of Gerald Weil.

Library of Congress Cataloging in Publication Data
Banchoff, Thomas.
Linear algebra through geometry.
(Undergraduate texts in mathematics)
Bibliography: p.
Includes index.
1. Algebras, Linear. I. Wermer, John. II. Title.
III. Series.
QA184.B36 1983 512'.5 82-19446

Typeset by Computype, Inc., St. Paul, MN.
Printed in the United States of America
Printed and bound by R. R. Donnelley & Sons, Harrisonburg, VA.

9 8 7 6 5 4 3 2 1

ISBN 0-387-**90787**-4 Springer-Verlag New York Heidelberg Berlin
ISBN 3-540-**90787**-4 Springer-Verlag Berlin Heidelberg New York

To our wives Lynore and Kerstin

Preface

In this book we lead the student to an understanding of elementary linear algebra by emphasizing the geometric significance of the subject.

Our experience in teaching beginning undergraduates over the years has convinced us that students learn the new ideas of linear algebra best when these ideas are grounded in the familiar geometry of two and three dimensions. Many important notions of linear algebra already occur in these dimensions in a non-trivial way, and a student with a confident grasp of these ideas will encounter little difficulty in extending them to higher dimensions and to more abstract algebraic systems. Moreover, we feel that this geometric approach provides a solid basis for the linear algebra needed in engineering, physics, biology, and chemistry, as well as in economics and statistics.

The great advantage of beginning with a thorough study of the linear algebra of the plane is that students are introduced quickly to the most important new concepts while they are still on the familiar ground of two-dimensional geometry. In short order, the student sees and uses the notions of dot product, linear transformations, determinants, eigenvalues, and quadratic forms. This is done in Chapters 2.0–2.7.

Then the very same outline is used in Chapters 3.0–3.7 to present the linear algebra of three-dimensional space, so that the former ideas are reinforced while new concepts are being introduced.

In Chapters 4.0–4.2 we deal with geometry in a space of four dimensions and introduce linear transformations and matrices in four variables.

Finally, in Chapters 5.1–5.3 we study systems of linear equations in n unknowns, and in conjunction with such systems we develop the notions of n-dimensional vector algebra and the ideas of subspace, basis, and dimension.

Except in a single chapter, the student need only know basic high school algebra and geometry and introductory trigonometry in order to read this book. The exception is Chapter 2.8, Differential Systems, where we assume a knowledge of elementary calculus.

Acknowledgments

We would like to thank the many students in our classes over the years whose interest and suggestions have helped in the development of this book. Our special thanks go to Dale Cavanaugh, and to Elaine Haste for the work of typing our manuscript.

Contents

Preface		vii
Acknowledgments		viii
1.0	Vectors in the Line	1
2.0	The Geometry of Vectors in the Plane	3
2.1	Transformations of the Plane	23
2.2	Linear Transformations and Matrices	29
2.3	Sums and Products of Linear Transformations	39
2.4	Inverses and Systems of Equations	49
2.5	Determinants	60
2.6	Eigenvalues	74
2.7	Classification of Conic Sections	84
2.8	Differential Systems	96
3.0	Vector Geometry in 3-Space	111
3.1	Transformations of 3-Space	126
3.2	Linear Transformations and Matrices	130
3.3	Sums and Products of Linear Transformations	135
3.4	Inverses and Systems of Equations	145
3.5	Determinants	163
3.6	Eigenvalues	175
3.7	Symmetric Matrices	190
3.8	Classification of Quadric Surfaces	202

4.0 Vector Geometry in 4-Space 207
4.1 Transformations of 4-Space 216
4.2 Linear Transformations and Matrices 224

5.1 Homogeneous Systems of Equations 228
5.2 Subspace, Linear Dependence, Dimension 236
5.3 Inhomogeneous Systems of Equations 244

Afterword 253

Index 255

CHAPTER 1.0

Vectors in the Line

Analytic geometry begins with the line. Every point on the line has a real number as its coordinate and every real number is the coordinate of exactly one point. A *vector in the line* is a directed line segment from the origin to a point with coordinate x. We denote this vector by a single capital letter $\mathbf{X}$. The collection of all vectors in the line is denoted by $\mathbb{R}^1$.

We add two vectors by adding their coordinates, so if $\mathbf{U}$ has coordinate u, then $\mathbf{X} + \mathbf{U}$ has coordinate $x + u$. To multiply a vector $\mathbf{X}$ by a real number r, we multiply the coordinate by r, so the coordinate of $r\mathbf{X}$ is rx. The vector with coordinate zero is denoted by $\mathbf{0}$. (See Fig. 1.1.)

The familiar properties of real numbers then lead to corresponding properties for vectors in 1-space. For any vectors $\mathbf{X}, \mathbf{U}, \mathbf{W}$ and any real numbers r and s we have:

$\mathbf{X} + \mathbf{U} = \mathbf{U} + \mathbf{X}$.
$(\mathbf{X} + \mathbf{U}) + \mathbf{W} = \mathbf{X} + (\mathbf{U} + \mathbf{W})$.
For all $\mathbf{X}$, $\mathbf{0} + \mathbf{X} = \mathbf{X} = \mathbf{X} + \mathbf{0}$.
For any $\mathbf{X}$, there is a vector $-\mathbf{X}$ such that $\mathbf{X} + (-\mathbf{X}) = \mathbf{0}$.
$r(\mathbf{X} + \mathbf{U}) = r\mathbf{X} + r\mathbf{U}$
$(r + s)\mathbf{X} = r\mathbf{X} + s\mathbf{X}$
$r(s\mathbf{X}) = (rs)\mathbf{X}$
$1\mathbf{X} = \mathbf{X}$

We can define the length of a vector $\mathbf{X}$ with coordinate x as the absolute value of x, i.e., the distance from the point labelled x to the origin. We denote this length by $|\mathbf{X}|$ and we may write $|\mathbf{X}| = \sqrt{x^2}$. (We always understand this symbol to stand for the non-negative square root.) Then $\mathbf{0}$ is the

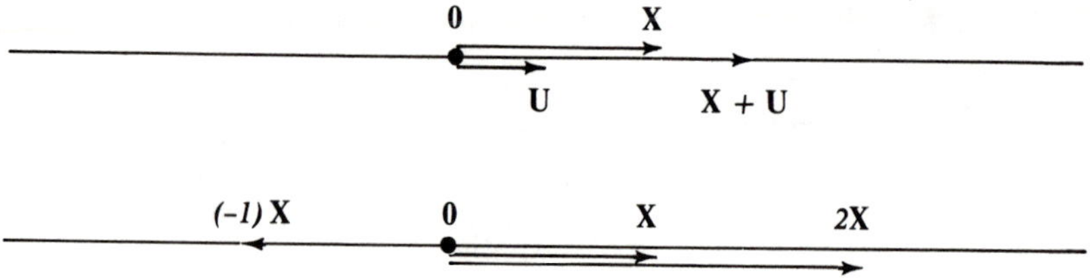

Figure 1.1

unique vector of the length 0 and there are just two vectors with length 1, with coordinates 1 and -1.

CHAPTER 2.0

The Geometry of Vectors in the Plane

Many of the familiar theorems of plane geometry appear in a new light when we rephrase them in the language of *vectors*. This is particularly true for theorems which are usually expressed in the language of analytic or coordinate geometry, because vector notation enables us to use a single symbol to refer to a pair of numbers which gives the coordinates of a point. Not only does this give us convenient notations for expressing important results, but it also allows us to concentrate on algebraic properties of vectors, and these enable us to apply the techniques used in plane geometry to study problems in space, in higher dimensions, and also in situations from calculus and differential equations which at first have little resemblance to plane geometry. Thus, we begin our study of linear algebra with the study of the geometry of vectors in the plane.

§1. The Algebra of Vectors

In vector geometry we define a *vector* in the plane as a pair of numbers $\begin{pmatrix} x \\ y \end{pmatrix}$ written in column form, with the *first coordinate* x written above the *second coordinate* y. We designate this vector by a single capital letter **X**, i.e., we write $\mathbf{X} = \begin{pmatrix} x \\ y \end{pmatrix}$. We can picture the vector **X** as an arrow, or directed line segment, starting at the origin in the coordinate plane and ending at the point with coordinates x and y. We illustrate the vectors $\mathbf{A} = \begin{pmatrix} 3 \\ 1 \end{pmatrix}$, $\mathbf{B} = \begin{pmatrix} 1 \\ 2 \end{pmatrix}$, $\mathbf{C} = \begin{pmatrix} 4 \\ 3 \end{pmatrix}$, and $\mathbf{D} = \begin{pmatrix} 2 \\ 4 \end{pmatrix}$ in Figure 2.1.

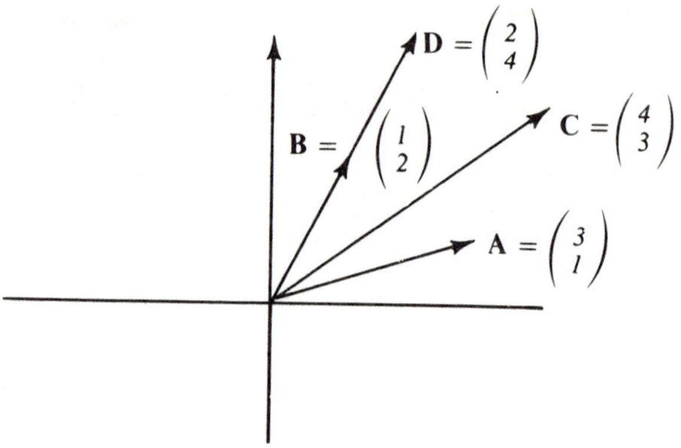

Figure 2.1

We *add* two vectors by adding their components, so if $\mathbf{X} = \begin{pmatrix} x \\ y \end{pmatrix}$ and $\mathbf{U} = \begin{pmatrix} u \\ v \end{pmatrix}$, we have

$$\mathbf{X} + \mathbf{U} = \begin{pmatrix} x + u \\ y + v \end{pmatrix}. \tag{1}$$

Thus, in the above diagram, we have $\mathbf{A} + \mathbf{B} = \mathbf{C}$, since

$$\mathbf{A} + \mathbf{B} = \begin{pmatrix} 3 \\ 1 \end{pmatrix} + \begin{pmatrix} 1 \\ 2 \end{pmatrix} = \begin{pmatrix} 3+1 \\ 1+2 \end{pmatrix} = \begin{pmatrix} 4 \\ 3 \end{pmatrix} = \mathbf{C}.$$

We *multiply* a vector $\mathbf{X}$ by a number r by multiplying each coordinate of $\mathbf{X}$ by r, i.e.,

$$r\mathbf{X} = r\begin{pmatrix} x \\ y \end{pmatrix} = \begin{pmatrix} rx \\ ry \end{pmatrix}. \tag{2}$$

In Fig. 2.1, $\mathbf{D} = \begin{pmatrix} 2 \\ 4 \end{pmatrix} = 2\begin{pmatrix} 1 \\ 2 \end{pmatrix} = 2\mathbf{B}$, and we also have $\mathbf{B} = \frac{1}{2}\mathbf{D}$.

Multiplying by a number r *scales* the vector $\mathbf{X}$ giving a longer vector $r\mathbf{X}$ if $r > 1$ and a shorter vector $r\mathbf{X}$ if $0 < r < 1$. Such multiplication of a vector by a number is called *scalar multiplication*, and the number r is called a *scalar*. If $r = 1$, then the result is the vector itself, so $1\mathbf{X} = \mathbf{X}$. If $r = 0$, then multiplication of any vector by $r = 0$ yields the *zero* vector $\begin{pmatrix} 0 \\ 0 \end{pmatrix}$, denoted by $\mathbf{0} = \begin{pmatrix} 0 \\ 0 \end{pmatrix}$. If $\mathbf{X}$ is not the zero vector, then the scalar multiples of $\mathbf{X}$ all lie on a line through the origin and the point at the endpoint of the arrow representing $\mathbf{X}$. We call this line the *line along* $\mathbf{X}$. If $r > 0$, we get the points on the ray from $\begin{pmatrix} 0 \\ 0 \end{pmatrix}$ through $\begin{pmatrix} x \\ y \end{pmatrix}$, while if $r < 0$, we get the points on the opposite ray. In particular, if $r = -1$, we get the vector $(-1)\mathbf{X} = (-1)\begin{pmatrix} x \\ y \end{pmatrix} = \begin{pmatrix} -x \\ -y \end{pmatrix}$ which has the same length as $\mathbf{X}$ but the opposite

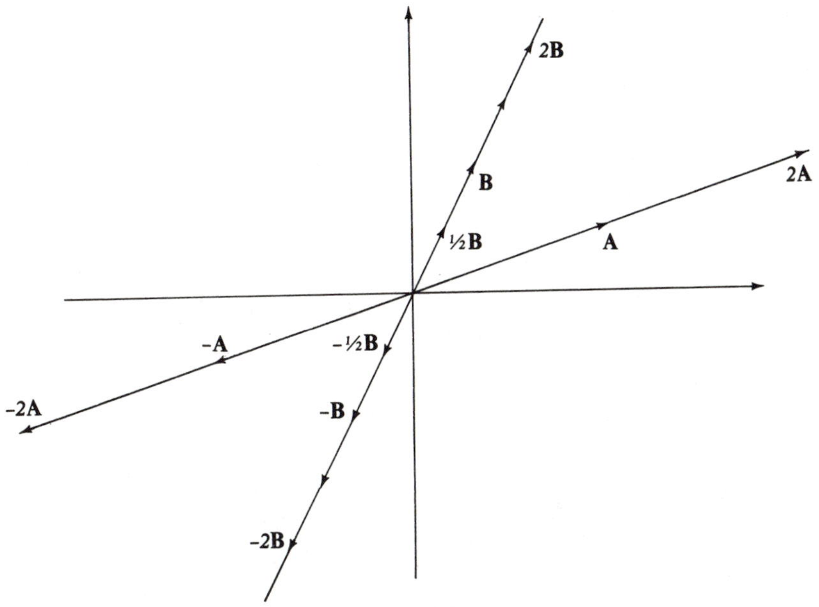

Figure 2.2

direction. We denote this vector by $-\mathbf{X} = \begin{pmatrix} -x \\ -y \end{pmatrix}$ and we note that

$$\mathbf{X} + (-\mathbf{X}) = \begin{pmatrix} x \\ y \end{pmatrix} + \begin{pmatrix} -x \\ -y \end{pmatrix} = \begin{pmatrix} x + (-x) \\ y + (-y) \end{pmatrix} = \begin{pmatrix} 0 \\ 0 \end{pmatrix} = \mathbf{0}. \qquad (3\ )$$

We say that the vector $-\mathbf{X}$ is the *negative* of $\mathbf{X}$ or the *additive inverse* of $\mathbf{X}$.

In Figure 2.2, we indicate some scalar multiples of the vectors $\mathbf{A} = \begin{pmatrix} 3 \\ 1 \end{pmatrix}$ and $\mathbf{B} = \begin{pmatrix} 1 \\ 2 \end{pmatrix}$.

Two particularly important vectors are $\mathbf{E}_1 = \begin{pmatrix} 1 \\ 0 \end{pmatrix}$ and $\mathbf{E}_2 = \begin{pmatrix} 0 \\ 1 \end{pmatrix}$, which we call the *basis vectors* of the plane. The collection of all scalar multiples $r\mathbf{E}_1 = r\begin{pmatrix} 1 \\ 0 \end{pmatrix} = \begin{pmatrix} r \\ 0 \end{pmatrix}$ of $\mathbf{E}_1$ then gives the *first coordinate axis*, and the *second coordinate axis* is given similarly by $s\mathbf{E}_2 = s\begin{pmatrix} 0 \\ 1 \end{pmatrix} = \begin{pmatrix} 0 \\ s \end{pmatrix}$. Since $\mathbf{X} = \begin{pmatrix} x \\ y \end{pmatrix} = \begin{pmatrix} x \\ 0 \end{pmatrix} + \begin{pmatrix} 0 \\ y \end{pmatrix} = x\begin{pmatrix} 1 \\ 0 \end{pmatrix} + y\begin{pmatrix} 0 \\ 1 \end{pmatrix} = x\mathbf{E}_1 + y\mathbf{E}_2$, we may express any vector $\mathbf{X}$ uniquely as a sum of one vector from the first coordinate axis and one vector from the second coordinate axis. Thus,

$$\mathbf{A} = \begin{pmatrix} 3 \\ 1 \end{pmatrix} = \begin{pmatrix} 3 \\ 0 \end{pmatrix} + \begin{pmatrix} 0 \\ 1 \end{pmatrix} = 3\begin{pmatrix} 1 \\ 0 \end{pmatrix} + 1\begin{pmatrix} 0 \\ 1 \end{pmatrix} = 3\mathbf{E}_1 + \mathbf{E}_2,$$

and, similarly, $\mathbf{D} = \begin{pmatrix} 2 \\ 4 \end{pmatrix} = 2\mathbf{E}_1 + 4\mathbf{E}_2$.

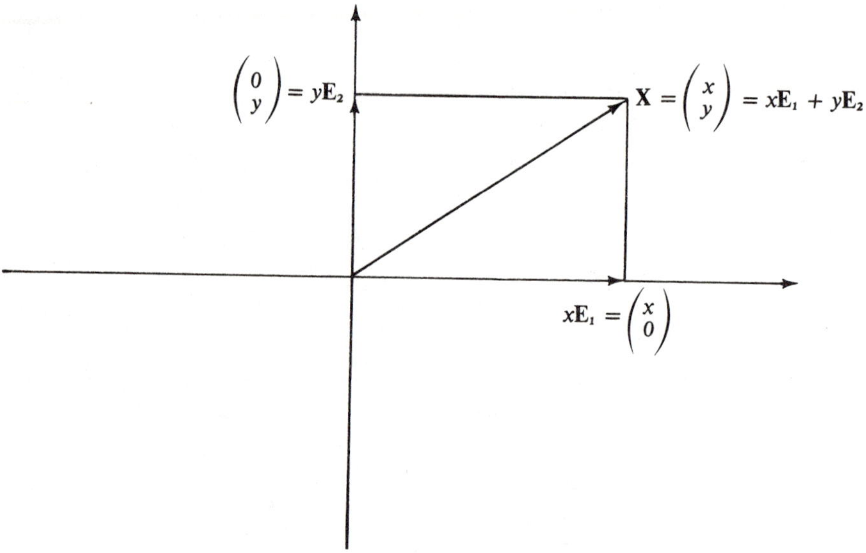

Figure 2.3

Writing a vector in this way expresses the point $\begin{pmatrix} x \\ y \end{pmatrix}$ as the fourth vertex of a rectangle whose other three coordinates are $\begin{pmatrix} x \\ 0 \end{pmatrix}$, $\begin{pmatrix} 0 \\ 0 \end{pmatrix}$, and $\begin{pmatrix} 0 \\ y \end{pmatrix}$. (See Fig. 2.3.)

More generally, we may obtain a geometric interpretation of vector addition as follows. If we start with the triangle with vertices $\begin{pmatrix} 0 \\ 0 \end{pmatrix}$, $\begin{pmatrix} x \\ 0 \end{pmatrix}$, $\begin{pmatrix} x \\ y \end{pmatrix}$

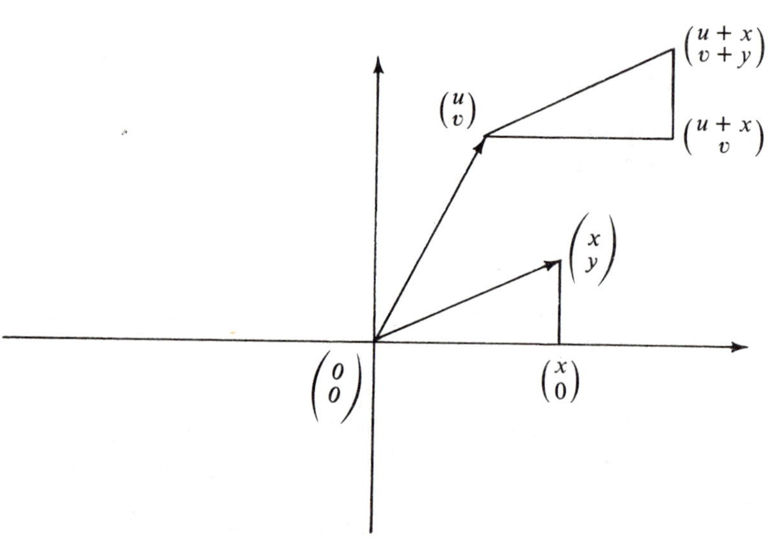

Figure 2.4

and move it parallel to itself so that its first vertex lies on $\begin{pmatrix} u \\ v \end{pmatrix}$, then the other two vertices lie on $\begin{pmatrix} u+x \\ v \end{pmatrix}$ and $\begin{pmatrix} u+x \\ v+y \end{pmatrix}$, respectively. (See Fig. 2.4.) Thus, the sum of the vectors $\mathbf{X} = \begin{pmatrix} x \\ y \end{pmatrix}$ and $\mathbf{U} = \begin{pmatrix} u \\ v \end{pmatrix}$ can be obtained by translating the directed segment from **0** to **X** parallel to itself until its beginning point lies at **U**. The new endpoint will represent **U** + **X**, and this will be the fourth coordinate of a parallelogram with **U**, **0**, and **X** as the other three vertices.

In our diagrams we have pictured addition of a vector **X** with positive coordinates, but a similar argument shows that the parallelogram interpretation is still valid if one or both coordinates are negative or zero.

By referring either to the coordinate description or the geometric description, we can establish the following algebraic properties of vector addition and scalar multiplication which are analogous to familiar facts about arithmetic of numbers:

(4)	$\mathbf{X} + \mathbf{U} = \mathbf{U} + \mathbf{X}$.	Commutative law for vectors
(5)	$(\mathbf{X} + \mathbf{U}) + \mathbf{A} = \mathbf{X} + (\mathbf{U} + \mathbf{A})$.	Associative law for vectors
(6)	There is a vector **0** such that $\mathbf{X} + \mathbf{0} = \mathbf{X} = \mathbf{0} + \mathbf{X}$ for all **X**.	Additive identity
(7)	For any **X** there is a vector $-\mathbf{X}$ such that $\mathbf{X} + (-\mathbf{X}) = \mathbf{0}$.	Additive inverse
(8)	$r(\mathbf{X} + \mathbf{U}) = r\mathbf{X} + r\mathbf{U}$.	Distributive law for vectors
(9)	$(r + s)(\mathbf{X}) = r\mathbf{X} + s\mathbf{X}$.	Distributive law for scalars
(10)	$r(s\mathbf{X}) = (rs)\mathbf{X}$.	Associative law for scalars
(11)	$1 \cdot \mathbf{X} = \mathbf{X}$ for each **X**.	

Note that it is possible for the parallelogram to collapse to a doubly covered line segment if we add two multiples of the same vector. In Fig. 2.5, we show the parallelograms for **B** + **B**, **A** + **B**, and **A** + (−**A**).

We can use the negative of a vector to help define the notion of *difference* **U** − **X** of the vectors **X** and **U**. (See Fig. 2.6.) We define

$$\mathbf{U} - \mathbf{X} = \mathbf{U} + (-\mathbf{X}),$$

so, in coordinates,

$$\begin{pmatrix} u \\ v \end{pmatrix} - \begin{pmatrix} x \\ y \end{pmatrix} = \mathbf{U} - \mathbf{X} = \mathbf{U} + (-\mathbf{X}) = \begin{pmatrix} u \\ v \end{pmatrix} + \begin{pmatrix} -x \\ -y \end{pmatrix} = \begin{pmatrix} u-x \\ v-y \end{pmatrix}.$$

Since $(\mathbf{U} - \mathbf{X}) + \mathbf{X} = \mathbf{U} + ((-\mathbf{X}) + \mathbf{X}) = \mathbf{U} + \mathbf{0} = \mathbf{U}$, we see that **U** − **X** is the vector we add to **X** to get **U**. Thus, if we move **U** − **X** parallel to itself until its beginning point lies on **X**, we get the directed line segment from **X** to **U**. Thus,

$$\mathbf{A} - \mathbf{B} = \begin{pmatrix} 3 \\ 1 \end{pmatrix} - \begin{pmatrix} 1 \\ 2 \end{pmatrix} = \begin{pmatrix} 2 \\ -1 \end{pmatrix} \quad \text{and} \quad \mathbf{B} - \mathbf{A} = \begin{pmatrix} 1 \\ 2 \end{pmatrix} - \begin{pmatrix} 3 \\ 1 \end{pmatrix} = \begin{pmatrix} -2 \\ 1 \end{pmatrix}.$$

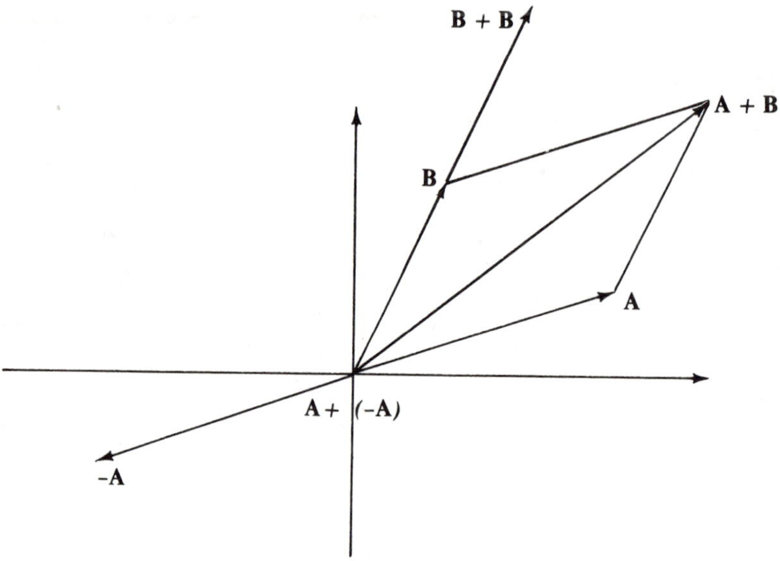

Figure 2.5

A pair of vectors $\mathbf{A}, \mathbf{B}$ is said to be *linearly dependent* if one of them is a multiple of the other. If $\mathbf{A} = \mathbf{0}$, then the pair $\mathbf{A}, \mathbf{B}$ is linearly dependent, since $\mathbf{0} = 0 \cdot \mathbf{B}$ no matter what $\mathbf{B}$ is. If $\mathbf{A} \neq \mathbf{0}$ and the pair $\mathbf{A}, \mathbf{B}$ is linearly dependent, then $\mathbf{B} = t\mathbf{A}$ for some t. If $\mathbf{B} = \mathbf{0}$, then we use $t = 0$, but if $\mathbf{A}$ and $\mathbf{B}$ are both nonzero, we have $\mathbf{B} = t\mathbf{A}$ and $(1/t)\mathbf{B} = \mathbf{A}$, so each of the vectors is a multiple of the other.

If $\mathbf{A}$, $\mathbf{B}$ is a linearly dependent pair of vectors and both $\mathbf{A}$ and $\mathbf{B}$ are nonzero, then the vectors $r\mathbf{A}$ for different values of r all lie on a line through the origin. The fact that $\mathbf{A}, \mathbf{B}$ is a linearly dependent pair means that $\mathbf{B}$ lies on the line determined by $\mathbf{A}$.

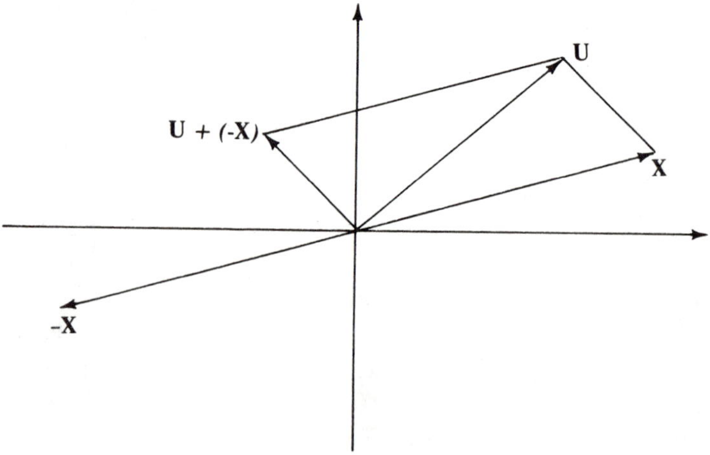

Figure 2.6

Exercise 1. For which choice of x will the following pairs be linearly dependent?

$$\text{(a) } \left(\begin{pmatrix} x \\ 1 \end{pmatrix}, \begin{pmatrix} 4 \\ 2 \end{pmatrix}\right), \quad \text{(b) } \left(\begin{pmatrix} x \\ x^2 \end{pmatrix}, \begin{pmatrix} -3 \\ 9 \end{pmatrix}\right),$$

$$\text{(c) } \left(\begin{pmatrix} x \\ 1 \end{pmatrix}, \begin{pmatrix} 9 \\ x \end{pmatrix}\right), \quad \text{(d) } \left(\begin{pmatrix} x \\ x^2 \end{pmatrix}, \begin{pmatrix} 1 \\ x \end{pmatrix}\right).$$

Exercise 2. True of false? If A is a scalar multiple of **B**, then **B** is a multiple of **A**.

Exercise 3. True or false? If **A** is a nonzero scalar multiple of **B**, then **B** is a nonzero scalar multiple of **A**.

Just as the multiples $t\mathbf{X}$ of a nonzero vector **X** give a description of a line through the origin, we may describe a line through a point **U** parallel to the vector **X** by taking the sum of **U** and all multiples of **X**. The line is then given by $\mathbf{U} + t\mathbf{X}$ for all real t. (See Fig. 2.7.)

For example, the line through $\mathbf{B} = \begin{pmatrix} 2 \\ 1 \end{pmatrix}$ parallel to the vector $\mathbf{A} = \begin{pmatrix} 3 \\ 1 \end{pmatrix}$ is given by $\mathbf{X} = \mathbf{B} + t\mathbf{A} = \begin{pmatrix} 2 \\ 1 \end{pmatrix} + t\begin{pmatrix} 3 \\ 1 \end{pmatrix} = \begin{pmatrix} 2 \\ 1 \end{pmatrix} + \begin{pmatrix} 3t \\ t \end{pmatrix} = \begin{pmatrix} 2 + 3t \\ 1 + t \end{pmatrix}$. This is called the *parametric representation* of a line in the plane, since the coordinates $x = 2 + 3t$ and $y = 1 + t$ are given linear functions of the *parameter* t. Similarly, the line given by the parametric equation in coordinates $\begin{pmatrix} x \\ y \end{pmatrix} = \begin{pmatrix} 3 + 4t \\ 1 + 2t \end{pmatrix}$ can be written in vector form as $\mathbf{X} = \begin{pmatrix} 3 \\ 1 \end{pmatrix} + \begin{pmatrix} 4t \\ 2t \end{pmatrix} = \begin{pmatrix} 3 \\ 1 \end{pmatrix} + t\begin{pmatrix} 4 \\ 2 \end{pmatrix} = \mathbf{A} + t\mathbf{D}$.

Exercise 4. Write an equation of the line through $\begin{pmatrix} 1 \\ 4 \end{pmatrix}$ parallel to the vector $\begin{pmatrix} 2 \\ 2 \end{pmatrix}$.

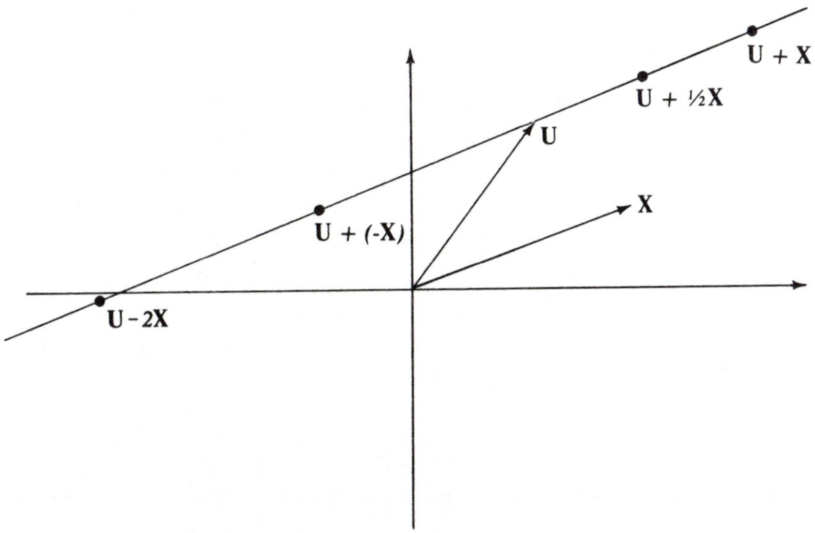

Figure 2.7

Exercise 5. Write an equation for the line through $\mathbf{A} = \begin{pmatrix} 3 \\ 1 \end{pmatrix}$ and $\mathbf{B} = \begin{pmatrix} 1 \\ 2 \end{pmatrix}$. *Hint*: This line will go through $\mathbf{B}$ and be parallel to the vector $\mathbf{B} - \mathbf{A}$.

Exercise 6. Show that the parametric equation $\mathbf{X} = \mathbf{V} + t(\mathbf{U} - \mathbf{V})$ represents the line through $\mathbf{U}$ and $\mathbf{V}$ if $\mathbf{U}$ and $\mathbf{V}$ are any two vectors which are not equal.

By the Pythagorean Theorem, the distance from a point $\begin{pmatrix} x \\ y \end{pmatrix}$ to the origin $\begin{pmatrix} 0 \\ 0 \end{pmatrix}$ is $\sqrt{x^2 + y^2}$, and we define this number to be the *length* of the vector $\mathbf{X} = \begin{pmatrix} x \\ y \end{pmatrix}$, written $|\mathbf{X}|$. For example, if $\mathbf{X} = \begin{pmatrix} 3 \\ 4 \end{pmatrix}$, then $|\mathbf{X}| = \sqrt{3^2 + 4^2} = 5$, while $|\mathbf{E}_1| = \left|\begin{pmatrix} 1 \\ 0 \end{pmatrix}\right| = 1$ and $|\mathbf{0}| = \sqrt{0^2 + 0^2} = 0$. Since the square root is always considered to be positive or zero, the length of a vector is never negative, and in fact $|\mathbf{X}|$ is positive unless $\mathbf{X} = \begin{pmatrix} 0 \\ 0 \end{pmatrix}$.

EXAMPLE 1. $|\mathbf{X} - \mathbf{U}| = \left|\begin{pmatrix} x \\ y \end{pmatrix} - \begin{pmatrix} u \\ v \end{pmatrix}\right| = \left|\begin{pmatrix} x - u \\ y - v \end{pmatrix}\right| = \sqrt{(x - u)^2 + (y - v)^2}$.

For any scalar r, we have

$$|r\mathbf{X}| = \left|\begin{pmatrix} rx \\ ry \end{pmatrix}\right| = \sqrt{(rx)^2 + (ry)^2} = \sqrt{r^2x^2 + r^2y^2} = |r|\sqrt{x^2 + y^2} = |r||\mathbf{X}|.$$

Thus, the length of a scalar multiple of a vector is the length of the vector multiplied by the absolute value of the scalar. For example, $|-5\mathbf{X}| = |-5|\,|\mathbf{X}| = 5|\mathbf{X}|$.

Exercise 7. Show that the midpoint of the segment joining points $\mathbf{X}$ and $\mathbf{U}$ is $\frac{1}{2}(\mathbf{X} + \mathbf{U})$.

If $\mathbf{X} \neq \begin{pmatrix} 0 \\ 0 \end{pmatrix}$, then $\mathbf{X} \neq \mathbf{0}$, so we may scale by the reciprocal $(1/|\mathbf{X}|)$ to get a vector $(1/|\mathbf{X}|)\mathbf{X}$. This vector lies along the ray from $\mathbf{0}$ to $\mathbf{X}$ and it has length equal to 1 since

$$\left|\left(\frac{1}{|\mathbf{X}|}\right)\mathbf{X}\right| = \left|\frac{1}{|\mathbf{X}|}\right||\mathbf{X}| = \frac{1}{|\mathbf{X}|}|\mathbf{X}| = 1.$$

The vectors of length 1 are called *unit vectors*, and they are represented by the points on the unit circle in the coordinate plane. The vector $(1/|\mathbf{X}|)\mathbf{X}$ is represented by the point where the ray from $\mathbf{0}$ to $\mathbf{X}$ intersects this unit circle. (See Fig. 2.8).

Any vector on the unit circle may be described by its angle θ from the ray along the positive x-axis to the ray along the unit vector. We call θ the *polar angle* of the vector. We may then write the unit vector using trigonometric functions as $\begin{pmatrix} \cos\theta \\ \sin\theta \end{pmatrix}$.

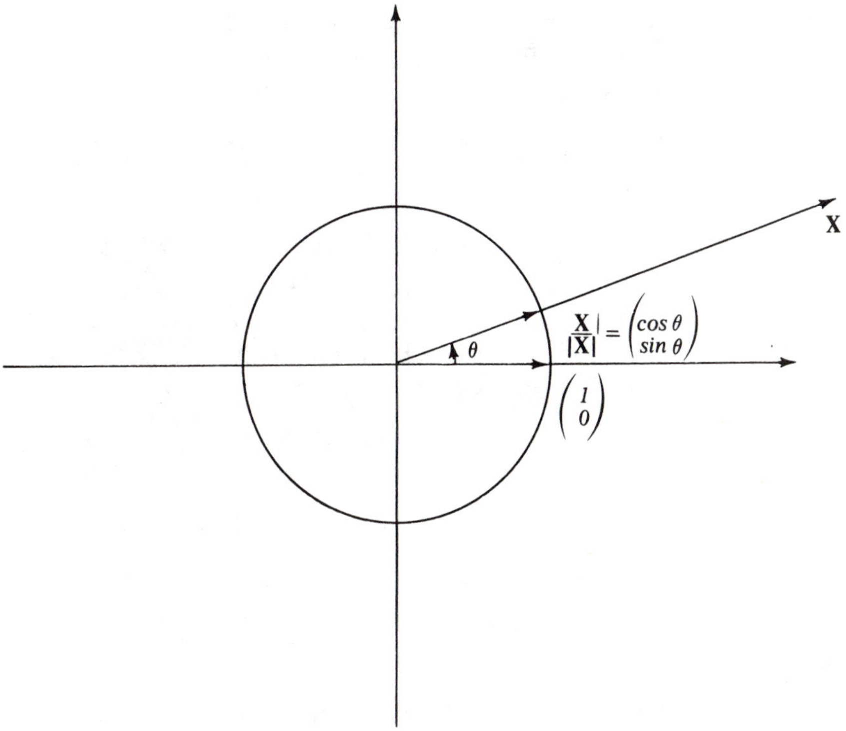

Figure 2.8

If **X** is any vector, we have

$$\mathbf{X} = |\mathbf{X}|\left(\frac{1}{|\mathbf{X}|}\mathbf{X}\right) = |\mathbf{X}|\begin{pmatrix}\cos\theta\\ \sin\theta\end{pmatrix} = \begin{pmatrix}|\mathbf{X}|\cos\theta\\ |\mathbf{X}|\sin\theta\end{pmatrix}$$

for some angle θ. This representation of X as a positive scalar multiple of a unit vector is called the *polar form* of the vector **X**, since we have written the coordinates of **X** in the form of polar coordinates.

EXAMPLE 2. If $\mathbf{X} = \begin{pmatrix}3\\0\end{pmatrix}$, we have $\mathbf{X} = 3\mathbf{E}_1$, where $\mathbf{E}_1$ is the unit vector $\begin{pmatrix}1\\0\end{pmatrix} = \begin{pmatrix}\cos(0)\\ \sin(0)\end{pmatrix}$. If $\mathbf{X} = \begin{pmatrix}1\\1\end{pmatrix}$, then $\mathbf{X} = \sqrt{2}\begin{bmatrix}1/\sqrt{2}\\ 1/\sqrt{2}\end{bmatrix} = \sqrt{2}\begin{pmatrix}\cos\theta\\ \sin\theta\end{pmatrix}$, where $\theta = 45° = \pi/4$.

§2. The Dot Product

An extremely useful notion in linear algebra is the *dot product* of two vectors **X** and **U** defined by

$$\mathbf{X}\cdot\mathbf{U} = \begin{pmatrix}x\\y\end{pmatrix}\cdot\begin{pmatrix}u\\v\end{pmatrix} = xu + yv. \tag{12}$$

The dot product of two vectors is then a number formed by adding the product of their first coordinates to the product of their second coordinates. For example, if $\mathbf{A} = \begin{pmatrix} 3 \\ 1 \end{pmatrix}$ and $\mathbf{B} = \begin{pmatrix} 1 \\ 2 \end{pmatrix}$ then $\mathbf{A} \cdot \mathbf{B} = 3 \cdot 1 + 1 \cdot 2 = 5$ and $\mathbf{A} \cdot \mathbf{A} = 3 \cdot 3 + 1 \cdot 1 = 10$. Note that $\mathbf{E}_1 \cdot \mathbf{E}_1 = 1 \cdot 1 + 0 \cdot 0 = 1$ while $\mathbf{E}_1 \cdot \mathbf{E}_2 = \begin{pmatrix} 1 \\ 0 \end{pmatrix} \cdot \begin{pmatrix} 0 \\ 1 \end{pmatrix} = 1 \cdot 0 + 0 \cdot 1 = 0$.

In general, $\mathbf{X} \cdot \mathbf{X} = \begin{pmatrix} x \\ y \end{pmatrix} \cdot \begin{pmatrix} x \\ y \end{pmatrix} = x^2 + y^2 = |\mathbf{X}|^2$, so the length of any vector is the square root of the dot product of the vector with itself. We therefore have $\mathbf{X} \cdot \mathbf{X} \geqslant 0$ for all $\mathbf{X}$, with equality if and only if $\mathbf{X} = \mathbf{0}$.

The dot product has certain algebraic properties that are similar to properties of ordinary multiplication of numbers:

(13)	$\mathbf{X} \cdot \mathbf{U} = \mathbf{U} \cdot \mathbf{X}$	Commutative law for dot product
(14)	$(r\mathbf{X}) \cdot \mathbf{U} = r(\mathbf{X} \cdot \mathbf{U})$	Associative law for scalar and dot product
(15)	$\mathbf{A} \cdot (\mathbf{X} + \mathbf{U}) = \mathbf{A} \cdot \mathbf{X} + \mathbf{A} \cdot \mathbf{U}$	Distributive law for dot product

Each of these properties can be established easily by referring to the coordinate definition. For example, if $\mathbf{A} = \begin{pmatrix} a \\ b \end{pmatrix}$, $\mathbf{X} = \begin{pmatrix} x \\ y \end{pmatrix}$, and $\mathbf{U} = \begin{pmatrix} u \\ v \end{pmatrix}$, we have

$$\begin{aligned} \mathbf{A} \cdot (\mathbf{X} + \mathbf{U}) &= \begin{pmatrix} a \\ b \end{pmatrix} \cdot \begin{pmatrix} x + u \\ y + v \end{pmatrix} = a(x + u) + b(y + v) \\ &= (ax + by) + (au + bv) = \begin{pmatrix} a \\ b \end{pmatrix} \cdot \begin{pmatrix} x \\ y \end{pmatrix} + \begin{pmatrix} a \\ b \end{pmatrix} \cdot \begin{pmatrix} u \\ v \end{pmatrix} \\ &= \mathbf{A} \cdot \mathbf{X} + \mathbf{A} \cdot \mathbf{U}. \end{aligned}$$

In ordinary multiplication of real numbers, the product ax equals zero only if either $a = 0$ or $x = 0$. Note, however, that it is possible for the dot product of two vectors to be zero even if neither vector is equal to zero. For example, $\begin{pmatrix} 2 \\ 2 \end{pmatrix} \cdot \begin{pmatrix} -1 \\ 1 \end{pmatrix} = 2(-1) + 2 \cdot 1 = 0$. (See Fig. 2.9.)

Exercise 8. Show that if $r\mathbf{X} = \mathbf{0}$, then either $r = 0$ or $\mathbf{X} = \mathbf{0}$.

We may ask under which circumstances the dot product of two nonzero vectors $\mathbf{X}$ and $\mathbf{U}$ will be zero, i.e., when do we have $\mathbf{X} \cdot \mathbf{U} = \begin{pmatrix} x \\ y \end{pmatrix} \cdot \begin{pmatrix} u \\ v \end{pmatrix} = xu + yv = 0$? One possibility is that one vector lies in the first coordinate axis and the other lies in the second coordinate axis, in which case the two vectors are perpendicular. If $\mathbf{X}$ does not lie in either coordinate axis, then $x \neq 0$ and $y \neq 0$. The slope of the line from the origin through $\begin{pmatrix} x \\ y \end{pmatrix}$ is y/x, and this is not equal to zero. Since $xu + yv = 0$, it follows that $yv = -ux$. If $u = 0$, then $v = 0$ as well. If $u \neq 0$, then $ux \neq 0$, so $-1 = yv/ux = (y/x)(v/u)$. Thus, either $\mathbf{U} = \mathbf{0}$ or the line from the origin to $\mathbf{U} = \begin{pmatrix} u \\ v \end{pmatrix}$ has

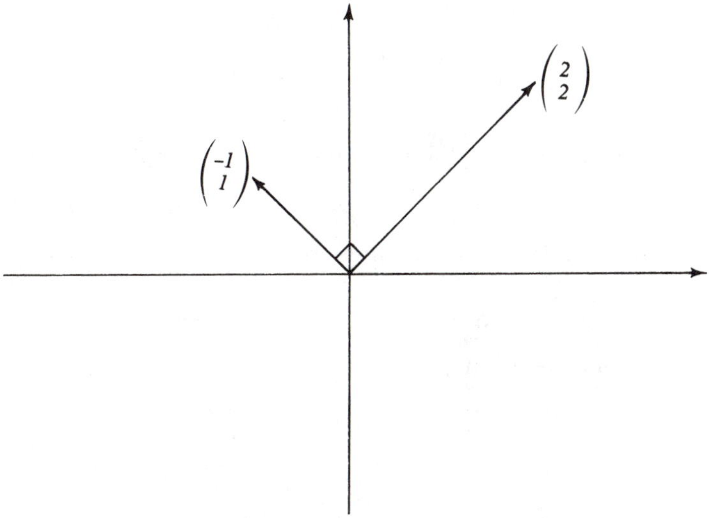

Figure 2.9

slope v/u equal to the negative reciprocal of y/x, the slope of the line from the origin to **X**. Therefore, these two lines must be perpendicular. It follows that in every case if the dot product of two nonzero vectors is zero, then the two vectors are perpendicular.

Retracing our steps, we easily see that, conversely, if **X** and **U** are any two perpendicular vectors, then $\mathbf{X} \cdot \mathbf{U} = 0$.

Exercise 9. Show that for any vector $\begin{pmatrix} x \\ y \end{pmatrix}$, we have $\begin{pmatrix} x \\ y \end{pmatrix}$ perpendicular to the vector $\begin{pmatrix} -y \\ x \end{pmatrix}$.

Note that the line with equation

$$ax + by = 0$$

may be described in two equivalent ways:

(i) The set of vectors $\mathbf{X} = \begin{pmatrix} x \\ y \end{pmatrix}$ which are perpendicular to the vector $\begin{pmatrix} a \\ b \end{pmatrix}$.

(ii) The line along the vector $\begin{pmatrix} -b \\ a \end{pmatrix}$ (by Exercise 9, the vector $\begin{pmatrix} -b \\ a \end{pmatrix}$ is perpendicular to $\begin{pmatrix} a \\ b \end{pmatrix}$).

Exercise 10. Find a vector **U** such that the line with equation $5x + 2y = 0$ lies along **U**.

Exercise 11. Find an equation of the form $ax + by = 0$ for the line along the vector $\begin{pmatrix} 3 \\ 1 \end{pmatrix}$.

Exercise 12. Find a vector **U** such that the line with equation $y = 2x$ lies along **U**.

For any vector $\mathbf{X} = \begin{pmatrix} x \\ y \end{pmatrix}$, we have $\mathbf{X} \cdot \mathbf{E}_1 = \begin{pmatrix} x \\ y \end{pmatrix} \cdot \begin{pmatrix} 1 \\ 0 \end{pmatrix} = (x \cdot 1) + (y \cdot 0) = x$. Thus, the dot product of $\mathbf{X}$ with the unit vector $\mathbf{E}_1$ is the coordinate of the *projection of* $\mathbf{X}$ *to the first coordinate axis*. Similarly, the dot product $\mathbf{X} \cdot \mathbf{E}_2 = \begin{pmatrix} x \\ y \end{pmatrix} \cdot \begin{pmatrix} 0 \\ 1 \end{pmatrix} = y$ of the vector $\mathbf{X}$ with the unit vector $\mathbf{E}_2$ is the coordinate of the *projection of* $\mathbf{X}$ *to the second coordinate axis.*

More generally, if we have any unit vector $\mathbf{W} = \begin{pmatrix} \cos\phi \\ \sin\phi \end{pmatrix}$, we may use the polar form of the vector $\mathbf{X} = |\mathbf{X}|\begin{pmatrix} \cos\theta \\ \sin\theta \end{pmatrix}$ to get a geometric interpretation of the dot product of $\mathbf{X}$ and $\mathbf{W}$ (see Fig. 2.10). We have

$$\mathbf{X} \cdot \mathbf{W} = \left(|\mathbf{X}|\begin{pmatrix} \cos\theta \\ \sin\theta \end{pmatrix}\right) \cdot \begin{pmatrix} \cos\phi \\ \sin\phi \end{pmatrix} = |\mathbf{X}|\left(\begin{pmatrix} \cos\theta \\ \sin\theta \end{pmatrix} \cdot \begin{pmatrix} \cos\phi \\ \sin\phi \end{pmatrix}\right)$$

$$= |\mathbf{X}|(\cos\theta\cos\phi + \sin\theta\sin\phi).$$

There is a basic trigonometric identity that states that

$$\cos\theta\cos\phi + \sin\theta\sin\phi = \cos(\theta - \phi) = \cos(\phi - \theta), \tag{16}$$

so we have $\mathbf{X} \cdot \mathbf{W} = |\mathbf{X}| \cdot \cos(\theta - \phi)$. Therefore, the dot product of a vector $\mathbf{X}$ with a unit vector $\mathbf{W}$ is the product of the length of $\mathbf{X}$ and the cosine of the angle between $\mathbf{X}$ and $\mathbf{W}$. If this angle $(\theta - \phi)$ is between 0 and $\pi/2$,

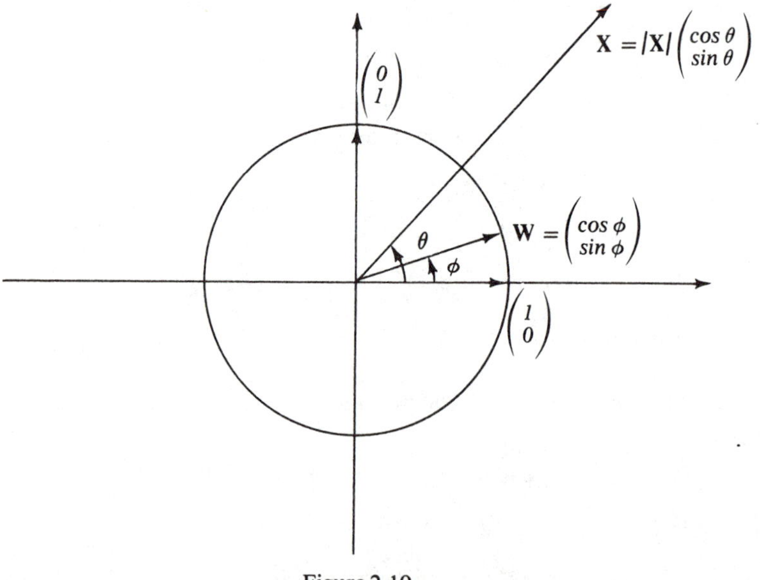

Figure 2.10

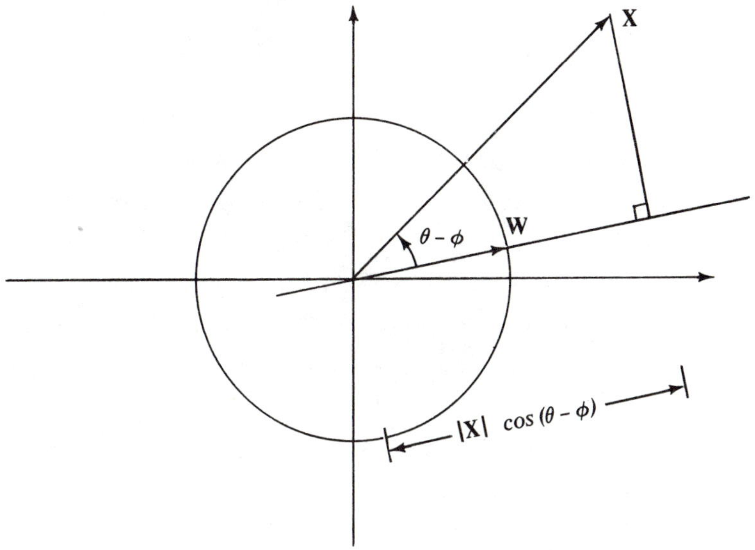

Figure 2.11

then this number $|\mathbf{X}|\cos(\theta - \phi)$ is the length of the adjacent side when the hypotenuse is $|\mathbf{X}|$. Thus, if $\mathbf{X}$ is a vector which makes an acute angle with the unit vector $\mathbf{W}$, then the dot product $\mathbf{X} \cdot \mathbf{W}$ is the *length of the projection of* $\mathbf{X}$ *to the line from the origin through* $\mathbf{W}$ (see Fig. 2.11).

If the angle between $\mathbf{X}$ and $\mathbf{W}$ is greater than $\pi/2$, then $\cos(\theta - \phi)$ is negative and the dot product $\mathbf{X} \cdot \mathbf{W}$ is a negative number. The projection of $\mathbf{X}$ to the line from $\mathbf{0}$ through $\mathbf{W}$ will lie on the ray opposite the ray from $\mathbf{0}$ through $\mathbf{W}$, and the length of this projection is the absolute value of the dot product of $\mathbf{X}$ and $\mathbf{W}$. (See Fig. 2.12). In all cases, then, we can say that the dot product $\mathbf{X} \cdot \mathbf{W}$ represents the coordinate of the projection of the vector $\mathbf{X}$ to the directed line from the origin through the unit vector $\mathbf{W}$.

In general, if we take the dot product of two nonzero vectors in polar form $\mathbf{X} = |\mathbf{X}|\begin{pmatrix}\cos\theta \\ \sin\theta\end{pmatrix}$ and $\mathbf{U} = |\mathbf{U}|\begin{pmatrix}\cos\phi \\ \sin\phi\end{pmatrix}$, we get

$$\mathbf{X} \cdot \mathbf{U} = |\mathbf{X}|\begin{pmatrix}\cos\theta \\ \sin\theta\end{pmatrix} \cdot |\mathbf{U}|\begin{pmatrix}\cos\phi \\ \sin\phi\end{pmatrix} = |\mathbf{X}|\,|\mathbf{U}|\begin{pmatrix}\cos\theta \\ \sin\theta\end{pmatrix} \cdot \begin{pmatrix}\cos\phi \\ \sin\phi\end{pmatrix} = |\mathbf{X}|\,|\mathbf{U}|\cos(\theta - \phi). \tag{17}$$

Thus, the dot product of two nonzero vectors is the product of their lengths multiplied by the cosine of the angle between them.

We may use the dot product to calculate the angle between two nonzero vectors just by writing

$$\cos(\phi - \theta) = \frac{\mathbf{X} \cdot \mathbf{U}}{|\mathbf{X}|\,|\mathbf{U}|} = \frac{xu + yv}{\sqrt{x^2 + y^2}\,\sqrt{u^2 + v^2}}. \tag{18}$$

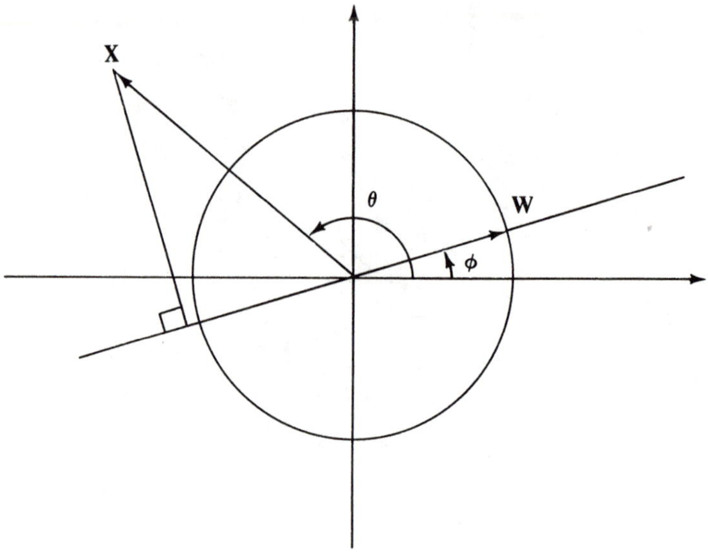

Figure 2.12

For example, if $\mathbf{A} = \begin{pmatrix} 3 \\ 1 \end{pmatrix}$, and $\mathbf{B} = \begin{pmatrix} 1 \\ 2 \end{pmatrix}$, then $|\mathbf{A}| = \sqrt{10}$, $|\mathbf{B}| = \sqrt{5}$, and $\mathbf{A} \cdot \mathbf{B} = 5$. Thus,

$$\cos(\phi - \theta) = \frac{5}{\sqrt{5} \cdot \sqrt{10}} = \frac{1}{\sqrt{2}}$$

and $\theta - \phi = \pi/4$.

Exercise 13. Find the angle between $\begin{pmatrix} 2 \\ 2 \end{pmatrix}$ and $\begin{pmatrix} 0 \\ 3 \end{pmatrix}$.

Exercise 14. Find the angle between $\begin{pmatrix} 3 \\ 1 \end{pmatrix}$ and $\begin{pmatrix} -1 \\ -2 \end{pmatrix}$.

Similarly, using the trigonometric relation

$$\sin(\phi - \theta) = \sin\phi\cos\theta - \cos\phi\sin\theta, \tag{19}$$

we obtain an expression for $\sin(\phi - \theta)$. Setting $\mathbf{X} = \begin{pmatrix} x \\ y \end{pmatrix}$, $\mathbf{U} = \begin{pmatrix} u \\ v \end{pmatrix}$, we have

$$\cos\theta = \frac{x}{\sqrt{x^2 + y^2}}, \qquad \sin\theta = \frac{y}{\sqrt{x^2 + y^2}}, \qquad \cos\phi = \frac{u}{\sqrt{u^2 + v^2}}$$

and

$$\sin\phi = \frac{v}{\sqrt{u^2 + v^2}},$$

and so

$$\sin(\phi - \theta) = \frac{vx - uy}{\sqrt{x^2 + y^2}\,\sqrt{u^2 + v^2}}. \tag{20}$$

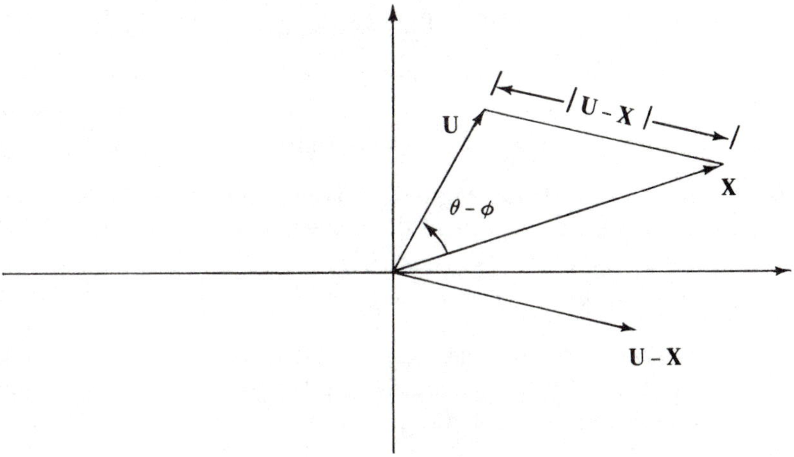

Figure 2.13

Note that the sine of the angle from $\binom{u}{v}$ to $\binom{x}{y}$ is the opposite of the sine of the angle from $\binom{x}{y}$ to $\binom{u}{v}$. We will return to this formula in Chapter 2.5.

If we apply this notion of dot product to the difference of two vectors, we obtain an important result from trigonometry. We calculate the square of the length of the segment from $\mathbf{X} = |\mathbf{X}|\begin{pmatrix}\cos\theta\\ \sin\theta\end{pmatrix}$ to $\mathbf{U} = |\mathbf{U}|\begin{pmatrix}\cos\phi\\ \sin\phi\end{pmatrix}$ (see Fig. 2.13):

$$\begin{aligned}|\mathbf{U}-\mathbf{X}|^2 &= (\mathbf{U}-\mathbf{X})\cdot(\mathbf{U}-\mathbf{X}) = \mathbf{U}\cdot\mathbf{U} - \mathbf{X}\cdot\mathbf{U} - \mathbf{U}\cdot\mathbf{X} + \mathbf{X}\cdot\mathbf{X}\\ &= |\mathbf{U}|^2 - 2\mathbf{U}\cdot\mathbf{X} + |\mathbf{X}|^2 = |\mathbf{U}|^2 + |\mathbf{X}|^2 - 2|\mathbf{U}|\,|\mathbf{X}|\cos(\theta-\phi). \qquad (21)\end{aligned}$$

Thus, the square of the length of one side of a triangle is the sum of the squares of the lengths of the other two sides minus twice the product of those lengths and the cosine of the angle between them. This result is known in trigonometry as the *law of cosines*.

In particular, if the vectors **U** and **X** are perpendicular, so that the angle between them is $\theta - \phi = \pi/2$, then $|\mathbf{U}-\mathbf{X}|^2 = |\mathbf{U}|^2 + |\mathbf{X}|^2$, so $\mathbf{X}\cdot\mathbf{U} = |\mathbf{X}|\,|\mathbf{U}|\cos(\theta-\phi) = 0$. We thus have another proof of the result that two nonzero vectors are *perpendicular* if and only if $\mathbf{X}\cdot\mathbf{U} = 0$. In linear algebra, we use the convention that the zero vector is perpendicular to every vector, and we frequently use the synonym *orthogonal* instead of perpendicular. We may thus say that two vectors **X** and **U** are *orthogonal* if and only if $\mathbf{X}\cdot\mathbf{U} = 0$.

We use the notion of dot product to solve some geometric problems which will be crucial in our further development of linear algebra:

(i) To find the projection of a given vector to a given line through the origin.

(ii) To compute the distance from a given point to the line through the origin with equation $ax + by = 0$.
(iii) To calculate the area of a parallelogram with one vertex at the origin.
(iv) To give a geometric interpretation of a system of two linear equations in two unknowns (where both lines go through the origin).

(i) We already know that if **W** is a unit vector, then the dot product of **X** and **W** represents the coordinate of the projection of the point **X** to the line from the origin through **W**. We set $P_{\mathbf{W}}(\mathbf{X})$ (read "P sub **W** of **X**") equal to the *vector* on this line which is the projection of **X** to the line. Thus, $P_{\mathbf{W}}(\mathbf{X}) = (\mathbf{X} \cdot \mathbf{W})\mathbf{W}$.

If **U** is an arbitrary vector, then we can find a formula for the projection $P_{\mathbf{U}}(\mathbf{X})$ of **X** to the line from the origin along **U** by using the above formula to find the projection of **X** to the line from the origin through the unit vector $\mathbf{U}/|\mathbf{U}|$, i.e.,

$$P_{\mathbf{U}}(\mathbf{X}) = \left(\mathbf{X} \cdot \frac{\mathbf{U}}{|\mathbf{U}|}\right)\frac{\mathbf{U}}{|\mathbf{U}|} = \frac{(\mathbf{X} \cdot \mathbf{U})\mathbf{U}}{|\mathbf{U}|^2} = \frac{(\mathbf{X} \cdot \mathbf{U})\mathbf{U}}{(\mathbf{U} \cdot \mathbf{U})}. \tag{22}$$

Hence the length of the projection of **X** to the line along **U** is given by

$$\left|\mathbf{X} \cdot \frac{\mathbf{U}}{|\mathbf{U}|}\right| = \frac{|\mathbf{X} \cdot \mathbf{U}|}{|\mathbf{U}|} \tag{23}$$

Alternatively, we may try to find the projection of **X** to the line along the nonzero vector **U** by finding a scalar t such that $\mathbf{X} - t\mathbf{U}$ is orthogonal to **U**. (See Fig. 2.14). In terms of the dot product, we obtain

$$0 = (\mathbf{X} - t\mathbf{U}) \cdot \mathbf{U} = (\mathbf{X} \cdot \mathbf{U}) - t(\mathbf{U} \cdot \mathbf{U}),$$

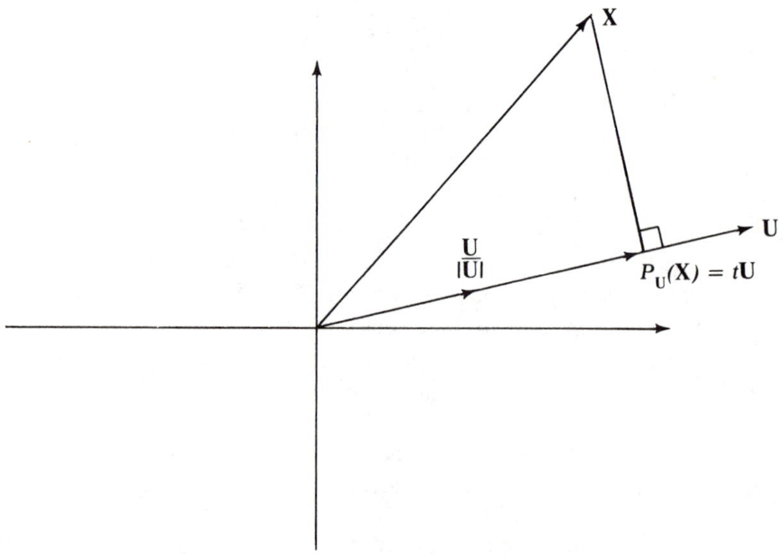

Figure 2.14

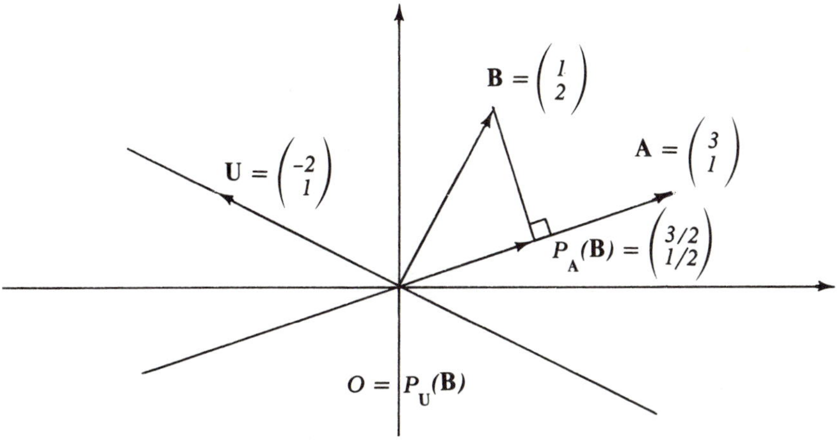

Figure 2.15

so $t = (\mathbf{X} \cdot \mathbf{U})/(\mathbf{U} \cdot \mathbf{U})$ and the projection of $\mathbf{X}$ to the line along $\mathbf{U}$ is

$$P_{\mathbf{U}}(\mathbf{X}) = t\mathbf{U} = \left(\frac{\mathbf{X} \cdot \mathbf{U}}{\mathbf{U} \cdot \mathbf{U}}\right)\mathbf{U}$$

which agrees with formula (22).

Example 3. To find the projection of $\mathbf{B} = \begin{pmatrix} 1 \\ 2 \end{pmatrix}$ to the line along $\mathbf{A} = \begin{pmatrix} 3 \\ 1 \end{pmatrix}$, we have (see Fig. 2.15)

$$P_{\mathbf{A}}(\mathbf{B}) = \left(\frac{\mathbf{A} \cdot \mathbf{B}}{\mathbf{A} \cdot \mathbf{A}}\right)\mathbf{A} = \frac{5}{10}\mathbf{A} = \frac{1}{2}\mathbf{A} = \begin{bmatrix} \frac{3}{2} \\ \frac{1}{2} \end{bmatrix}.$$

To find the projection of $\mathbf{B} = \begin{pmatrix} 1 \\ 2 \end{pmatrix}$ to the line along $\mathbf{U} = \begin{pmatrix} -2 \\ 1 \end{pmatrix}$, we have

$$P_{\mathbf{U}}(\mathbf{B}) = \left(\frac{\mathbf{B} \cdot \mathbf{U}}{\mathbf{U} \cdot \mathbf{U}}\right)\mathbf{U} = \left(\frac{0}{2}\right)\mathbf{U} = 0.$$

(This fits with our intuition that if $\mathbf{X}$ is orthogonal to $\mathbf{U}$, the projection of $\mathbf{X}$ to the line along $\mathbf{U}$ will be the origin itself.)

Exercise 15. Find the projection of $\begin{pmatrix} -1 \\ 2 \end{pmatrix}$ to the line along $\begin{pmatrix} 3 \\ 1 \end{pmatrix}$.

Exercise 16. Find the projection of $\begin{pmatrix} 3 \\ 1 \end{pmatrix}$ to the line along $\begin{pmatrix} 1 \\ 2 \end{pmatrix}$.

Exercise 17. Find the projection of $\begin{pmatrix} 3 \\ 1 \end{pmatrix}$ to the line along $\begin{pmatrix} -2 \\ 6 \end{pmatrix}$.

Exercise 18. Find the distance from $\begin{pmatrix} 3 \\ 1 \end{pmatrix}$ to the line along $\begin{pmatrix} 1 \\ 2 \end{pmatrix}$.

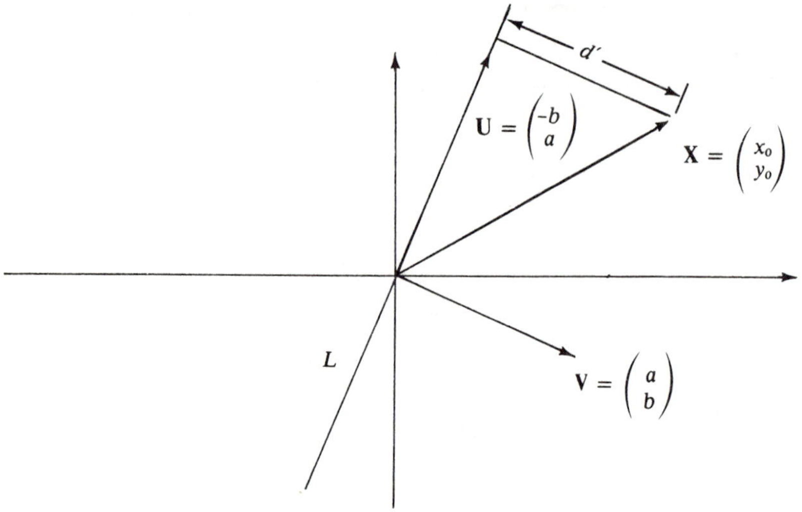

Figure 2.16

(ii) Now we want to find the distance d' from a point $\begin{pmatrix}x_0\\ y_0\end{pmatrix}$ to a line L with equation $ax + by = 0$, where a and b are not both 0 (see Fig. 2.16). The vector $\mathbf{U} = \begin{pmatrix}-b\\ a\end{pmatrix}$ is a nonzero vector on this line and the vector $\mathbf{V} = \begin{pmatrix}a\\ b\end{pmatrix}$ is a non-zero vector perpendicular to this line. The distance from $\mathbf{X} = \begin{pmatrix}x_0\\ y_0\end{pmatrix}$ to the line L is then given by the length of the projection of $\mathbf{X}$ to the line along $\mathbf{V}$. By (23) we obtain

$$d' = \frac{|\mathbf{X}\cdot\mathbf{V}|}{|\mathbf{V}|} = \frac{\left|\begin{pmatrix}x_0\\ y_0\end{pmatrix}\cdot\begin{pmatrix}a\\ b\end{pmatrix}\right|}{\left|\begin{pmatrix}a\\ b\end{pmatrix}\right|} = \frac{|ax_0 + by_0|}{\sqrt{a^2+b^2}} \tag{24}$$

If $\begin{pmatrix}x_0\\ y_0\end{pmatrix}$ is a point on L, then the expression $ax_0 + by_0 = 0$, so by (24), $d' = 0$, as we expected.

Exercise 19. Find the distance from $\begin{pmatrix}3\\ 1\end{pmatrix}$ to the line $y = 2x$.

Exercise 20. Verify that the sum of the squares of the distances from a point $\mathbf{X}$ to the perpendicular lines $ax + by = 0$ and $bx - ay = 0$ is equal to the square of the length of $\mathbf{X}$.

(iii) Once we have the formula for the distance from a point to the line along a given vector, it is an easy matter to find a formula for the area of a parallelogram with one vertex at the origin. If the other three vertices are $\mathbf{A} = \begin{pmatrix}a\\ c\end{pmatrix}$, $\mathbf{B} = \begin{pmatrix}b\\ d\end{pmatrix}$, and $\mathbf{A} + \mathbf{B} = \begin{pmatrix}a+b\\ c+d\end{pmatrix}$, then the distance from $\mathbf{A}$ to the

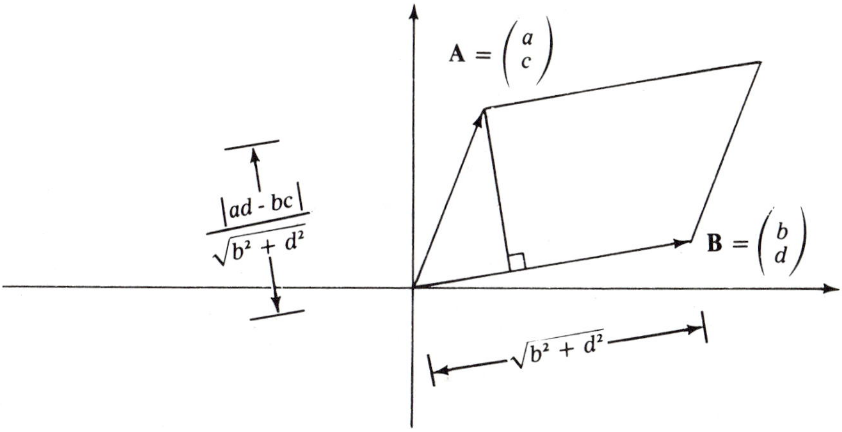

Figure 2.17

line along **B** is given by formula (24) by the expression

$$\frac{|ad - bc|}{\sqrt{b^2 + d^2}}.$$

Multiplying this distance by the length $\sqrt{b^2 + d^2}$ of the base **B**, we get the formula

$$|ad - bc| = \text{area of the parallelogram with sides } \begin{pmatrix} a \\ c \end{pmatrix} \text{ and } \begin{pmatrix} b \\ d \end{pmatrix}. \qquad (25)$$

(See Figure 2.17.)

(iv) Let us try to solve the system of equations

$$\begin{aligned} 2x + 3y &= 0, \\ 4x - y &= 0 \end{aligned} \qquad (26)$$

for the unknowns x and y.

Suppose x, y is a solution. We define the vector $\mathbf{X} = \begin{pmatrix} x \\ y \end{pmatrix}$. The first equation then says that $\mathbf{X} \cdot \begin{pmatrix} 2 \\ 3 \end{pmatrix} = 0$ and the second that $\mathbf{X} \cdot \begin{pmatrix} 4 \\ -1 \end{pmatrix} = 0$. Thus, the vector **X** is orthogonal to both the vectors $\begin{pmatrix} 2 \\ 3 \end{pmatrix}$ and $\begin{pmatrix} 4 \\ -1 \end{pmatrix}$. This is possible only if $\mathbf{X} = \mathbf{0}$.

Hence, $\begin{pmatrix} x \\ y \end{pmatrix} = \begin{pmatrix} 0 \\ 0 \end{pmatrix}$, so the only solution to (26) is $x = 0$, $y = 0$.

Now look at the general case of a system

$$\begin{aligned} ax + by &= 0, \\ cx + dy &= 0, \end{aligned} \qquad (27)$$

where a, b, c, d are given constants such that not both a and b are zero and not both c and d are zero.

Of course, $x = 0$, $y = 0$ is a solution of (27). Are there others, and if so what are they?

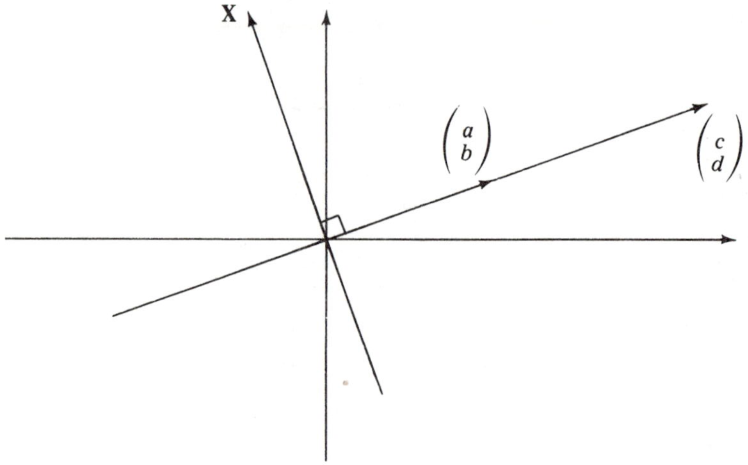

Figure 2.18

Let x, y be a solution other than 0, 0. Set $\mathbf{X} = \binom{x}{y}$. Then $\mathbf{X} \neq \mathbf{0}$ and

$$\mathbf{X} \cdot \binom{a}{b} = \mathbf{0} \quad \text{and} \quad \mathbf{X} \cdot \binom{c}{d} = 0.$$

Thus, there is a nonzero vector orthogonal to both vectors $\binom{a}{b}$ and $\binom{c}{d}$. (See Fig. 2.18.) This can only happen where $\binom{a}{b}$ and $\binom{c}{d}$ lie on the same line through the origin. Since $\binom{c}{d} \neq \mathbf{0}$, there is some scalar t with $\binom{a}{b} = t\binom{c}{d}$, and so $a = tc$ and $b = td$. Also $t \neq 0$. Every vector $\mathbf{X}$ on the line perpendicular to $\binom{a}{b}$ is then orthogonal to both $\binom{a}{b}$ and $\binom{c}{d}$. Our result is this:

(27) has a solution other than $x = 0, y = 0$ only when there is some scalar t such that

$$a = tc, \qquad b = td.$$

In that case, there is a line consisting of solutions x, y, namely the line orthogonal to $\binom{a}{b}$.

Exercise 21. Find all solutions to the system of equations

$$3x + 2y = 0,$$
$$4x - y = 0.$$

Exercise 22. Find all solutions to the system of equations

$$5x + y = 0,$$
$$-10x - 2y = 0.$$

CHAPTER 2.1

Transformations of the Plane

Recall the notion of a "function." A function is a rule which assigns to each number some number. This suggests the following definition: *A transformation of the plane* is a rule which assigns to each vector in the plane some vector in the plane.

We denote transformation by letters A, B, R, S, T, etc.

EXAMPLE 1. Let P be the transformation which assigns to each vector $\mathbf{X}$ the projection of $\mathbf{X}$ on the line along the vector $\mathbf{U} = \begin{pmatrix} 1 \\ 2 \end{pmatrix}$.

We write $P(\mathbf{X})$ for the vector which P assigns to $\mathbf{X}$ and we call $P(\mathbf{X})$ the *image* of $\mathbf{X}$ (see Fig. 2.19).

Let $\mathbf{X} = \begin{pmatrix} x \\ y \end{pmatrix}$ and let us calculate $P(\mathbf{X})$. By formula (22) of Chapter 2.0,

$$P(\mathbf{X}) = \left(\frac{\mathbf{X}\cdot\mathbf{U}}{\mathbf{U}\cdot\mathbf{U}}\right)\mathbf{U} = \frac{x+2y}{1+4}\begin{bmatrix} 1 \\ 2 \end{bmatrix} = \begin{bmatrix} \dfrac{x+2y}{5} \\ \dfrac{2}{5}(x+2y) \end{bmatrix}. \tag{1}$$

EXAMPLE 2. Let S be the transformation which assigns to each vector $\mathbf{X}$ the reflection of $\mathbf{X}$ in the line along the vector $\begin{pmatrix} 1 \\ 2 \end{pmatrix}$.

Given $\mathbf{X} = \begin{pmatrix} x \\ y \end{pmatrix}$, we want to find the coordinates of the point $S(\mathbf{X})$ such that the midpoint of the segment from $\mathbf{X}$ to $S(\mathbf{X})$ is the projection of $\mathbf{X}$ to the line along $\mathbf{U} = \begin{pmatrix} 1 \\ 2 \end{pmatrix}$. Denote the coordinates of $S(\mathbf{X})$ by x', y'. Then

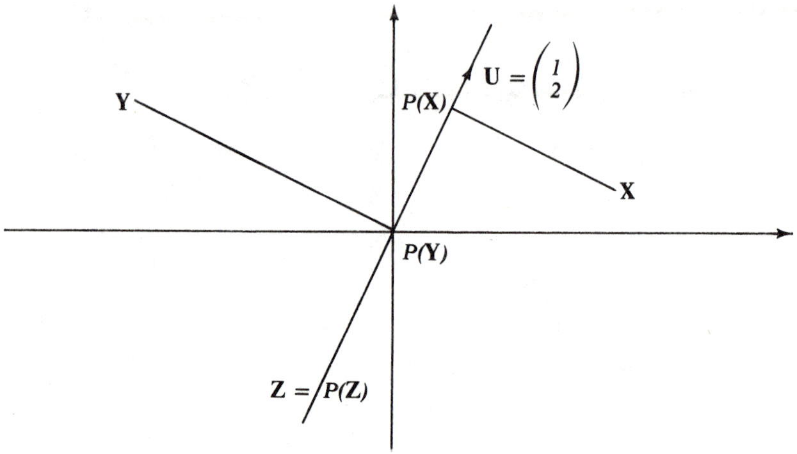

Figure 2.19

$\frac{1}{2}(\mathbf{X} + S(\mathbf{X})) = P(\mathbf{X})$, where P was defined in the preceding example. So,

$$\mathbf{X} + S(\mathbf{X}) = 2P(\mathbf{X})$$

and

$$S(\mathbf{X}) = 2P(\mathbf{X}) - \mathbf{X}.$$

From formula (1), we then obtain

$$S(\mathbf{X}) = \begin{pmatrix} x' \\ y' \end{pmatrix} = 2\begin{pmatrix} \dfrac{x+2y}{5} \\ \dfrac{2(x+2y)}{5} \end{pmatrix} - \begin{pmatrix} x \\ y \end{pmatrix}$$

$$= \begin{pmatrix} 2\dfrac{(x+2y)}{5} - x \\ 4\dfrac{(x+2y)}{5} - y \end{pmatrix}.$$

Thus

$$\begin{pmatrix} x' \\ y' \end{pmatrix} = \begin{pmatrix} -\frac{3}{5}x + \frac{4}{5}y \\ \frac{4}{5}x + \frac{3}{5}y \end{pmatrix}.$$

For example, if $\mathbf{X} = \begin{pmatrix} 0 \\ 10 \end{pmatrix}$, then $S(\mathbf{X}) = \begin{pmatrix} x' \\ y' \end{pmatrix} = \begin{pmatrix} 8 \\ 6 \end{pmatrix}$.

Exercise 1. In each of the following problems, $\mathbf{U}$ is a nonzero vector and P denotes the transformation which projects each vector $\mathbf{X}$ to the line along $\mathbf{U}$. Let $\mathbf{X} = \begin{pmatrix} x \\ y \end{pmatrix}$ and $P(\mathbf{X}) = \begin{pmatrix} x' \\ y' \end{pmatrix}$, and calculate x' and y' in terms of x and y.

(a) $\mathbf{U} = \begin{pmatrix} 1 \\ 0 \end{pmatrix}$,
(b) $\mathbf{U} = \begin{pmatrix} 0 \\ 1 \end{pmatrix}$,
(c) $\mathbf{U} = \begin{pmatrix} 1 \\ -1 \end{pmatrix}$,
(d) $\mathbf{U} = \begin{pmatrix} 3 \\ 1 \end{pmatrix}$.

In each case draw a diagram and indicate several vectors and their images.

Exercise 2. Consider the line $5x - 2y = 0$ and let P denote projection on this line. If $\begin{pmatrix} a \\ b \end{pmatrix}$ is a given vector and $\begin{pmatrix} a' \\ b' \end{pmatrix} = P\begin{pmatrix} a \\ b \end{pmatrix}$, express a' and b' in terms of a and b.

Exercise 3. For each of the vectors $\mathbf{U}$ in Exercise 1, let $S(\mathbf{X}) = \begin{pmatrix} x' \\ y' \end{pmatrix}$ denote the reflection of $\begin{pmatrix} x \\ y \end{pmatrix}$ in the line along $\mathbf{U}$. Calculate the coordinates x' and y' in terms of x and y. In each case draw a diagram and indicate several vectors and their images.

Exercise 4. Let L be the line $5x - 2y = 0$, and let S denote reflection in L. If $\begin{pmatrix} a \\ b \end{pmatrix}$ is a given vector and $\begin{pmatrix} a' \\ b' \end{pmatrix} = S\begin{pmatrix} a \\ b \end{pmatrix}$, express a' and b' in terms of a and b.

Example 3. Let D_2 be the transformation which sends each vector into twice itself:

$$D_2(\mathbf{X}) = 2\mathbf{X}.$$

If $x = \begin{pmatrix} x \\ y \end{pmatrix}$ and $D_2(\mathbf{X}) = \begin{pmatrix} x' \\ y' \end{pmatrix}$, let us calculate x' and y'.

$$\begin{pmatrix} x' \\ y' \end{pmatrix} = D_2(\mathbf{X}) = 2\mathbf{X} = 2\begin{pmatrix} x \\ y \end{pmatrix} = \begin{pmatrix} 2x \\ 2y \end{pmatrix}.$$

So

$$\begin{cases} x' = 2x, \\ y' = 2y. \end{cases} \tag{2}$$

An obvious generalization of this example consists in replacing the number 2 by the number r and defining the transformation D_r by $D_r(\mathbf{X}) = r\mathbf{X}$. For $\mathbf{X} = \begin{pmatrix} x \\ y \end{pmatrix}$, $D_r(\mathbf{X}) = \begin{pmatrix} x' \\ y' \end{pmatrix}$, then, we find that $x' = rx$, $y' = ry$. We call D_r the transformation of *stretching by r*.

Fix a scalar θ with $0 \leqslant \theta \leqslant 2\pi$. We define the transformation R_θ of *rotation by θ radians* as follows (see Fig. 2.20): Let $\mathbf{X}$ be a vector. Rotate the segment from $\mathbf{0}$ to $\mathbf{X}$ around $\mathbf{0}$ counterclockwise through an angle of θ radians. The endpoint of the new segment is $R_\theta(\mathbf{X})$.

Example 4. Let $\mathbf{X} = \begin{pmatrix} x \\ y \end{pmatrix}$ and set $\begin{pmatrix} x' \\ y' \end{pmatrix} = R_{\pi/2}(\mathbf{X})$. Find x', y'. (See Fig. 2.21.) By rotating the triangle with vertices $\begin{pmatrix} 0 \\ 0 \end{pmatrix}, \begin{pmatrix} x \\ y \end{pmatrix}, \begin{pmatrix} x \\ 0 \end{pmatrix}$ through a right

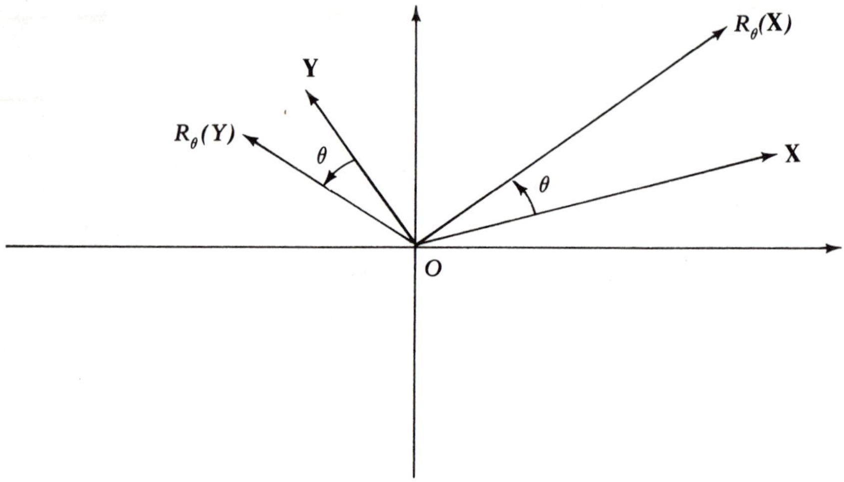

Figure 2.20

angle at the origin, we see that

$$x' = -y, \qquad y' = x.$$

Exercise 5. Let $\mathbf{X} = \binom{x}{y}$. Calculate

(a) $R_{3\pi/2}(\mathbf{X})$,
(b) $R_\pi(\mathbf{X})$,
(c) $R_{2\pi}(\mathbf{X})$,
(d) $R_{\pi/4}(\mathbf{X})$.

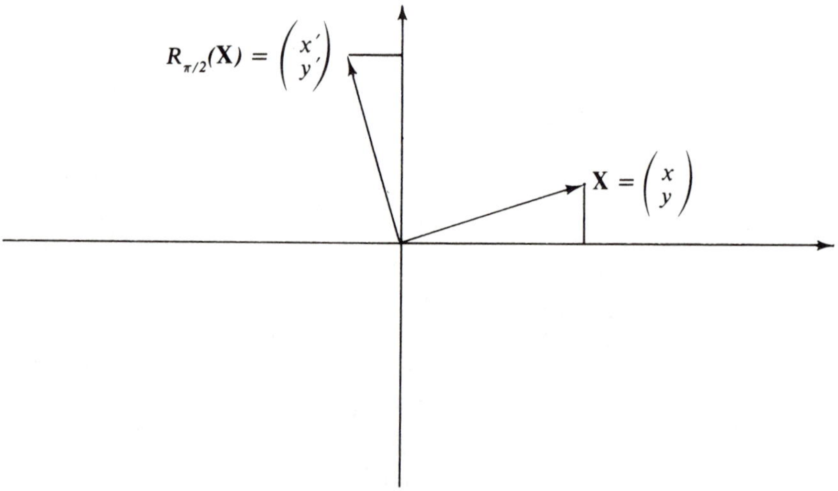

Figure 2.21

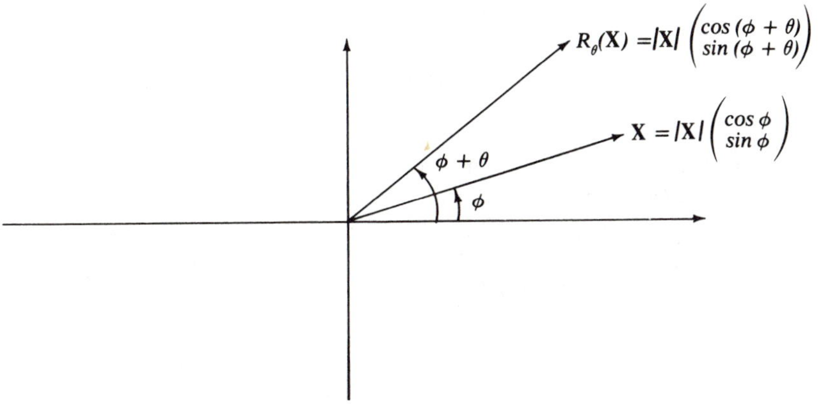

Figure 2.22

You may have found part (d) of Exercise 5 a bit difficult. Here is a method that lets us calculate $R_\theta(\mathbf{X})$ for any θ:

Set $\mathbf{X} = \begin{pmatrix} x \\ y \end{pmatrix}$ and $\begin{pmatrix} x' \\ y' \end{pmatrix} = R_\theta(\mathbf{X})$. We can write $\mathbf{X}$ and $R_\theta(\mathbf{X})$ in the form

$$\mathbf{X} = |\mathbf{X}| \begin{pmatrix} \cos\phi \\ \sin\phi \end{pmatrix}, \qquad R_\theta(\mathbf{X}) = |R_\theta(\mathbf{X})| \begin{pmatrix} \cos\phi' \\ \sin\phi' \end{pmatrix},$$

where ϕ is the polar angle of $\mathbf{X}$, and ϕ' is the polar angle of $R_\theta(\mathbf{X})$. Then $\phi' = \phi + \theta$ and $|R_\theta(\mathbf{X})| = |\mathbf{X}|$. (See Fig. 2.22.) So

$$R_\theta(\mathbf{X}) = |\mathbf{X}| \begin{pmatrix} \cos(\phi+\theta) \\ \sin(\phi+\theta) \end{pmatrix} = |\mathbf{X}| \begin{pmatrix} \cos\phi\cos\theta - \sin\phi\sin\theta \\ \sin\phi\cos\theta + \cos\phi\sin\theta \end{pmatrix}$$

$$= \begin{pmatrix} |\mathbf{X}|\cos\phi\cos\theta - |\mathbf{X}|\sin\phi\sin\theta \\ |\mathbf{X}|\sin\phi\cos\theta + |\mathbf{X}|\cos\phi\sin\theta \end{pmatrix}.$$

Now

$$|\mathbf{X}|\cos\phi = x, \qquad |\mathbf{X}|\sin\phi = y.$$

So

$$\begin{pmatrix} x' \\ y' \end{pmatrix} = R_\theta(\mathbf{X}) = \begin{pmatrix} x\cos\theta - y\sin\theta \\ y\cos\theta + x\sin\theta \end{pmatrix},$$

or

$$\begin{cases} x' = (\cos\theta)x - (\sin\theta)y, \\ y' = (\sin\theta)x + (\cos\theta)y. \end{cases} \tag{3}$$

Exercise 6. Interpret the results you obtained for Exercise 5 as corollaries of formula (3), by giving θ suitable values.

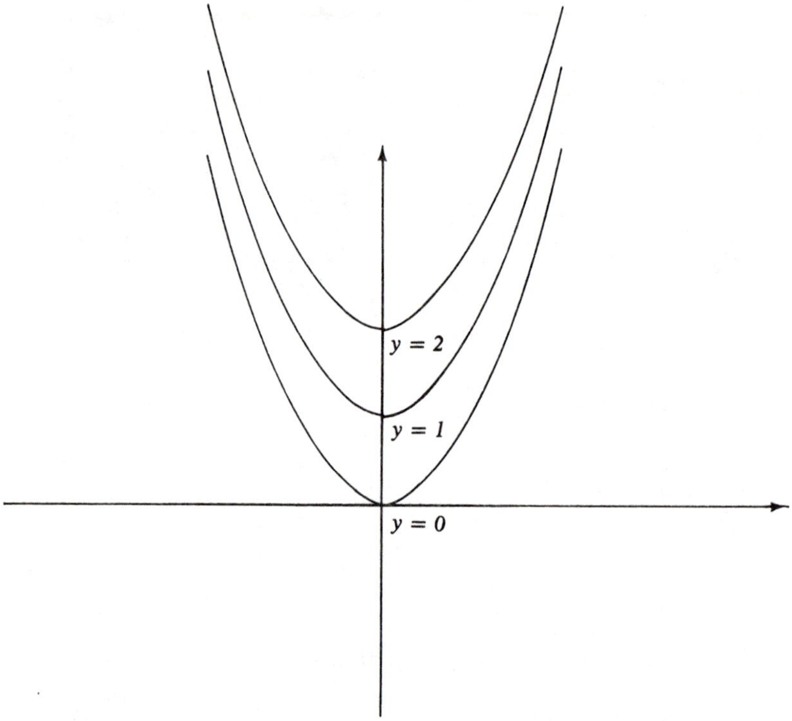

Figure 2.23

EXAMPLE 5. Let T_0 be the following transformation: T_0 sends every horizontal line $y = c$ into the parabola $y = x^2 + c$ by sending $\begin{pmatrix} x \\ y \end{pmatrix}$ into $\begin{pmatrix} x \\ x^2 + y \end{pmatrix}$. (See Fig. 2.23.) In other words, if $\mathbf{X} = \begin{pmatrix} x \\ y \end{pmatrix}$ and $\begin{pmatrix} x' \\ y' \end{pmatrix} = \mathbf{X}' = T_0(\mathbf{X})$, then

$$\begin{cases} x' = x, \\ y' = x^2 + y. \end{cases} \tag{4}$$

Let S, T be two transformations. When do we say that they are *equal*, i.e., $S = T$? Recall that two functions f, g were called equal if $f(x) = g(x)$ for every number x. In a similar spirit, we say $S = T$ provided

$$S(\mathbf{X}) = T(\mathbf{X}) \qquad \text{for every vector } \mathbf{X}.$$

EXAMPLE 6. The transformation $R_{-\pi/2}$, which rotates each vector clockwise by $\pi/2$ radians, and the transformation $R_{3\pi/2}$, which rotates each vector counterclockwise by $3\pi/2$ radians, are equal, i.e.,

$$R_{-\pi/2} = R_{3\pi/2}.$$

CHAPTER 2.2

Linear Transformations and Matrices

In Chapter 2.1, we looked at a number of transformations of the plane. Let us list the results we obtained for each transformation T, giving x', y' in terms of x, y, where $\begin{pmatrix} x' \\ y' \end{pmatrix} = T\begin{pmatrix} x \\ y \end{pmatrix}$.

(i) P = projection to the line along $\begin{pmatrix} 1 \\ 2 \end{pmatrix}$.

$$x' = \frac{x + 2y}{5},$$
$$y' = \frac{2(x + 2y)}{5}.$$

(ii) S is reflection about the line along $\begin{pmatrix} 1 \\ 2 \end{pmatrix}$.

$$x' = -\frac{3}{5}x + \frac{4}{5}y,$$
$$y' = \frac{4}{5}x + \frac{3}{5}y.$$

(iii) D_r is stretching by r.

$$x' = rx,$$
$$y' = ry.$$

(iv) R_θ is rotation by θ radians.

$$x' = (\cos\theta)x - (\sin\theta)y,$$
$$y' = (\sin\theta)x + (\cos\theta)y.$$

(v) T_0 is the transformation of Example 5.

$$x' = x,$$
$$y' = x^2 + y.$$

Can we describe some single general type of transformation, expressing x' and y' each in terms of x and y, which includes the above examples as special cases?

Let a, b, c, d be scalars. Denote by A the transformation which sends each vector $\mathbf{X} = \begin{pmatrix} x \\ y \end{pmatrix}$ into the vector $\mathbf{X}' = \begin{pmatrix} x' \\ y' \end{pmatrix}$, where

$$\begin{cases} x' = ax + by, \\ y' = cx + dy. \end{cases} \tag{1}$$

Setting $a = \frac{1}{5}$, $b = \frac{2}{5}$, $c = \frac{2}{5}$, $d = \frac{4}{5}$, we re-obtain example (i) above. Setting $a = -\frac{3}{5}$, $b = \frac{4}{5}$, $c = \frac{4}{5}$, $d = \frac{3}{5}$, we get (ii). Setting $a = r$, $b = 0$, $c = 0$, $d = r$, we get (iii). If we take $a = \cos\theta$, $b = -\sin\theta$, $c = \sin\theta$, $d = \cos\theta$, we obtain (iv). However, no choice of a, b, c, d will give us (v).

A transformation A given by a system (1) is called a *linear transformation of the plane* and the symbol

$$\begin{pmatrix} a & b \\ c & d \end{pmatrix}$$

is called the *matrix* of A, denoted $m(A)$. The plural of "matrix" is "matrices." Reflection through a line and projection to a line are linear transformations, provided the line goes through the origin. Stretchings D_r and rotations are also linear transformations. It is not possible to describe *all* linear transformations in simple geometrical terms. However, Equations (1) provide a simple algebraic description.

Let us list the matrices of the linear transformations (i)–(iv) considered above.

(i) $$m(P) = \begin{pmatrix} \frac{1}{5} & \frac{2}{5} \\ \frac{2}{5} & \frac{4}{5} \end{pmatrix},$$

(ii) $$m(S) = \begin{pmatrix} -\frac{3}{5} & \frac{4}{5} \\ \frac{4}{5} & \frac{3}{5} \end{pmatrix},$$

(iii) $$m(D_r) = \begin{pmatrix} r & 0 \\ 0 & r \end{pmatrix},$$

(iv) $$m(R_\theta) = \begin{pmatrix} \cos\theta & -\sin\theta \\ \sin\theta & \cos\theta \end{pmatrix},$$

(v) T_0 is *not* a linear transformation.

We need the linear transformation which is the analogue of the function $f(x) = x$. That function sends every number into itself. The *identity transformation*, denoted I, sends every vector into itself:

$$I(\mathbf{X}) = \mathbf{X}, \qquad \text{for every vector } \mathbf{X}.$$

Since I sends $\mathbf{X}=\begin{pmatrix}x\\y\end{pmatrix}$ into $I(\mathbf{X})=\begin{pmatrix}x\\y\end{pmatrix}$, the system

$$x' = x,$$
$$y' = y$$

describes I. Thus the matrix of I is

(vi) $$m(I)=\begin{pmatrix}1 & 0\\0 & 1\end{pmatrix}.$$

Finally, we need the linear transformation *zero*, denoted 0, which sends every vector into the zero vector:

$$0(\mathbf{X}) = \mathbf{0}, \qquad \text{for all } \mathbf{X}.$$

Evidently, the matrix of 0 is

(vii) $$m(0)=\begin{pmatrix}0 & 0\\0 & 0\end{pmatrix}.$$

Next we introduce some useful notation. Let A be the linear transformation with matrix $\begin{pmatrix}p & q\\r & s\end{pmatrix}$ and let $\mathbf{X}=\begin{pmatrix}x\\y\end{pmatrix}$ be a vector. We shall write

$$\begin{pmatrix}p & q\\r & s\end{pmatrix}\begin{pmatrix}x\\y\end{pmatrix}=A\begin{pmatrix}x\\y\end{pmatrix}=A(\mathbf{X}). \tag{2}$$

For instance, if D is stretching by 2, then

$$\begin{pmatrix}2 & 0\\0 & 2\end{pmatrix}\begin{pmatrix}x\\y\end{pmatrix}=D\begin{pmatrix}x\\y\end{pmatrix}=\begin{pmatrix}2x\\2y\end{pmatrix},$$

or if P is projection on the line along $\begin{pmatrix}1\\2\end{pmatrix}$, then

$$\begin{pmatrix}\frac{1}{5} & \frac{2}{5}\\ \frac{2}{5} & \frac{4}{5}\end{pmatrix}\begin{pmatrix}x\\y\end{pmatrix}=P\begin{pmatrix}x\\y\end{pmatrix}=\begin{pmatrix}\frac{1}{5}x+\frac{2}{5}y\\ \frac{2}{5}x+\frac{4}{5}y\end{pmatrix}.$$

Let A have the matrix $\begin{pmatrix}p & q\\r & s\end{pmatrix}$, $\mathbf{X}=\begin{pmatrix}x\\y\end{pmatrix}$ and $A(\mathbf{X})=\begin{pmatrix}x'\\y'\end{pmatrix}$. Then

$$x' = px + qy,$$
$$y' = rx + sy.$$

By definition (2),

$$\begin{pmatrix}p & q\\r & s\end{pmatrix}\begin{pmatrix}x\\y\end{pmatrix}=A\begin{pmatrix}x\\y\end{pmatrix}=\begin{pmatrix}x'\\y'\end{pmatrix},$$

and so

$$\begin{pmatrix}p & q\\r & s\end{pmatrix}\begin{pmatrix}x\\y\end{pmatrix}=\begin{pmatrix}px+qy\\rx+sy\end{pmatrix}. \tag{3}$$

Formula (3) is basic. We interpret (3) as saying that the matrix $\begin{pmatrix}p & q\\r & s\end{pmatrix}$ *acts on the vector* $\begin{pmatrix}x\\y\end{pmatrix}$ to yield the vector $\begin{pmatrix}px+qy\\rx+sy\end{pmatrix}$.

EXAMPLE 1.

$$\begin{pmatrix} 5 & 6 \\ 7 & 8 \end{pmatrix}\begin{pmatrix} x \\ y \end{pmatrix} = \begin{pmatrix} 5x+6y \\ 7x+8y \end{pmatrix},$$

$$\begin{pmatrix} 5 & 6 \\ 7 & 9 \end{pmatrix}\begin{pmatrix} 1 \\ 1 \end{pmatrix} = \begin{pmatrix} 11 \\ 16 \end{pmatrix},$$

$$\begin{pmatrix} 0 & 0 \\ 2 & 2 \end{pmatrix}\begin{pmatrix} x \\ y \end{pmatrix} = \begin{pmatrix} 0 \\ 2x+2y \end{pmatrix},$$

$$\begin{pmatrix} 1 & 0 \\ 0 & 1 \end{pmatrix}\begin{pmatrix} 5 \\ \pi \end{pmatrix} = \begin{pmatrix} 5 \\ \pi \end{pmatrix}.$$

Let A be an arbitrary linear transformation. We claim that A sends the origin into the origin, i.e.,

$$A(\mathbf{0}) = \mathbf{0},$$

for if $\begin{pmatrix} a & b \\ c & d \end{pmatrix}$ is the matrix of A, then

$$A(\mathbf{0}) = \begin{pmatrix} a & b \\ c & d \end{pmatrix}\begin{pmatrix} 0 \\ 0 \end{pmatrix} = \begin{pmatrix} 0 \\ 0 \end{pmatrix} = \mathbf{0}.$$

A basic reason why linear transformations are interesting is that a linear transformation acts in a simple way on the sum of two vectors. Let A be the linear transformation with matrix $\begin{pmatrix} a & b \\ c & d \end{pmatrix}$ and let $\mathbf{X} = \begin{pmatrix} x \\ y \end{pmatrix}$, $\overline{\mathbf{X}} = \begin{pmatrix} \bar{x} \\ \bar{y} \end{pmatrix}$ be two vectors.

$$\begin{aligned} A(\mathbf{X}+\overline{\mathbf{X}}) &= A\begin{pmatrix} x+\bar{x} \\ y+\bar{y} \end{pmatrix} = \begin{pmatrix} a & b \\ c & d \end{pmatrix}\begin{pmatrix} x+\bar{x} \\ y+\bar{y} \end{pmatrix} \\ &= \begin{pmatrix} a(x+\bar{x})+b(y+\bar{y}) \\ c(x+\bar{x})+d(y+\bar{y}) \end{pmatrix} = \begin{pmatrix} (ax+by)+(a\bar{x}+b\bar{y}) \\ (cx+dy)+(c\bar{x}+d\bar{y}) \end{pmatrix} \\ &= \begin{pmatrix} a & b \\ c & d \end{pmatrix}\begin{pmatrix} x \\ y \end{pmatrix} + \begin{pmatrix} a & b \\ c & d \end{pmatrix}\begin{pmatrix} \bar{x} \\ \bar{y} \end{pmatrix} = A(\mathbf{X}) + A(\overline{\mathbf{X}}). \end{aligned}$$

Thus, we have found

$$A(\mathbf{X}+\overline{\mathbf{X}}) = A(\mathbf{X}) + A(\overline{\mathbf{X}}) \tag{4}$$

for every pair of vectors $\mathbf{X}$, $\overline{\mathbf{X}}$.

A similar calculation shows

$$A(t\mathbf{X}) = tA(\mathbf{X}), \tag{5}$$

if $\mathbf{X}$ is a vector and t is a scalar.

Exercise 1. Verify that formula (5) is true.

Conversely, let B be a transformation of the plane. Let us not assume that B is linear, but instead let us suppose that (4) and (5) are valid for B, i.e., suppose

$$B(\mathbf{X}+\overline{\mathbf{X}}) = B(\mathbf{X}) + B(\overline{\mathbf{X}}), \qquad B(t\mathbf{X}) = tB(\mathbf{X}), \tag{6}$$

whenever $\mathbf{X}$ and $\overline{\mathbf{X}}$ are vectors and t is a scalar. We claim that it follows that B is a linear transformation, i.e., B is given by a system (1) for suitable a, b, c, d.

To see this, set $\mathbf{E}_1 = \begin{pmatrix} 1 \\ 0 \end{pmatrix}$ and $\mathbf{E}_2 = \begin{pmatrix} 0 \\ 1 \end{pmatrix}$. Then an arbitrary vector $\mathbf{X} = \begin{pmatrix} x \\ y \end{pmatrix}$ can be expressed as

$$\mathbf{X} = x\mathbf{E}_1 + y\mathbf{E}_2.$$

Set $B(\mathbf{X}) = \begin{pmatrix} x' \\ y' \end{pmatrix}$. By hypothesis,

$$B(\mathbf{X}) = B(x\mathbf{E}_1) + B(y\mathbf{E}_2) = xB(\mathbf{E}_1) + yB(\mathbf{E}_2).$$

$B(\mathbf{E}_1)$ can be written

$$B(\mathbf{E}_1) = \begin{pmatrix} u \\ v \end{pmatrix},$$

and similarly $B(\mathbf{E}_2) = \begin{pmatrix} w \\ z \end{pmatrix}$. Thus

$$\begin{pmatrix} x' \\ y' \end{pmatrix} = B(\mathbf{X}) = x\begin{pmatrix} u \\ v \end{pmatrix} + y\begin{pmatrix} w \\ z \end{pmatrix} = \begin{pmatrix} ux + wy \\ vx + zy \end{pmatrix}.$$

So

$$x' = ux + wy,$$
$$y' = vx + zy.$$

Thus, x', y' have the form of Eq. (1) of this chapter. Hence B is a linear transformation, by definition. The matrix of B is $\begin{pmatrix} u & w \\ v & z \end{pmatrix}$. Thus we have proved that if B is a transformation satisfying (6), then B is a linear transformation. Summing up, we have shown:

Theorem 2.1. *Let A be a transformation of the plane. Then A is a linear transformation if and only if for every pair of vectors $\mathbf{X}$ and $\overline{\mathbf{X}}$ and every scalar t:*

$$A(\mathbf{X}+\overline{\mathbf{X}}) = A(\mathbf{X}) + A(\overline{\mathbf{X}}), \tag{7a}$$

and

$$A(t\mathbf{X}) = tA(\mathbf{X}). \tag{7b}$$

Note: (7a) and (7b) together imply

$$A(t\mathbf{X} + s\overline{\mathbf{X}}) = tA(\mathbf{X}) + sA(\overline{\mathbf{X}}) \tag{8}$$

for every pair of scalars t, s and every pair of vectors $\mathbf{X}$, $\overline{\mathbf{X}}$. This is so, since by (7a),

$$A(t\mathbf{X} + s\overline{\mathbf{X}}) = A(t\mathbf{X}) + A(s\overline{\mathbf{X}}),$$

while by (7b), $A(t\mathbf{X}) = tA(\mathbf{X})$ and $A(s\overline{\mathbf{X}}) = sA(\overline{\mathbf{X}})$. On the other hand, (8) clearly implies both (7a) and (7b). Thus, in Theorem 2.1, we may replace the two conditions (7a) and (7b) by the single condition (8). From now on, when presented with a transformation T, if we wish to show that T is a linear transformation, we can do either of the following: Show that T satisfies (7a) and (7b) (or, equivalently, (8)), or show that there is some matrix $\begin{pmatrix} a & b \\ c & d \end{pmatrix}$ such that for every vector $\mathbf{X} = \begin{pmatrix} x \\ y \end{pmatrix}$, the vector $T(\mathbf{X}) = \begin{pmatrix} x' \\ y' \end{pmatrix}$ is given by:

$$x' = ax + by,$$
$$y' = cx + dy.$$

If A is the linear transformation with matrix $\begin{pmatrix} a & b \\ c & d \end{pmatrix}$, then $A(\mathbf{E}_1) = \begin{pmatrix} a & b \\ c & d \end{pmatrix}\begin{pmatrix} 1 \\ 0 \end{pmatrix} = \begin{pmatrix} a \\ c \end{pmatrix}$ and $A(\mathbf{E}_2) = \begin{pmatrix} a & b \\ c & d \end{pmatrix}\begin{pmatrix} 0 \\ 1 \end{pmatrix} = \begin{pmatrix} b \\ d \end{pmatrix}$. Thus we can describe the matrix of A by saying that its first column is the image of the first basis vector $\mathbf{E}_1$ and the second column is the image of $\mathbf{E}_2$.

EXAMPLE 2. Let P denote projection on the line along $\mathbf{U} = \begin{pmatrix} u \\ v \end{pmatrix}$, so $P(\mathbf{X}) = \left(\frac{\mathbf{X} \cdot \mathbf{U}}{\mathbf{U} \cdot \mathbf{U}}\right)\mathbf{U}$. Then

$$P(\mathbf{E}_1) = \left(\frac{u}{u^2 + v^2}\right)\begin{pmatrix} u \\ v \end{pmatrix} = \begin{bmatrix} \dfrac{u^2}{u^2 + v^2} \\ \dfrac{uv}{u^2 + v^2} \end{bmatrix},$$

$$P(\mathbf{E}_2) = \left(\frac{v}{u^2 + v^2}\right)\begin{pmatrix} u \\ v \end{pmatrix} = \begin{bmatrix} \dfrac{uv}{u^2 + v^2} \\ \dfrac{v^2}{u^2 + v^2} \end{bmatrix}.$$

Thus the matrix of P is given by

$$m(P) = \begin{bmatrix} \dfrac{u^2}{u^2 + v^2} & \dfrac{vu}{u^2 + v^2} \\ \dfrac{uv}{u^2 + v^2} & \dfrac{v^2}{u^2 + v^2} \end{bmatrix}. \tag{9}$$

For example, if $\mathbf{U}$ is the unit vector $\mathbf{U}=\begin{pmatrix}\cos\theta\\ \sin\theta\end{pmatrix}$, then $u^2+v^2=1$, so

$$m(P)=\begin{pmatrix}\cos^2\theta & \sin\theta\cos\theta\\ \cos\theta\sin\theta & \sin^2\theta\end{pmatrix}. \tag{10}$$

Theorem 2.1 allows us to give a simple solution of the following geometric problem: Let A be a linear transformation and let L be a straight line. By the *image of L under A* we mean the collection of all vectors $A(\mathbf{X})$ when $\mathbf{X}$ is a vector whose endpoint lies on L. We denote this image by $A(L)$.

Example 3. The image of the x-axis under the transformation $R_{\pi/2}$, which is rotation by $\pi/2$ radians, is the y-axis.

What kind of geometric object does the image of L under A turn out to be? The answer is given by:

Theorem 2.2. *Let A be a linear transformation and let L be a straight line. Then the image of L under A is either a straight line or a single point.*

Proof. Remember from Chapter 2.0 that we can choose vectors $\mathbf{X}_0$ and $\mathbf{U}$ in such a way that L is described by

$$\mathbf{X}=\mathbf{X}_0+t\mathbf{U},\qquad t \text{ a real scalar.}$$

Thus, for each $\mathbf{X}$ on L,

$$\mathbf{X}=\mathbf{X}_0+t\mathbf{U}.$$

Hence, by (7a) and (7b),

$$A(\mathbf{X})=A(\mathbf{X}_0+t\mathbf{U})=A(\mathbf{X}_0)+tA(\mathbf{U}).$$

If $A(\mathbf{U})\neq 0$, then as $\mathbf{X}$ runs through all vectors with endpoint on L, $A(\mathbf{X})$ runs through the collection of points

$$A(\mathbf{X}_0)+tA(\mathbf{U}),\qquad t \text{ real.}$$

This is a straight line, and so the image of L under A is this line. (See Fig. 2.24.) If $A(\mathbf{U})=0$, then $A(\mathbf{X})=A(\mathbf{X}_0)$ for each $\mathbf{X}$ with endpoint on L. So the image of L under A is the single point $A(\mathbf{X}_0)$.

Example 4. Let A denote reflection in the x-axis and let L be the line along $\begin{pmatrix}1\\2\end{pmatrix}$. Find the image of L under A. The point $\begin{pmatrix}t\\2t\end{pmatrix}$ has image $\begin{pmatrix}t\\-2t\end{pmatrix}$ so the line along $\begin{pmatrix}1\\-2\end{pmatrix}$ is the image of L under A. (See Fig. 2.25.)

Example 5. Let P denote projection on the y-axis and let L be the x-axis. Find the image of L under P. (See Fig. 2.26.) If $\mathbf{X}$ lies on L, then $P(\mathbf{X})=\mathbf{0}$. Hence, the image of L under P is a single point, the origin.

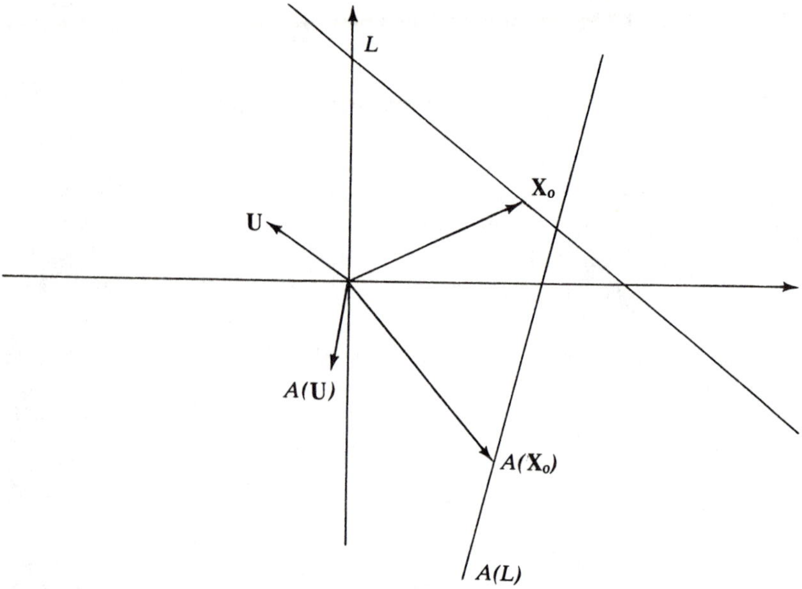

Figure 2.24

Exercise 2. For each of the following transformation T calculate the image of the x-axis.

(a) T is rotation by 45°.
(b) T is reflection in the line $y = 2x$.
(c) T is projection on the line $y = x$.

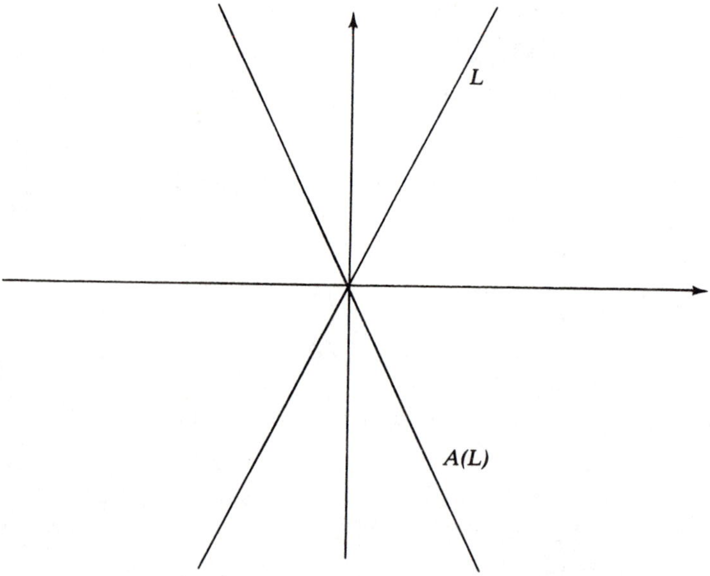

Figure 2.25

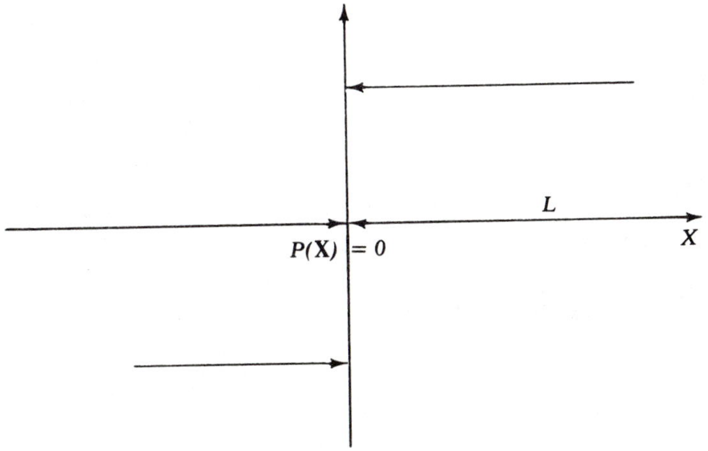

Figure 2.26

Exercise 3. A is the transformation with matrix $\begin{pmatrix} 2 & 3 \\ -1 & 0 \end{pmatrix}$.

(a) Find the image of the line along $\begin{pmatrix} 3 \\ 1 \end{pmatrix}$ under A.

(b) Find the image of the line along $\begin{pmatrix} a \\ b \end{pmatrix}$ under A.

Exercise 4. Let B be a linear transformation such that whenever $\mathbf{X}$ is a nonzero vector, then $B(\mathbf{X}) \neq \mathbf{0}$. Show that for every straight line L through the origin, the image of L under B is a straight line through the origin.

If $\mathbf{X}$ is a nonzero vector then the set of vectors $\{r\mathbf{X} \mid 0 \leqslant r \leqslant 1\}$ is the segment from $\mathbf{0}$ to the point $\mathbf{X}$. If $\mathbf{X} = \mathbf{0}$, then the collection $\{r\mathbf{X} \mid 0 \leqslant r \leqslant 1\}$ contains only the zero vector, and in this case we say that the segment *degenerates to a point*.

If T is any linear transformation, then $T(r\mathbf{X}) = rT(\mathbf{X})$, so the image of the segment $\{r\mathbf{X} \mid 0 \leqslant r \leqslant 1\}$ is the segment $\{r(T(\mathbf{X})) \mid 0 \leqslant r \leqslant 1\}$, possibly degenerate if $T(\mathbf{X}) = \mathbf{0}$.

The set of points $\{\mathbf{U} + r\mathbf{X} \mid 0 \leqslant r \leqslant 1\}$ is also a segment, from $\mathbf{U}$ to $\mathbf{U} + \mathbf{X}$.

If $\mathbf{X}$ and $\mathbf{U}$ are linearly independent vectors, then the set of vectors $\{r\mathbf{X} + s\mathbf{U} \mid 0 \leqslant r \leqslant 1, 0 \leqslant s \leqslant 1\}$ describes the *parallelogram* determined by $\mathbf{X}$ and $\mathbf{U}$ (see Fig. 2.27). The sets $\{r\mathbf{X} \mid 0 \leqslant r \leqslant 1\}$ and $\{s\mathbf{U} \mid 0 \leqslant s \leqslant 1\}$ form two edges of the parallelogram and the other two edges are $\{r\mathbf{X} + \mathbf{U} \mid 0 \leqslant r \leqslant 1\}$ and $\{\mathbf{X} + s\mathbf{U} \mid 0 \leqslant s \leqslant 1\}$. The four corners of the parallelogram are, in order: $\mathbf{U}$, $\mathbf{0}$, $\mathbf{X}$, $\mathbf{U} + \mathbf{X}$.

If $\mathbf{X}$ and $\mathbf{U}$ are linearly dependent, but not both $\mathbf{0}$, then the four points $\mathbf{U}$, $\mathbf{0}$, $\mathbf{X}$, and $\mathbf{U} + \mathbf{X}$ all lie on the same line and the set $\{r\mathbf{X} + s\mathbf{U} \mid 0 \leqslant r \leqslant 1, 0 \leqslant s \leqslant 1\}$ is then a *degenerate* or *collapsed* parallelogram.

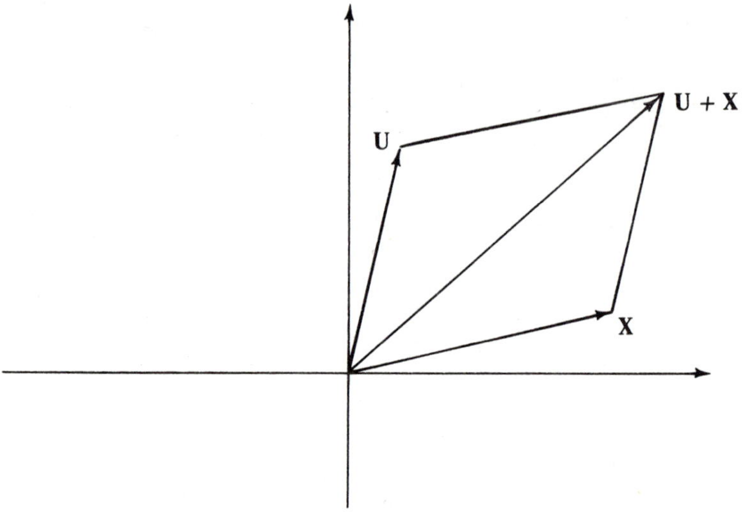

Figure 2.27

If **X** and **U** are both **0**, then $\{r\mathbf{X} + s\mathbf{U} \mid 0 \leqslant r, s \leqslant 1\}$ is also just the point **0**, so the parallelogram degenerates to a single point.

If T is a linear transformation, then $T(r\mathbf{X} + s\mathbf{U}) = rT(\mathbf{X}) + sT(\mathbf{U})$, so the image of the parallelogram $\Pi = \{r\mathbf{X} + s\mathbf{U} \mid 0 \leqslant r, s \leqslant 1\}$ is the parallelogram $T(\Pi) = \{rT(\mathbf{X}) + sT(\mathbf{U}) \mid 0 \leqslant r \leqslant 1,\ 0 \leqslant s \leqslant 1\}$. Even if $\mathbf{X}, \mathbf{U}$ is a linearly independent pair, the parallelogram $T(\Pi)$ might be degenerate.

Exercise 5. Describe the parallelograms determined by the following pairs of vectors:

(a) $\begin{pmatrix} 2 \\ 1 \end{pmatrix}, \begin{pmatrix} 1 \\ 1 \end{pmatrix}$,

(b) $\begin{pmatrix} 2 \\ 1 \end{pmatrix}, \begin{pmatrix} 4 \\ 2 \end{pmatrix}$,

(c) $\begin{pmatrix} 2 \\ 1 \end{pmatrix}, \begin{pmatrix} -2 \\ -1 \end{pmatrix}$,

(d) $\begin{pmatrix} 0 \\ 0 \end{pmatrix}, \begin{pmatrix} 2 \\ 1 \end{pmatrix}$,

(e) $\begin{pmatrix} 0 \\ 0 \end{pmatrix}, \begin{pmatrix} 0 \\ 0 \end{pmatrix}$.

Exercise 6. Describe the images of each of the preceding parallelograms under the projection to the first coordinate axis:

$$P\begin{pmatrix} x \\ y \end{pmatrix} = \begin{pmatrix} x \\ 0 \end{pmatrix}.$$

Exercise 7. Do the same for the linear transformation with matrix $\begin{pmatrix} -1 & -2 \\ 2 & 4 \end{pmatrix}$.

CHAPTER 2.3

Products of Linear Transformations

Let A and B be two linear transformations. We define the transformation C which consists of A followed by B, i.e., if $\mathbf{X}$ is any vector

$$C(\mathbf{X}) = B(A(\mathbf{X})).$$

We write $C = BA$ and we call C *the product B times A*.

Associating to A and B their product BA is in some ways analogous to multiplying two numbers, and we shall pursue this analogy later on.

EXAMPLE 1. B is reflection in the x-axis and A is reflection in the y-axis (see Fig. 2.28). Find BA.

Choose $\mathbf{X} = \begin{pmatrix} x \\ y \end{pmatrix}$. Then

$$A(\mathbf{X}) = \begin{pmatrix} -x \\ y \end{pmatrix} \quad \text{and} \quad B(A(\mathbf{X})) = \begin{pmatrix} -x \\ -y \end{pmatrix}.$$

So

$$(BA)(\mathbf{X}) = \begin{pmatrix} -x \\ -y \end{pmatrix} = -\mathbf{X}.$$

Thus, BA sends each vector into its negative. In other words, $BA = R_\pi$, rotation by π radians.

Exercise 1. Show that if S, T are linear transformations, then ST and TS are linear transformations. (Use (7a) and (7b) or (8) in Chapter 2.2.)

Exercise 2. Let A, B have the same meaning as in Example 1. Show that $AB = R_\pi$.

EXAMPLE 2. Let P be projection on the x-axis and Q projection on the y-axis. Find QP. (See Fig. 2.29.)

If $\mathbf{X} = \begin{pmatrix} x \\ y \end{pmatrix}$, then $P(\mathbf{X}) = \begin{pmatrix} x \\ 0 \end{pmatrix}$ and so $(QP)(\mathbf{X}) = Q(P(\mathbf{X})) = Q\begin{pmatrix} x \\ 0 \end{pmatrix} = \begin{pmatrix} 0 \\ 0 \end{pmatrix}$.

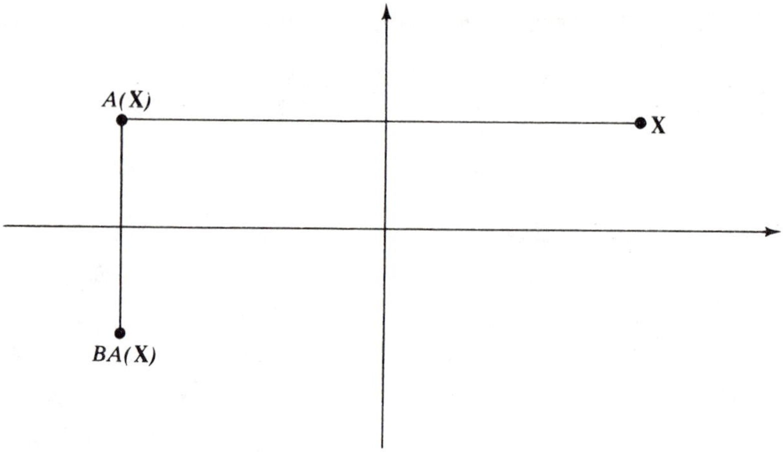

Figure 2.28

Thus QP is the transformation which sends every vector into the origin, i.e., $QP = 0$.

EXAMPLE 3. Let A be a linear transformation and let I denote the identity transformation. Let us find AI and IA.

Fix a vector **X**

$$(AI)(\mathbf{X}) = A(I(\mathbf{X})) = A(\mathbf{X})$$

and

$$(IA)(\mathbf{X}) = I(A(\mathbf{X})) = A(\mathbf{X}).$$

Hence,

$$AI = A \quad \text{and} \quad IA = A. \tag{1}$$

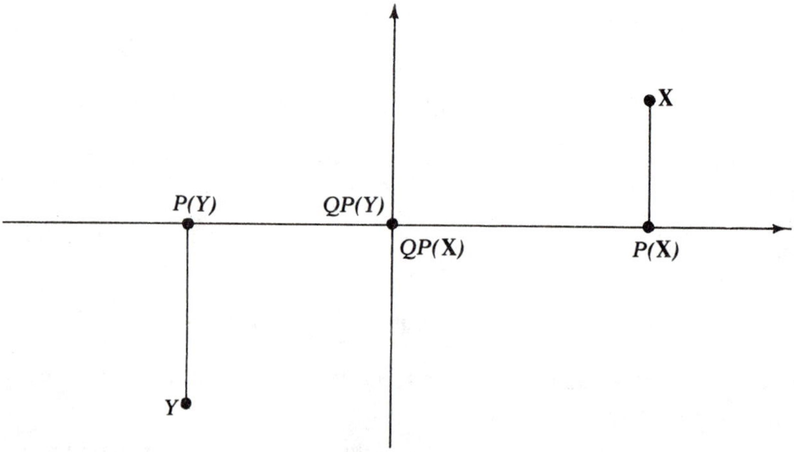

Figure 2.29

Note: The number 1 has the property that

$$a1 = 1a = a$$

for every number a. In view of (1), the identity transformation I plays the same role in multiplying linear transformations as the number 1 does in multiplying numbers.

Exercise 3. Let P be projection on the x-axis and let $R_{\pi/2}$ be rotation by $\pi/2$ radians.

(a) Calculate $PR_{\pi/2}$.
(b) Calculate $(R_{\pi/2})P$.

Observe that your answers for (a) and (b) in Exercise 3 are different. Thus $PR_{\pi/2} \neq R_{\pi/2}P$. The commutative law of multiplication, i.e., the law that $ab = ba$, which is valid for every pair of numbers a, b is false for the product of linear transformations. That is, if A, B are linear transformations, then sometimes $AB = BA$ and sometimes $AB \neq BA$. If $AB = BA$, we say that A and B *commute*. For instance, if A is any linear transformation and I is the identity, then A and I commute.

Now suppose that A and B are two linear transformations having matrices $\begin{pmatrix} a & b \\ c & d \end{pmatrix}$ and $\begin{pmatrix} \bar{a} & \bar{b} \\ \bar{c} & \bar{d} \end{pmatrix}$, respectively. What is the matrix of the transformation AB?

Let $\mathbf{X} = \begin{pmatrix} x \\ y \end{pmatrix}$. Then

$$B(\mathbf{X}) = \begin{pmatrix} \bar{a} & \bar{b} \\ \bar{c} & \bar{d} \end{pmatrix}\begin{pmatrix} x \\ y \end{pmatrix} = \begin{pmatrix} \bar{a}x + \bar{b}y \\ \bar{c}x + \bar{d}y \end{pmatrix}.$$

So

$$AB(\mathbf{X}) = A(B(\mathbf{X})) = \begin{pmatrix} a & b \\ c & d \end{pmatrix}\begin{pmatrix} \bar{a}x + \bar{b}y \\ \bar{c}x + \bar{d}y \end{pmatrix}$$

$$= \begin{pmatrix} a(\bar{a}x + \bar{b}y) + b(\bar{c}x + \bar{d}y) \\ c(\bar{a}x + \bar{b}y) + d(\bar{c}x + \bar{d}y) \end{pmatrix}$$

$$= \begin{pmatrix} (a\bar{a} + b\bar{c})x + (a\bar{b} + b\bar{d})y \\ (c\bar{a} + d\bar{c})x + (c\bar{b} + d\bar{d})y \end{pmatrix}$$

$$= \begin{pmatrix} a\bar{a} + b\bar{c} & a\bar{b} + b\bar{d} \\ c\bar{a} + d\bar{c} & c\bar{b} + d\bar{d} \end{pmatrix}\begin{pmatrix} x \\ y \end{pmatrix}.$$

So the matrix of AB is

$$\begin{pmatrix} a\bar{a} + b\bar{c} & a\bar{b} + b\bar{d} \\ c\bar{a} + d\bar{c} & c\bar{b} + d\bar{d} \end{pmatrix}. \tag{2}$$

We define the *product of matrices* $m(A)$ *and* $m(B)$ to be the matrix $m(AB)$ of AB. Thus

$$m(A)m(B) = m(AB). \tag{3}$$

In other words,

$$\begin{pmatrix} a & b \\ c & d \end{pmatrix}\begin{pmatrix} \bar{a} & \bar{b} \\ \bar{c} & \bar{d} \end{pmatrix} = \begin{pmatrix} a\bar{a} + b\bar{c} & a\bar{b} + b\bar{d} \\ c\bar{a} + d\bar{c} & c\bar{b} + d\bar{d} \end{pmatrix}. \tag{4}$$

Note that on the right-hand side of (4), the upper left-hand entry is the dot product

$$\begin{pmatrix} a \\ b \end{pmatrix} \cdot \begin{pmatrix} \bar{a} \\ \bar{c} \end{pmatrix},$$

the upper right-hand entry is

$$\begin{pmatrix} a \\ b \end{pmatrix} \cdot \begin{pmatrix} \bar{b} \\ \bar{d} \end{pmatrix},$$

the lower left-hand entry is

$$\begin{pmatrix} c \\ d \end{pmatrix} \cdot \begin{pmatrix} \bar{a} \\ \bar{c} \end{pmatrix},$$

and the lower right-hand entry is

$$\begin{pmatrix} c \\ d \end{pmatrix} \cdot \begin{pmatrix} \bar{b} \\ \bar{d} \end{pmatrix}.$$

EXAMPLE 4. Find the product $\begin{pmatrix} 1 & 0 \\ 2 & 3 \end{pmatrix}\begin{pmatrix} 5 & 4 \\ -1 & 7 \end{pmatrix}$. By formula (4),

$$\begin{pmatrix} 1 & 0 \\ 2 & 3 \end{pmatrix}\begin{pmatrix} 5 & 4 \\ -1 & 7 \end{pmatrix} = \begin{pmatrix} 1 \cdot 5 + 0 \cdot -1 & 1 \cdot 4 + 0 \cdot 7 \\ 2 \cdot 5 + 3 \cdot -1 & 2 \cdot 4 + 3 \cdot 7 \end{pmatrix} = \begin{pmatrix} 5 & 4 \\ 7 & 29 \end{pmatrix}.$$

Exercise 4. In each case, calculate the indicated product of two matrices:

(a) $\begin{pmatrix} -1 & 0 \\ 0 & 0 \end{pmatrix}\begin{pmatrix} -1 & 0 \\ 0 & 0 \end{pmatrix}$,

(b) $\begin{pmatrix} 0 & 1 \\ 0 & 0 \end{pmatrix}\begin{pmatrix} 0 & 1 \\ 0 & 0 \end{pmatrix}$,

(c) $\begin{pmatrix} 1 & 2 \\ 3 & 4 \end{pmatrix}\begin{pmatrix} 5 & 6 \\ 7 & 8 \end{pmatrix}$.

Exercise 5. Let U be the linear transformation having matrix $\begin{pmatrix} 0 & 1 \\ 1 & 0 \end{pmatrix}$.

(a) Interpret U geometrically.
(b) Show that $UU = I$.

Exercise 6. Let V be the linear transformation having matrix $\begin{pmatrix} 0 & -1 \\ 1 & 0 \end{pmatrix}$. Show that $VV = R_\pi$, rotation by π radians.

Exercise 7. Let R_θ and R_ϕ be rotation by angles of θ radians and Q radians, respectively. Show that

$$R_\theta R_\phi = R_{\theta+\phi},$$

rotation by $\theta + \phi$ radians.

Exercise 8. Exhibit a linear transformation N such that $N \neq 0$, while $NN = 0$.

Exercise 9. Let A be the linear transformation with matrix $\begin{pmatrix} 1 & -1 \\ 2 & -2 \end{pmatrix}$.

(a) Show that if $\mathbf{X}$ is any vector which lies on the line along $\begin{pmatrix} 1 \\ 1 \end{pmatrix}$, then $A(\mathbf{X}) = 0$.

(b) Show that if $\mathbf{X}$ is any vector, then $A(\mathbf{X})$ lies on the line along $\begin{pmatrix} 1 \\ 2 \end{pmatrix}$.

(c) Find a linear transformation B with $B \neq 0$ such that $BA = 0$.

(d) Find a linear transformation C with $C \neq 0$ such that $AC = 0$.

If a, b, c are three numbers, then the *associative law* holds, i.e., $(ab)c = a(bc)$.

If A, B, C are three linear transformations, then

$$(AB)C = A(BC), \tag{5a}$$

and

$$(m(A)m(B))m(C) = m(A)(m(B)m(C)). \tag{5b}$$

Note: (5a) says that the associative law holds for multiplication of linear transformations, while (5b) says that it holds for multiplication of matrices.

Proof of (5a): Let $\mathbf{X}$ be any vector. Then

$$\begin{aligned}((AB)C)(\mathbf{X}) &= AB(C(\mathbf{X})) = A(B(C(\mathbf{X}))) \\ &= A((BC)(\mathbf{X})) = (A(BC))(\mathbf{X}).\end{aligned}$$

Hence

$$(AB)C = A(BC).$$

So (5a) holds.

Proof of (5b): By definition of multiplication of matrices, if S and T are linear transformation, then $m(S)m(T) = m(ST)$. Hence, using (5a), we get

$$\begin{aligned}(m(A)m(B))m(C) &= m(AB)m(C) \\ &= m((AB)C) = m(A(BC)) \\ &= m(A)m(BC) = m(A)(m(B)m(C)).\end{aligned}$$

Thus (5b) holds.

Note: If $\begin{pmatrix} a & b \\ c & d \end{pmatrix}$, $\begin{pmatrix} x & y \\ u & v \end{pmatrix}$, and $\begin{pmatrix} p & q \\ r & s \end{pmatrix}$ are three matrices, (5b) yields

$$\left(\begin{pmatrix} a & b \\ c & d \end{pmatrix}\begin{pmatrix} x & y \\ u & v \end{pmatrix}\right)\begin{pmatrix} p & q \\ r & s \end{pmatrix} = \begin{pmatrix} a & b \\ c & d \end{pmatrix}\left(\begin{pmatrix} x & y \\ u & v \end{pmatrix}\begin{pmatrix} p & q \\ r & s \end{pmatrix}\right).$$

We could obtain this equation directly, using formula (4), but that would require more effort.

Let A and B be two linear transformations. By the *sum of A and B*, $A + B$, we mean the transformation which assigns to each vector $\mathbf{X}$ the vector $A(\mathbf{X}) + B(\mathbf{X})$ so

$$(A + B)(\mathbf{X}) = A(\mathbf{X}) + B(\mathbf{X}), \qquad \text{for each } \mathbf{X}.$$

If the matrix $m(A) = \begin{pmatrix} a & b \\ c & d \end{pmatrix}$ and the matrix $m(B) = \begin{pmatrix} \bar{a} & \bar{b} \\ \bar{c} & \bar{d} \end{pmatrix}$, then

$$(A + B)\left[\begin{pmatrix} x \\ y \end{pmatrix}\right] = A\begin{pmatrix} x \\ y \end{pmatrix} + B\begin{pmatrix} x \\ y \end{pmatrix} = \begin{pmatrix} a & b \\ c & d \end{pmatrix}\begin{pmatrix} x \\ y \end{pmatrix} + \begin{pmatrix} \bar{a} & \bar{b} \\ \bar{c} & \bar{d} \end{pmatrix}\begin{pmatrix} x \\ y \end{pmatrix}$$

$$= \begin{pmatrix} ax + by \\ cx + dy \end{pmatrix} + \begin{pmatrix} \bar{a}x + \bar{b}y \\ \bar{c}x + \bar{d}y \end{pmatrix} = \begin{bmatrix} (a + \bar{a})x + (b + \bar{b})y \\ (c + \bar{c})x + (d + \bar{d})y \end{bmatrix}$$

$$= \begin{pmatrix} a + \bar{a} & b + \bar{b} \\ c + \bar{c} & d + \bar{d} \end{pmatrix}\begin{pmatrix} x \\ y \end{pmatrix}.$$

Thus, $A + B$ is a linear transformation and its matrix is $\begin{pmatrix} a + \bar{a} & b + \bar{b} \\ c + \bar{d} & d + \bar{d} \end{pmatrix}$. We define the *sum of the matrices* $m(A)$ and $m(B)$, denoted $m(A) + m(B)$, as $m(A + B)$. Thus

$$\begin{pmatrix} a & b \\ c & d \end{pmatrix} + \begin{pmatrix} \bar{a} & \bar{b} \\ \bar{c} & \bar{d} \end{pmatrix} = \begin{pmatrix} a + \bar{a} & b + \bar{b} \\ c + \bar{c} & d + \bar{d} \end{pmatrix}.$$

Similarly, if A is as above and t is a scalar, we denote by tA the transformation defined by

$$(tA)(\mathbf{X}) = tA(\mathbf{X}), \qquad \text{for every vector } \mathbf{X},$$

and we define

$$tm(A) = m(tA).$$

It follows that

$$t\begin{pmatrix} a & b \\ c & d \end{pmatrix} = \begin{pmatrix} ta & tb \\ tc & td \end{pmatrix}.$$

EXAMPLE 5.

$$\begin{pmatrix} 1 & 2 \\ 3 & 4 \end{pmatrix} + 2\begin{pmatrix} -1 & 0 \\ 0 & 1 \end{pmatrix} = \begin{pmatrix} 1 + 2(-1) & 2 \\ 3 & 4 + 2(1) \end{pmatrix} = \begin{pmatrix} -1 & 2 \\ 3 & 6 \end{pmatrix}.$$

As an application of the notion of the sum of two linear transformations, let us do the following example.

EXAMPLE 6. Let L be a straight line through the origin and denote by S the transformation which reflects each vector in $\mathbf{X}$. Let P denote the transformation of projection to L. If $\mathbf{X}$ is any vector, then $\frac{1}{2}(\mathbf{X} + S(\mathbf{X})) = P(\mathbf{X})$. Hence

$$\mathbf{X} + S(\mathbf{X}) = 2P(\mathbf{X}),$$

and so

$$S(\mathbf{X}) = 2P(\mathbf{X}) - \mathbf{X} = (2P - I)(\mathbf{X}).$$

Since this holds for every vector $\mathbf{X}$, we get

$$S = 2P - I. \tag{6}$$

EXAMPLE 7. Let L be a straight line through the origin and denote by θ the angle from the positive x-axis to L. Find the matrix of the transformation S which reflects each vector in L. (See Fig. 2.30.) The vector $\mathbf{U} = \begin{pmatrix} \cos\theta \\ \sin\theta \end{pmatrix}$ is a unit vector and lies on L. By (10) of Chapter 2.2, if P is the transformation which projects to L, then

$$m(P) = \begin{pmatrix} \cos^2\theta & \sin\theta\cos\theta \\ \cos\theta\sin\theta & \sin^2\theta \end{pmatrix}.$$

By Example 6, $S = 2P - I$, so

$$m(S) = 2m(P) - m(I) = \begin{pmatrix} 2\cos^2\theta & 2\cos\theta\sin\theta \\ 2\cos\theta\sin\theta & 2\sin^2\theta \end{pmatrix} - \begin{pmatrix} 1 & 0 \\ 0 & 1 \end{pmatrix}$$

$$= \begin{pmatrix} 2\cos^2\theta - 1 & 2\cos\theta\sin\theta \\ 2\cos\theta\sin\theta & 2\sin^2\theta - 1 \end{pmatrix}.$$

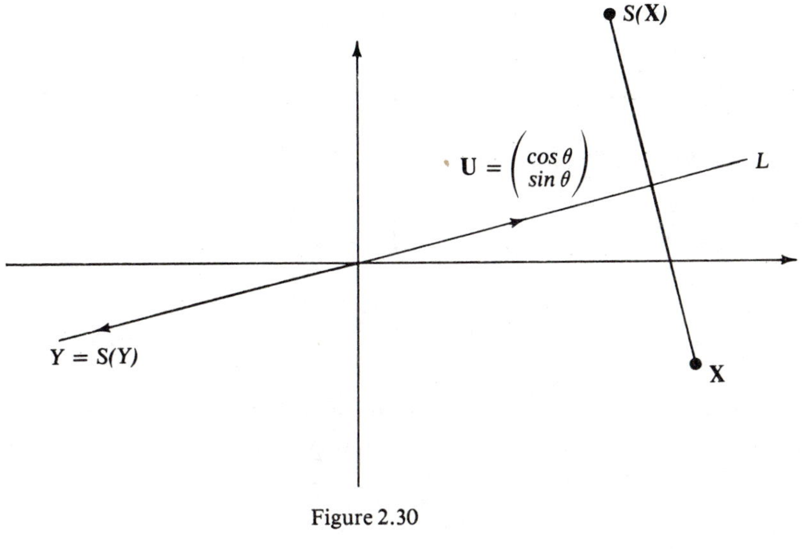

Figure 2.30

By the double-angle formulae from trigonometry, we have

$$\cos 2\theta = 2\cos^2\theta - 1, \qquad \sin 2\theta = 2\sin\theta\cos\theta.$$

So

$$m(S) = \begin{pmatrix} \cos 2\theta & \sin 2\theta \\ \sin 2\theta & -\cos 2\theta \end{pmatrix}. \tag{7}$$

Exercise 10.

(a) Using (7) show that $(m(S))^2 = m(I)$.
(b) Give a geometric explanation of the result of part (a).

Exercise 11. Using formula (7), find the matrix of the transformation which reflects each vector in the line along $\begin{pmatrix} 1 \\ 2 \end{pmatrix}$.

Exercise 12. Let H_k be the transformation with matrix $\begin{pmatrix} 1 & 0 \\ k & 1 \end{pmatrix}$. Show $\begin{pmatrix} 1 & 0 \\ k & 1 \end{pmatrix}\begin{pmatrix} 1 & 0 \\ m & 1 \end{pmatrix} = \begin{pmatrix} 1 & 0 \\ k+m & 1 \end{pmatrix}$, and conclude that $H_k H_m = H_{k+m}$.

Exercise 13. Let J_k be the transformation with matrix $\begin{pmatrix} 1 & k \\ 0 & 1 \end{pmatrix}$. Show that $J_k J_m = J_{k+m}$.

Exercise 14. Conclude $H_k J_m$ and $J_m H_k$ for a given pair of scalars k, m.

Exercise 15. Describe the images of the unit square under the transformations H_1, H_2, H_{-1}.

Exercise 16. Describe the images of the unit square under the transformations J_1, J_2, J_{-1}.

Note: The transformations H_k and J_m are called *shear transformations*. The transformation K with matrix

$$m(K) = \begin{pmatrix} 0 & 1 \\ 1 & 0 \end{pmatrix}$$

is called a *permutation* and its matrix is called a *permutation matrix*.

Exercise 17. Let $\begin{pmatrix} a & b \\ c & d \end{pmatrix}$ be a matrix. Show that

$$\begin{pmatrix} 0 & 1 \\ 1 & 0 \end{pmatrix}\begin{pmatrix} a & b \\ c & d \end{pmatrix} = \begin{pmatrix} c & d \\ a & b \end{pmatrix}$$

and

$$\begin{pmatrix} a & b \\ c & d \end{pmatrix}\begin{pmatrix} 0 & 1 \\ 1 & 0 \end{pmatrix} = \begin{pmatrix} b & a \\ d & c \end{pmatrix}.$$

Note that multiplying a matrix on the left by a permutation matrix *interchanges the rows*, while the corresponding multiplication on the right *interchanges the columns*.

Permutation matrices and shear matrices, as well as the identity matrix, are called *elementary matrices*. A matrix

$$\begin{pmatrix} d_1 & 0 \\ 0 & d_2 \end{pmatrix}$$

with entries 0 except on the diagonal is called a *diagonal matrix*.

Theorem 2.3. *Let* $\begin{pmatrix} a & b \\ c & d \end{pmatrix}$ *be an arbitrary matrix. We can find elementary matrices* e_1, e_2, e_3, e_4 *and a diagonal matrix* $\begin{pmatrix} d_1 & 0 \\ 0 & d_2 \end{pmatrix}$, *such that*

$$e_1 e_2 \begin{pmatrix} a & b \\ c & d \end{pmatrix} e_3 e_4 = \begin{pmatrix} d_1 & 0 \\ 0 & d_2 \end{pmatrix}. \tag{8}$$

PROOF. Suppose $a \neq 0$. We have

$$\begin{pmatrix} 1 & 0 \\ k & 1 \end{pmatrix}\begin{pmatrix} a & b \\ c & d \end{pmatrix} = \begin{pmatrix} a & b \\ ka + c & kb + d \end{pmatrix}.$$

Taking $k = -c/a$, we get

$$\begin{pmatrix} 1 & 0 \\ k & 1 \end{pmatrix}\begin{pmatrix} a & b \\ c & d \end{pmatrix} = \begin{pmatrix} a & b \\ 0 & x \end{pmatrix},$$

where $x = -(c/a)b + d$. Also,

$$\begin{pmatrix} a & b \\ 0 & x \end{pmatrix}\begin{pmatrix} 1 & m \\ 0 & 1 \end{pmatrix} = \begin{pmatrix} a & ma + b \\ 0 & x \end{pmatrix}.$$

Taking $m = -b/a$, we get

$$\begin{pmatrix} a & b \\ 0 & x \end{pmatrix}\begin{pmatrix} 1 & m \\ 0 & 1 \end{pmatrix} = \begin{pmatrix} a & 0 \\ 0 & x \end{pmatrix}.$$

Thus we have

$$\begin{pmatrix} 1 & 0 \\ k & 1 \end{pmatrix}\begin{pmatrix} a & b \\ c & d \end{pmatrix}\begin{pmatrix} 1 & m \\ 0 & 1 \end{pmatrix} = \begin{pmatrix} a & 0 \\ 0 & x \end{pmatrix},$$

and so (8) holds.

What if $a = 0$? Either $\begin{pmatrix} a & b \\ c & d \end{pmatrix}$ is the zero matrix, or some entry is nonzero, say $c \neq 0$. Then

$$\begin{pmatrix} 0 & 1 \\ 1 & 0 \end{pmatrix}\begin{pmatrix} a & b \\ c & d \end{pmatrix} = \begin{pmatrix} c & d \\ a & b \end{pmatrix}.$$

Since $c \neq 0$, the preceding reasoning applies to $\begin{pmatrix} c & d \\ a & b \end{pmatrix}$ and we can choose shear matrices e_1 and e_3 such that

$$e_1 \begin{pmatrix} c & d \\ a & b \end{pmatrix} e_3 = e_1 e_2 \begin{pmatrix} a & b \\ c & d \end{pmatrix} e_3$$

is a diagonal matrix, where $e_2 = \begin{pmatrix} 0 & 1 \\ 1 & 0 \end{pmatrix}$. So again (8) holds. If $b \neq 0$ or $d \neq 0$, we proceed in a similar way to obtain (8).

EXAMPLE 8. Let us find formula (8) for the matrix $\begin{pmatrix} 0 & 5 \\ 2 & 3 \end{pmatrix}$.

$$\begin{pmatrix} 0 & 1 \\ 1 & 0 \end{pmatrix}\begin{pmatrix} 0 & 5 \\ 2 & 3 \end{pmatrix} = \begin{pmatrix} 2 & 3 \\ 0 & 5 \end{pmatrix},$$

$$\begin{pmatrix} 0 & 1 \\ 1 & 0 \end{pmatrix}\begin{pmatrix} 0 & 5 \\ 2 & 3 \end{pmatrix}\begin{pmatrix} 1 & -\frac{3}{2} \\ 0 & 1 \end{pmatrix} = \begin{pmatrix} 2 & 3 \\ 0 & 5 \end{pmatrix}\begin{pmatrix} 1 & -\frac{3}{2} \\ 0 & 1 \end{pmatrix} = \begin{pmatrix} 2 & 0 \\ 0 & 5 \end{pmatrix}.$$

Thus $\begin{pmatrix} 2 & 0 \\ 0 & 5 \end{pmatrix}$ is the diagonal matrix of formula (8) here.

Exercise 18. Find elementary matrices e_1 and e_3 and a diagonal matrix $\begin{pmatrix} d_1 & 0 \\ 0 & d_2 \end{pmatrix}$ such that

$$e_1\begin{pmatrix} 2 & 2 \\ 3 & 3 \end{pmatrix}e_2 = \begin{pmatrix} d_1 & 0 \\ 0 & d_2 \end{pmatrix}.$$

CHAPTER 2.4

Inverses and Systems of Equations

§1 Inverses

If a, x, y are numbers, then

$$a(x+y) = ax + ay.$$

If A is a linear transformation and **X** and **Y** are vectors, then by Theorem 2.1 of Chapter 2.2,

$$A(\mathbf{X}+\mathbf{Y}) = A(\mathbf{X}) + A(\mathbf{Y}).$$

Thus we see that the operation which takes a number x into the number ax is somehow similar to the operation which takes a vector **X** into the vector $A(\mathbf{X})$, where A is a linear transformation.

Next, consider the equation:

$$ax = y \tag{1}$$

where a and y are given numbers, $a \neq 0$, and x is an unknown number. We solve (1) by taking the *reciprocal* $1/a$ of a, and multiplying both sides by it, arriving at

$$\frac{1}{a}(ax) = \frac{1}{a}y, \quad \text{and so} \quad x = \frac{1}{a}y.$$

As an analogue of equation (1) for vectors, we may consider a linear transformation A and a vector **Y** and look for a vector **X** such that

$$A(\mathbf{X}) = \mathbf{Y}. \tag{2}$$

To solve (2) we should like to have an analogue of the reciprocal for the transformation A. Now the reciprocal $1/a$ satisfies

$$\frac{1}{a}\cdot a = 1 \quad \text{and} \quad a\cdot\frac{1}{a} = 1.$$

A reasonable analogue would be a linear transformation B such that

$$BA = I \quad \text{and} \quad AB = I. \tag{3}$$

Suppose we have found such a B. Then we can solve Eq. (2) by applying B to both sides. This gives

$$B(A(\mathbf{X})) = B(\mathbf{Y}).$$

But

$$B(A(\mathbf{X})) = (BA)(\mathbf{X}) = I(\mathbf{X}) = \mathbf{X},$$

so we get

$$\mathbf{X} = B(\mathbf{Y}). \tag{4}$$

We can verify that (4) really gives a solution to (2) by applying A to both sides of (4). This gives

$$A(\mathbf{X}) = A(B(\mathbf{Y})) = (AB)(\mathbf{Y}) = I(\mathbf{Y}) = \mathbf{Y},$$

and so (2) is valid.

The problem of solving Eq. (2) will thus be resolved, provided we can find a linear transformation B satisfying $BA = I$ and $AB = I$. Such a linear transformation B is called an *inverse* of A.

Note that there is exactly one number which fails to have a reciprocal, namely the number 0. It turns out that there are many linear transformations which have no inverse, and later on in this chapter we shall see how we can decide whether or not a given linear transformation has an inverse.

Let A be a linear transformation. If b is an inverse of A, then B *undoes* the effect of A on a vector in the following sense: if A sends the vector $\mathbf{X}$ to the vector $\mathbf{Y}$, then B sends the vector $\mathbf{Y}$ to the vector $\mathbf{X}$.

To see that this is so, consider a vector $\mathbf{X}$. Define $\mathbf{Y} = A(\mathbf{X})$. By (3),

$$(BA)(\mathbf{X}) = I(\mathbf{X}) \quad \text{or} \quad B(A(\mathbf{X})) = \mathbf{X}.$$

So $B(\mathbf{Y}) = \mathbf{X}$, as we have claimed. (See Fig. 2.31.)

Example 1. Fix $r \neq 0$. Find the inverse of D_r, i.e., of stretching by r.

D_r takes the vector $\mathbf{X}$ into the vector $r\mathbf{X}$. To undo this, we must multiply by the scalar $1/r$. Thus, we set $B = D_{1/r}$. Then if $\mathbf{X}$ is any vector,

$$(BD_r)(\mathbf{X}) = B(D_r(\mathbf{X})) = B(r\mathbf{X}) = \frac{1}{r}(r\mathbf{X}) = \mathbf{X}.$$

Hence, $BD_r = I$. Also,

$$(D_rB)(\mathbf{X}) = D_r(B(\mathbf{X})) = D_r\left(\frac{1}{r}\mathbf{X}\right) = r\left(\frac{1}{r}\mathbf{X}\right) = \mathbf{X},$$

Thus, B satisfies (3) and so $D_{1/r} = B$ is an inverse of D.

Example 2. R_θ denotes rotation by θ radians. Find the inverse of $R_{\pi/2}$.

Let $\mathbf{X}$ be a vector. $R_{\pi/2}$ rotates $\mathbf{X}$ by $\pi/2$ radians counterclockwise around $\mathbf{0}$. To undo the effect of $R_{\pi/2}$, we can rotate by $-\pi/2$ radians.

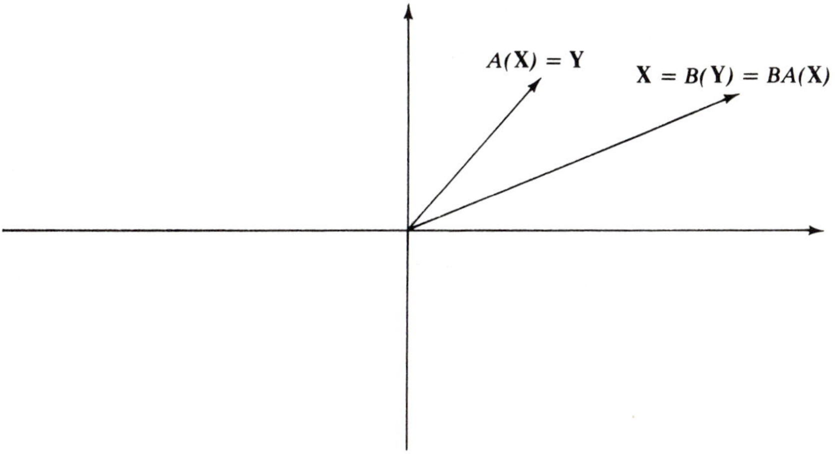

Figure 2.31

Thus, we set $B = R_{-\pi/2}$. If one prefers, we can write $B = R_{3\pi/2}$, since rotation by $-\pi/2$ radians and rotation by $3\pi/2$ radians have the same effect on each vector, and so $R_{-\pi/2} = R_{3\pi/2}$. Then, if $\mathbf{X}$ is a vector,

$$(BR_{\pi/2})(\mathbf{X}) = B(R_{\pi/2}(\mathbf{X})) = R_{-\pi/2}(R_{\pi/2}(\mathbf{X})) = \mathbf{X}.$$

Thus, $BR_{\pi/2} = I$. Also,

$$(R_{\pi/2}B)(\mathbf{X}) = R_{\pi/2}(B(\mathbf{X})) = R_{\pi/2}(R_{-\pi/2}(\mathbf{X})) = \mathbf{X}.$$

Thus, $R_{\pi/2}B = I$. So B satisfies (3), and $R_{-\pi/2} = B$ is an inverse of $R_{\pi/2}$.

Exercise 1. Find the inverse of $R_{3\pi/4}$.

EXAMPLE 3. Let L be a straight line through the origin. Let S be reflection in the line L. Find an inverse to S. (See Fig. 2.32.)

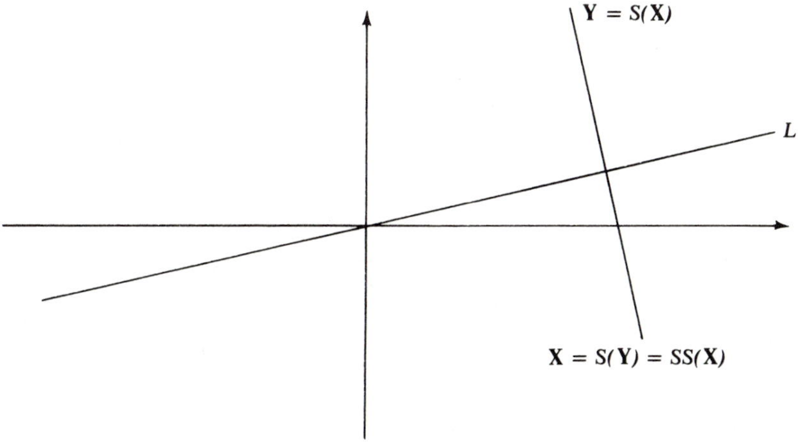

Figure 2.32

If we start with a vector $\mathbf{X}$, then reflect $\mathbf{X}$ in L and then reflect in L again, and we return to $\mathbf{X}$. In other words,

$$S(S(\mathbf{X})) = \mathbf{X} \quad \text{or} \quad (SS)(\mathbf{X}) = \mathbf{X}.$$

Thus, $SS = I$. Hence, if we take $A = S$ and $B = S$, then (3) is satisfied. So S is an inverse of itself.

Note: For numbers, the analogous situation occurs when a number a is its own reciprocal, as in case $a = 1$ or $a = -1$.

Now let A be an arbitrary linear transformation. Can A have *more than one* inverse? Assume that B and C are two linear transformations each of which satisfies (3), i.e., assume

$$AB = I \quad \text{and} \quad BA = I \tag{5}$$

and, also,

$$AC = I \quad \text{and} \quad CA = I. \tag{6}$$

By (5), $BA = I$.
Hence, $(BA)C = IC = C$.
By the associative property, $(BA)C = B(AC)$. So

$$B(AC) = C.$$

By (6), $AC = I$, so $B(AC) = BI = B$. Hence,

$$B = C.$$

We have seen, then, that if B and C each is an inverse of A, then $B = C$. In other words, A can have only *one* inverse. Thus we can speak of *the* inverse of A, and we denote this inverse, provided it exists, by A^{-1}. Thus, $A \cdot A^{-1} = I$ and $A^{-1} \cdot A = I$. Examples 1, 2, and 3 can then be expressed as follows:

$$(D_r)^{-1} = D_{1/r},$$

$$(R_{\pi/2})^{-1} = R_{-\pi/2},$$

$$S^{-1} = S.$$

Exercise 2. Let T be the transformation with matrix $\begin{pmatrix} 2 & 0 \\ 0 & 5 \end{pmatrix}$, so that $T\begin{pmatrix} x \\ y \end{pmatrix} = \begin{pmatrix} 2x \\ 5y \end{pmatrix}$ for every vector $\begin{pmatrix} x \\ y \end{pmatrix}$. Find the matrix of T^{-1}.

Exercise 3. Let T be the transformation with matrix $\begin{pmatrix} 2 & 1 \\ 0 & 5 \end{pmatrix}$. Find the matrix of T^{-1}.

EXAMPLE 4. Let P be projection on the x-axis. Suppose B is an inverse of P. Then, $BP = I$, so $(BP)(\mathbf{X}) = \mathbf{X}$ for every vector $\mathbf{X}$.

Choose a vector $\mathbf{X} = \begin{pmatrix} 0 \\ y \end{pmatrix}$ with $y \neq 0$. Then $P(\mathbf{X}) = \begin{pmatrix} 0 \\ 0 \end{pmatrix}$, and so

$$(BP)(\mathbf{X}) = B(P(\mathbf{X})) = B\begin{pmatrix} 0 \\ 0 \end{pmatrix} = \mathbf{0}.$$

Hence

$$\mathbf{X} = (BP)(\mathbf{X}) = \mathbf{0}.$$

But $\mathbf{X}$ was not the zero vector, and so we have reached a contradiction. From this we are forced to conclude that there is no linear transformation B satisfying $BP = I$. So P has no inverse.

Next we observe that if A is a linear transformation which has an inverse A^{-1}, then A satisfies the following condition:

$$\text{The only vector } \mathbf{X} \text{ with } A(\mathbf{X}) = \mathbf{0} \text{ is the vector } \mathbf{X} = \mathbf{0}. \tag{7}$$

To see this, choose $\mathbf{X}$ with $A(\mathbf{X}) = \mathbf{0}$. Then $A^{-1}(A(\mathbf{X})) = A^{-1}(\mathbf{0}) = \mathbf{0}$, and also $A^{-1}(A(\mathbf{X})) = (A^{-1}A)(\mathbf{X}) = I(\mathbf{X}) = \mathbf{X}$. So $\mathbf{X} = \mathbf{0}$, and so (7) is true.

Now let A be a linear transformation with matrix $\begin{pmatrix} a & b \\ c & d \end{pmatrix}$. We shall prove:

Proposition 1. *Condition* (7) *holds if and only if* $ad - bc \neq 0$.

Proof. Suppose $ad - bc = 0$. Then we have

$$A\begin{pmatrix} -b \\ a \end{pmatrix} = \begin{pmatrix} a & b \\ c & d \end{pmatrix}\begin{pmatrix} -b \\ a \end{pmatrix} = \begin{pmatrix} 0 \\ -cb + ad \end{pmatrix} = \begin{pmatrix} 0 \\ 0 \end{pmatrix},$$

$$A\begin{pmatrix} d \\ -c \end{pmatrix} = \begin{pmatrix} a & b \\ c & d \end{pmatrix}\begin{pmatrix} d \\ -c \end{pmatrix} = \begin{pmatrix} ad - bc \\ 0 \end{pmatrix} = \begin{pmatrix} 0 \\ 0 \end{pmatrix}.$$

If (7) holds, it follows that $\begin{pmatrix} -b \\ a \end{pmatrix} = \begin{pmatrix} 0 \\ 0 \end{pmatrix}$ and $\begin{pmatrix} d \\ -c \end{pmatrix} = \begin{pmatrix} 0 \\ 0 \end{pmatrix}$. Hence a, b, c, d are all zero, so $A(\mathbf{X}) = \mathbf{0}$ for every $\mathbf{X}$. But then (7) is false, so we have a contradiction. Hence if $ad - bc = 0$, then (7) does not hold.

Conversely, suppose $ad - bc \neq 0$. Let $\mathbf{X} = \begin{pmatrix} x \\ y \end{pmatrix}$ be a vector with $A(\mathbf{X}) = \mathbf{0}$. Then $\begin{pmatrix} 0 \\ 0 \end{pmatrix} = A(\mathbf{X}) = \begin{pmatrix} a & b \\ c & d \end{pmatrix}\begin{pmatrix} x \\ y \end{pmatrix} = \begin{pmatrix} ax + by \\ cx + dy \end{pmatrix}$. So

$$ax + by = 0,$$

$$cx + dy - 0.$$

Multiplying the first equation by d and the second by b and subtracting, we get

$$(ad - bc)x = 0,$$

and hence $x = 0$. Similarly, we get $y = 0$. Hence $\mathbf{X} = \begin{pmatrix} x \\ y \end{pmatrix} = \begin{pmatrix} 0 \\ 0 \end{pmatrix}$. Thus

$\mathbf{X} = \mathbf{0}$ is the only vector with $A(\mathbf{X}) = \mathbf{0}$, so (7) holds. The proposition is proved.

We saw earlier that if A has an inverse, then (7) holds and so $ad - bc \neq 0$. Let us now proceed in the converse direction.

Consider a linear transformation A with matrix $\begin{pmatrix} a & b \\ c & d \end{pmatrix}$. Assume

$$ad - bc \neq 0. \tag{8}$$

We seek an inverse B for A. Set $m(B) = \begin{pmatrix} p & q \\ r & s \end{pmatrix}$, where p, q, r, s are unknown numbers. We must have

$$\begin{pmatrix} a & b \\ c & d \end{pmatrix}\begin{pmatrix} p & q \\ r & s \end{pmatrix} = \begin{pmatrix} 1 & 0 \\ 0 & 1 \end{pmatrix},$$

so

$$\begin{aligned} ap + br &= 1, \\ cp + dr &= 0, \end{aligned} \tag{9}$$

and

$$\begin{aligned} aq + bs &= 0, \\ cq + ds &= 1. \end{aligned}$$

Hence

$$\begin{aligned} dap + dbr &= d, \\ bcp + bdr &= 0, \end{aligned}$$

and so

$$(ad - bc)p = d,$$

and, since (8) holds, we get

$$p = \frac{d}{ad - bc}.$$

To simplify the notation, we set $\Delta = ad - bc$.

Exercise 4. Using the system (9), show that

$$r = \frac{-c}{\Delta}. \tag{10}$$

Exercise 5. Using the relations

$$\begin{aligned} aq + bs &= 0, \\ cq + ds &= 1, \end{aligned}$$

show that

$$q = \frac{-b}{\Delta} \quad \text{and} \quad s = \frac{a}{\Delta}. \tag{11}$$

We have obtained

$$\begin{pmatrix} p & q \\ r & s \end{pmatrix} = \begin{pmatrix} \dfrac{d}{\Delta} & \dfrac{-b}{\Delta} \\ \dfrac{-c}{\Delta} & \dfrac{a}{\Delta} \end{pmatrix}.$$

Exercise 6. Calculate

$$\begin{pmatrix} \dfrac{d}{\Delta} & \dfrac{-b}{\Delta} \\ \dfrac{-c}{\Delta} & \dfrac{a}{\Delta} \end{pmatrix}\begin{pmatrix} a & b \\ c & d \end{pmatrix} \quad \text{and} \quad \begin{pmatrix} a & b \\ c & d \end{pmatrix}\begin{pmatrix} \dfrac{d}{\Delta} & \dfrac{-b}{\Delta} \\ \dfrac{-c}{\Delta} & \dfrac{a}{\Delta} \end{pmatrix}.$$

Show that both products equal $\begin{pmatrix} 1 & 0 \\ 0 & 1 \end{pmatrix}$.

Earlier we saw that if A has an inverse, then $ad - bc \neq 0$. Combining this with Exercise 6, we have

Theorem 2.4. *Let A be a linear transformation with matrix $\begin{pmatrix} a & b \\ c & d \end{pmatrix}$.*

(i) *If A has an inverse, then $ad - bc \neq 0$;*
(ii) *If $ad - bc \neq 0$, then A has an inverse B and*

$$m(B) = \begin{pmatrix} \dfrac{d}{\Delta} & \dfrac{-b}{\Delta} \\ \dfrac{-c}{\Delta} & \dfrac{a}{\Delta} \end{pmatrix}, \tag{12}$$

where Δ denotes $ad - bc$.

Example 5. Let A have the matrix $\begin{pmatrix} 1 & 2 \\ 3 & 4 \end{pmatrix}$. Since $1 \cdot 4 - 2 \cdot 3 = -2 \neq 0$, A has an inverse A^{-1}. The matrix of A^{-1} is

$$\begin{pmatrix} \dfrac{4}{-2} & \dfrac{-2}{-2} \\ \dfrac{-3}{-2} & \dfrac{1}{-2} \end{pmatrix} = \begin{pmatrix} -2 & 1 \\ \dfrac{3}{2} & -\dfrac{1}{2} \end{pmatrix}.$$

Example 6. Solve the system

$$\begin{aligned} x + 2y &= 4, \\ 3x + 4y &= 0 \end{aligned} \tag{13}$$

for x and y.

We write the system in the form

$$A\begin{pmatrix} x \\ y \end{pmatrix} = \begin{pmatrix} 1 & 2 \\ 3 & 4 \end{pmatrix}\begin{pmatrix} x \\ y \end{pmatrix} = \begin{pmatrix} 4 \\ 0 \end{pmatrix}.$$

By the preceding example, A^{-1} has the matrix $\begin{pmatrix} -2 & 1 \\ 3/2 & -1/2 \end{pmatrix}$. Hence,

$$\begin{pmatrix} x \\ y \end{pmatrix} = A^{-1}\begin{pmatrix} 4 \\ 0 \end{pmatrix} = \begin{pmatrix} -2 & 1 \\ 3/2 & -1/2 \end{pmatrix}\begin{pmatrix} 4 \\ 0 \end{pmatrix} = \begin{pmatrix} -8 \\ 6 \end{pmatrix}.$$

Hence, the solution is

$$x = -8, \qquad y = 6.$$

§2. Systems of Linear Equations

We consider the following system of two equations in two unknowns:

$$\begin{aligned} ax + by &= u, \\ cx + dy &= v. \end{aligned} \tag{14}$$

For each choice of numbers u, v, we may ask: Does the system (14) have a solution x, y? And if (14) has a solution, is this solution unique?

We may write the above system in matrix form by introducing the linear transformation A with matrix $\begin{pmatrix} a & b \\ c & d \end{pmatrix}$. Then the system (14) may be written

$$A(\mathbf{X}) = \mathbf{U}, \tag{15}$$

where $\mathbf{X}$ is the vector $\begin{pmatrix} x \\ y \end{pmatrix}$ and $\mathbf{U} = \begin{pmatrix} u \\ v \end{pmatrix}$.

Suppose that the transformation A has an inverse A^{-1}. For given vector $\mathbf{U}$,

$$A\big(A^{-1}(\mathbf{U})\big) = (AA^{-1})(\mathbf{U}) = \mathbf{U},$$

so $\mathbf{X} = A^{-1}(\mathbf{U})$ is a solution of (15). Conversely, if $\mathbf{X}$ is a solution of (15), then

$$\mathbf{X} = A^{-1}(A(\mathbf{X})) = A^{-1}(\mathbf{U}).$$

So (15) has the unique solution $\mathbf{X} = A^{-1}(\mathbf{U})$.

In particular, if $\mathbf{U} = 0$, we find that $\mathbf{X} = \begin{pmatrix} 0 \\ 0 \end{pmatrix} = A^{-1}\begin{pmatrix} 0 \\ 0 \end{pmatrix}$ is the unique solution of the system

$$\begin{aligned} ax + by &= 0, \\ cx + dy &= 0. \end{aligned} \tag{16}$$

This system, with zero on the right-hand side, is called the *homogeneous system* associated with the system (14).

No matter what the matrix $\begin{pmatrix} a & b \\ c & d \end{pmatrix}$ is, the homogeneous system has at least one solution, the solution $\mathbf{X} = \begin{pmatrix} 0 \\ 0 \end{pmatrix}$. This is called the *trivial solution* of the homogeneous system, and we have seen above that if A has an inverse, then the trivial solution is the only solution of the homogeneous system. If A does not have an inverse, then by Theorem 2.4, $ad - bc = 0$, and so by Proposition 1 in §1 of this chapter, there is a nonzero vector $\mathbf{X} = \binom{x}{y}$ with $A(\mathbf{X}) = \mathbf{0}$. Then x, y is a nontrivial solution of the homogeneous system (16).

What is the totality of solutions of (16)? If a, b, c, d are all 0, then every vector $\binom{x}{y}$ in the plane is a solution. If a and b are both 0, but c and d are not both 0, then the solutions are all $\binom{x}{y}$ with $cx + dy = 0$, and so the totality of solutions is the line $cx + dy = 0$. A similar statement holds if c and d are both 0, but a and b are not both 0.

Finally, if A does not have an inverse and either $a \neq 0$ or $b \neq 0$ and, also, either $c \neq 0$ or $d \neq 0$, we may conclude that the totality of solutions of (16) is the line through the origin orthogonal to $\binom{a}{b}$.

We can summarize what we have found so far in the following two propositions.

Proposition 2. *The system* (14) *has a unique solution for every vector* $\binom{u}{v}$ *if and only if the transformation A has an inverse.*

Proposition 3. *The homogeneous system* (16) *has a nontrivial solution if and only if A fails to have an inverse. In this case the totality of solutions of* (16) *is either the whole plane or a line through the origin.*

Now suppose that A fails to have an inverse, and $A \neq 0$. Then the solutions of (16) form a line through the origin, or, in other words, if we fix one nonzero solution $\mathbf{X}^h$ of (16), then every solution of (16) equals $t\mathbf{X}^h$ for some scalar t. If $\mathbf{X}$ and $\overline{\mathbf{X}}$ are two solutions of the system $A(\mathbf{X}) = \mathbf{U}$, then

$$A(\mathbf{X} - \overline{\mathbf{X}}) = A(\mathbf{X}) - A(\overline{\mathbf{X}}) = \mathbf{U} - \mathbf{U} = \mathbf{0},$$

so $\mathbf{X} - \overline{\mathbf{X}}$ is a solution of (3). Hence $\mathbf{X} - \overline{\mathbf{X}} = t\mathbf{X}^h$ and so for some t,

$$\mathbf{X} = \overline{\mathbf{X}} + t\mathbf{X}^h.$$

We can therefore describe all solutions of the nonhomogeneous system (15) in the following way:

Proposition 4. *Assume A is not the zero transformation. If A does not have an inverse, then if $\overline{\mathbf{X}}$ is a particular solution of* (15), *so that $A(\overline{\mathbf{X}}) = \mathbf{U}$, we may express every solution of* (15) *in the form $\overline{\mathbf{X}} + t\mathbf{X}^h$, where $\mathbf{X}^h$ is a non-trivial solution of the homogeneous system* (16).

EXAMPLE 7. Find all solutions of the system

$$\begin{aligned} x + 2y &= 3, \\ -2x - 4y &= -6. \end{aligned} \tag{17}$$

The corresponding homogeneous system is

$$\begin{aligned} x + 2y &= 0, \\ -2x - 4y &= 0. \end{aligned}$$

and the solutions to this system are the multiples $t\begin{pmatrix} -2 \\ 1 \end{pmatrix}$ of a vector $\mathbf{X}^h$ perpendicular to $\begin{pmatrix} 1 \\ 2 \end{pmatrix}$ and $\begin{pmatrix} -2 \\ -4 \end{pmatrix}$. We observe that $\begin{pmatrix} x \\ y \end{pmatrix} = \begin{pmatrix} 1 \\ 1 \end{pmatrix}$ is a particular solution of the system (17), so it follows, by the above proposition, that the set of all solutions is given by

$$\mathbf{X} = \begin{pmatrix} 1 \\ 1 \end{pmatrix} + t\begin{pmatrix} -2 \\ 1 \end{pmatrix} = \begin{pmatrix} 1 - 2t \\ 1 + t \end{pmatrix}.$$

EXAMPLE 8. Find all solutions of the system

$$\begin{aligned} x + 2y &= 3, \\ -2x + 4y &= -6. \end{aligned} \tag{18}$$

In this case the matrix $\begin{pmatrix} 1 & 2 \\ -2 & 4 \end{pmatrix}$ has an inverse, $\frac{1}{8}\begin{pmatrix} 4 & -2 \\ 2 & 1 \end{pmatrix}$, so the unique solution to the system (18) is given by

$$\frac{1}{8}\begin{pmatrix} 4 & -2 \\ 2 & 1 \end{pmatrix}\begin{pmatrix} 3 \\ -6 \end{pmatrix} = \begin{pmatrix} 3 \\ 0 \end{pmatrix}.$$

EXAMPLE 9. Find all solutions of the system

$$\begin{aligned} x + 2y &= 3, \\ -2x - 4y &= 5. \end{aligned} \tag{19}$$

In this case the system (19) has no solution. If we had a solution to the first equation, we could multiply both sides of the equation by -2, to get

$$-2x - 4y = -6,$$

and this is inconsistent with the second equation,

$$-2x - 4y = 5.$$

More generally, we can get a solution of the system

$$\begin{aligned} x + 2y &= u, \\ -2x - 4y &= v \end{aligned}$$

if and only if

$$-2x - 4y = -2u$$

and

$$-2x - 4y = v$$

are consistent, i.e., if $-2u = v$. For example, in the system (17), we have $u = 3$, $v = -6$.

Exercise 7. Find all solutions of the following systems.

(a) $2x + y = 0,$
$3x - y = 0.$

(b) $2x + y = 0,$
$-4x - 2y = 0.$

Exercise 8. Find all solutions of the following systems.

(a) $2x + y = 1,$
$3x - y = 1.$

(b) $2x + y = 1,$
$-4x - 2y = 1.$

Exercise 9. Find all solutions of the system

$$2x + y = 1,$$
$$-4x - 2y = -2.$$

Exercise 10. Find all solutions of the system

$$x + y = 10,$$
$$5x + 5y = 50.$$

Exercise 11. For what choices of the numbers u, v does the system

$$x + y = u,$$
$$5x + 5y = v$$

have a solution?

§3. Inverses of Shears and Permutations

Recall the elementary matrices, H_k, J_k, and K which we discussed at the end of Chapter 2.3. We had

$$m(H_k) = \begin{pmatrix} 1 & 0 \\ k & 1 \end{pmatrix}, \qquad m(J_k) = \begin{pmatrix} 1 & k \\ 0 & 1 \end{pmatrix}, \qquad m(K) = \begin{pmatrix} 0 & 1 \\ 1 & 0 \end{pmatrix}.$$

Exercise 12. Show that

$$H_k^{-1} = H_{-k}, \qquad J_k^{-1} = J_{-k}, \qquad K^{-1} = K.$$

CHAPTER 2.5

Determinants

Let A be a linear transformation with matrix $\begin{pmatrix} a & b \\ c & d \end{pmatrix}$. The quantity

$$ad - bc$$

is called the *determinant* of the matrix $\begin{pmatrix} a & b \\ c & d \end{pmatrix}$ and is denoted

$$\begin{vmatrix} a & b \\ c & d \end{vmatrix}. \tag{1}$$

Expressed in these terms, Theorem 2.4 states that A has an inverse if and only if $\begin{vmatrix} a & b \\ c & d \end{vmatrix} \neq 0$. We shall see that the determinant gives us further information about the behavior of A.

Consider a pair of vectors $\mathbf{X}_1, \mathbf{X}_2$ regarded as an *ordered* pair with $\mathbf{X}_1$ first and $\mathbf{X}_2$ second. Denote by α the angle from $\mathbf{X}_1$ to $\mathbf{X}_2$, measured counter-clockwise, and assume that $\alpha \neq 0$ and $\alpha \neq \pi$.

If $\sin\alpha > 0$, we say that the pair $\mathbf{X}_1, \mathbf{X}_2$ is *positively oriented*. This holds exactly when α lies between 0 and π (see Fig 2.33). If $\sin\alpha < 0$, we say the pair $\mathbf{X}_1, \mathbf{X}_2$ is *negatively oriented*. This holds if α is between π and 2π (see Fig. 2.34).

Example 1. The pair $\mathbf{E}_1, \mathbf{E}_2$ is positively oriented (see Fig. 2.35). The pair $\mathbf{E}_1, -\mathbf{E}_2$ is negatively oriented (see Fig. 2.36). The pair $\mathbf{E}_2, \mathbf{E}_1$ is negatively oriented (see Fig. 2.37). The pair $\begin{pmatrix} 1 \\ 2 \end{pmatrix}, \begin{pmatrix} -1 \\ 2 \end{pmatrix}$ is positively oriented (see Fig. 2.38).

We saw in (20), Chapter 2.0, that if $\begin{pmatrix} x \\ y \end{pmatrix}$ and $\begin{pmatrix} u \\ v \end{pmatrix}$ are two vectors and if α

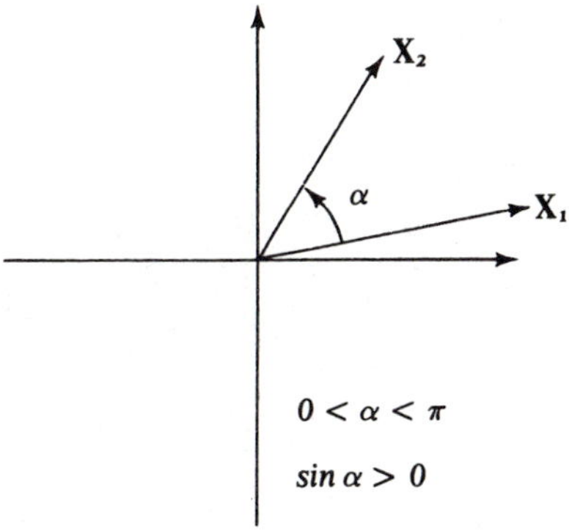

Figure 2.33

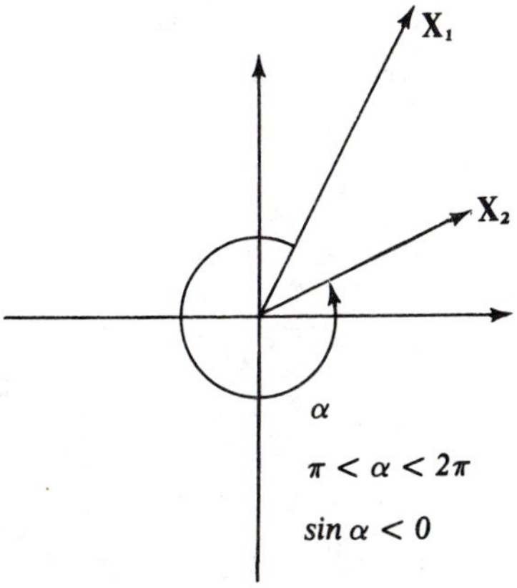

Figure 2.34

is the angle from $\begin{pmatrix} x \\ y \end{pmatrix}$ to $\binom{u}{v}$, then

$$\sin\alpha = \frac{xv - yu}{\sqrt{x^2 + y^2}\,\sqrt{u^2 + v^2}}.$$

Now let $\mathbf{X}_1 = \begin{pmatrix} x_1 \\ y_1 \end{pmatrix}$, $\mathbf{X}_2 = \begin{pmatrix} x_2 \\ y_2 \end{pmatrix}$ be a given pair of vectors. How can we tell from the numbers x_1, y_1, x_2, y_2 whether or not the pair $\mathbf{X}_1, \mathbf{X}_2$ is positively oriented? Let α denote the angle from $\mathbf{X}_1$ to $\mathbf{X}_2$, measured counterclockwise. By the preceding,

$$\sin\alpha = \frac{x_1 y_2 - y_1 x_2}{\sqrt{x_1^2 + y_1^2}\,\sqrt{x_2^2 + y_2^2}}. \tag{2}$$

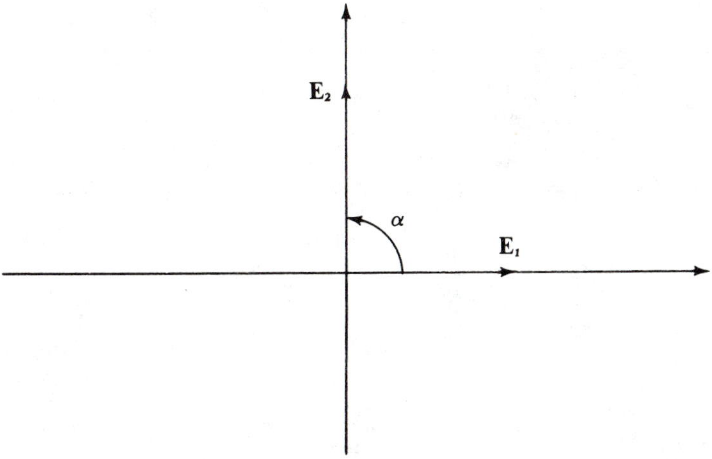

Figure 2.35

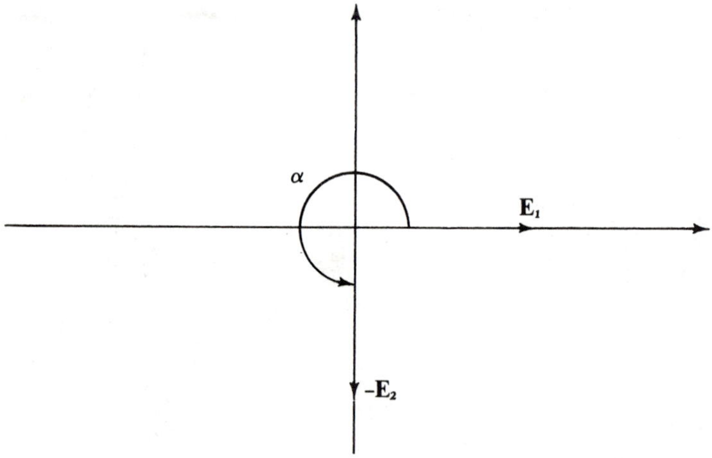

Figure 2.36

Hence, $\sin\alpha > 0$ if and only if $x_1y_2 - y_1x_2 > 0$. But $x_1y_2 - y_1x_2 = \begin{vmatrix} x_1 & x_2 \\ y_1 & y_2 \end{vmatrix}$. So, we conclude:

The pair $\mathbf{X}_1, \mathbf{X}_2$ is positively oriented if

and only if the determinant $\begin{vmatrix} x_1 & x_2 \\ y_1 & y_2 \end{vmatrix} > 0$. (3)

Next let A be a linear transformation which has an inverse. We say that *A preserves orientation* if whenever $\mathbf{X}_1, \mathbf{X}_2$ is a positively oriented pair of

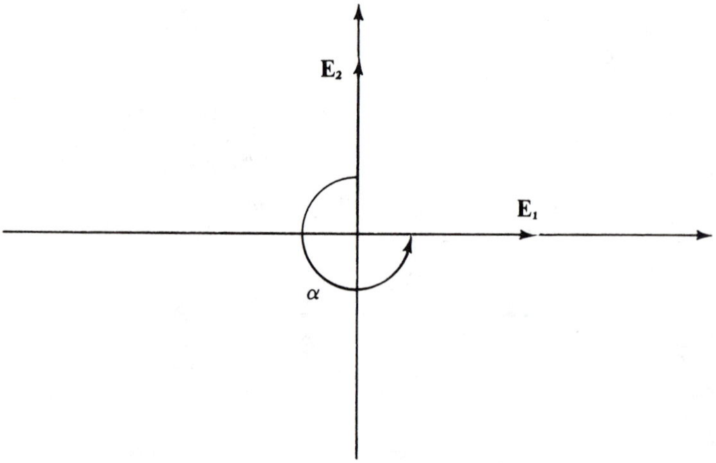

Figure 2.37

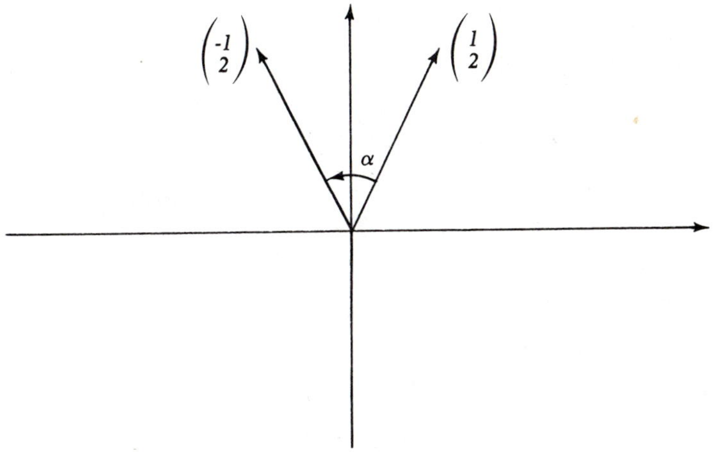

Figure 2.38

vectors, then the pair $A(\mathbf{X}_1), A(\mathbf{X}_2)$ of image-vectors is again positively oriented.

EXAMPLE 2.

(a) Rotation by $R_{\pi/2}$ preserves orientation;
(b) D_3, stretching by 3, preserves orientation;
(c) reflection in the x-axis does not preserve orientation.

Let A be a linear transformation which preserves orientation and let $\begin{pmatrix} a & b \\ c & d \end{pmatrix}$ be its matrix. Set $\mathbf{E}_1 = \begin{pmatrix} 1 \\ 0 \end{pmatrix}$, $\mathbf{E}_2 = \begin{pmatrix} 0 \\ 1 \end{pmatrix}$. The pair $\mathbf{E}_1, \mathbf{E}_2$ is positively oriented. Hence, the pair $A(\mathbf{E}_1), A(\mathbf{E}_2)$ is positively oriented. $A(\mathbf{E}_1) = \begin{pmatrix} a \\ c \end{pmatrix}$, $A(\mathbf{E}_2) = \begin{pmatrix} b \\ d \end{pmatrix}$. So by (3), we have

$$\begin{vmatrix} a & b \\ c & d \end{vmatrix} > 0.$$

Thus, if A preserves orientation, then the determinant is positive. Conversely, suppose $\begin{pmatrix} a & b \\ c & d \end{pmatrix} > 0$ and let us see whether it follows that A preserves orientation. Let $\mathbf{X}_1 = \begin{pmatrix} x_1 \\ y_1 \end{pmatrix}$, $\mathbf{X}_2 = \begin{pmatrix} x_2 \\ y_2 \end{pmatrix}$ be a positively oriented pair of vectors. Then

$$A(\mathbf{X}_1) = \begin{pmatrix} a & b \\ c & d \end{pmatrix}\begin{pmatrix} x_1 \\ y_1 \end{pmatrix} = \begin{pmatrix} ax_1 + by_1 \\ cx_1 + dy_1 \end{pmatrix}$$

and

$$A(\mathbf{X}_2) = \begin{pmatrix} a & b \\ c & d \end{pmatrix}\begin{pmatrix} x_2 \\ y_2 \end{pmatrix} = \begin{pmatrix} ax_2 + by_2 \\ cx_2 + dy_2 \end{pmatrix}.$$

The pair $A(\mathbf{X}_1), A(\mathbf{X}_2)$ is positively oriented, by (3), if and only if the determinant

$$\begin{vmatrix} ax_1 + by_1 & ax_2 + by_2 \\ cx_1 + dy_1 & cx_2 + dy_2 \end{vmatrix} > 0.$$

This determinant equals

$$\begin{aligned} &(ax_1 + by_1)(cx_2 + dy_2) - (ax_2 + by_2)(cx_1 + dy_1) \\ &\quad = acx_1x_2 + bdy_1y_2 + adx_1y_2 + bcy_1x_2 \\ &\qquad - acx_2x_1 - bdy_2y_1 - adx_2y_1 - bcy_2x_1 \\ &\quad = ad(x_1y_2 - x_2y_1) - bc(x_1y_2 - x_2y_1) \\ &\quad = (ad - bc)(x_1y_2 - x_2y_1). \end{aligned}$$

So we have found

$$\begin{vmatrix} ax_1 + by_1 & ax_2 + by_2 \\ cx_1 + dy_1 & cx_2 + dy_2 \end{vmatrix} = \begin{vmatrix} a & b \\ c & d \end{vmatrix} \cdot \begin{vmatrix} x_1 & x_2 \\ y_1 & y_2 \end{vmatrix}. \tag{4}$$

Since $\mathbf{X}_1, \mathbf{X}_2$ is positively oriented, the determinant $\begin{vmatrix} x_1 & x_2 \\ y_1 & y_2 \end{vmatrix} > 0$. By hypothesis, $\begin{vmatrix} a & b \\ c & d \end{vmatrix} > 0$. So

$$\begin{vmatrix} ax_1 + by_1 & ax_2 + by_2 \\ cx_1 + dy_1 & cx_2 + dy_2 \end{vmatrix} > 0,$$

and so the pair $A(\mathbf{X}_1), A(\mathbf{X}_2)$ is positively oriented. If $\begin{vmatrix} a & b \\ c & d \end{vmatrix} < 0$, the same calculation shows that $A(\mathbf{X}_1), A(\mathbf{X}_2)$ is negatively oriented. In the preceding, the pair $\mathbf{X}_1, \mathbf{X}_2$ could be any given positively oriented pair of vectors. So we have proved the following:

Theorem 2.5. *Let A be a linear transformation with matrix $\begin{pmatrix} a & b \\ c & d \end{pmatrix}$. If $\begin{vmatrix} a & b \\ c & d \end{vmatrix} > 0$, then A preserves orientation. If $\begin{vmatrix} a & b \\ c & d \end{vmatrix} < 0$, then A does not preserve orientation.*

Let us say that A *reverses orientation* if whenever $\mathbf{X}_1, \mathbf{X}_2$ is a positively oriented pair, then $A(\mathbf{X}_1), A(\mathbf{X}_2)$ is a negatively oriented pair. If we look back over our preceding argument, we see that, in fact, we have shown: if $\begin{vmatrix} a & b \\ c & d \end{vmatrix} < 0$, then A reverses orientation.

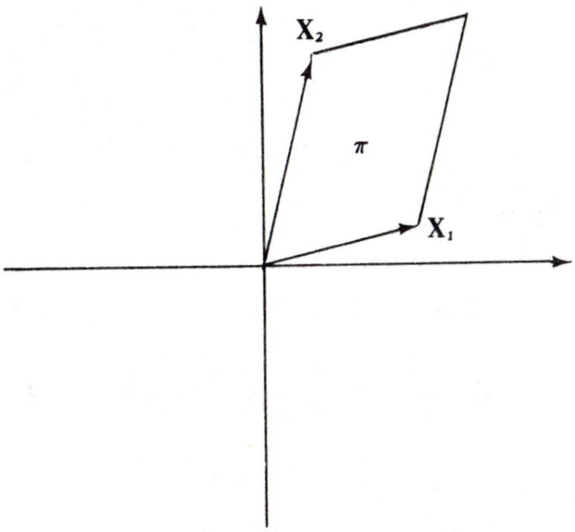

Figure 2.39

Next, we shall calculate the effect of a transformation A on *area*. Let A have the matrix $\begin{pmatrix} a & b \\ c & d \end{pmatrix}$ and assume $\begin{vmatrix} a & b \\ c & d \end{vmatrix} > 0$. Let π be a parallelogram, two of whose sides are the vectors $\mathbf{X}_1 = \begin{pmatrix} x_1 \\ y_1 \end{pmatrix}$ and $\mathbf{X}_2 = \begin{pmatrix} x_2 \\ y_2 \end{pmatrix}$, such that the pair $\mathbf{X}_1, \mathbf{X}_2$ is positively oriented.

Let $A(\pi)$ be the image of π under A, i.e., $A(\pi) = \{A(\mathbf{X}) \mid \mathbf{X}$ is a vector in $\pi\}$. (See Figs. 2.39 and 2.40) By (3), $\begin{vmatrix} x_1 & x_2 \\ y_1 & y_2 \end{vmatrix} > 0$. By (25), Chapter 2.0,

$$\text{area}(\pi) = \begin{vmatrix} x_1 & x_2 \\ y_1 & y_2 \end{vmatrix}. \tag{5}$$

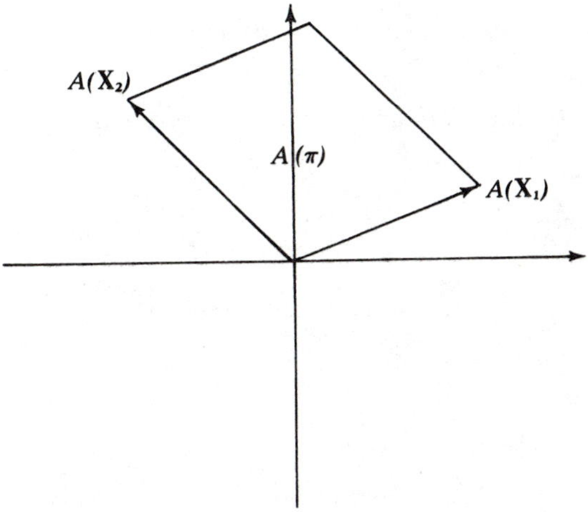

Figure 2.40

By Theorem 2.5 $A(\mathbf{X}_1), A(\mathbf{X}_2)$ is a positively oriented pair. By (5), with $A(\pi)$ replacing π,

$$\operatorname{area}(A(\pi)) = \begin{vmatrix} ax_1 + by_1 & ax_2 + by_2 \\ cx_1 + dy_1 & cx_2 + dy_2 \end{vmatrix}.$$

By calculation (4), this determinant equals $\begin{vmatrix} a & b \\ c & d \end{vmatrix} \cdot \begin{vmatrix} x_1 & x_2 \\ y_1 & y_2 \end{vmatrix}$. Thus, we have

$$\operatorname{area}(A(\pi)) = \begin{vmatrix} a & b \\ c & d \end{vmatrix} \operatorname{area}(\pi). \tag{6}$$

If we instead assume that $\begin{vmatrix} a & b \\ c & d \end{vmatrix} < 0$ and perform the corresponding calculation, we get

$$\operatorname{area}(A(\pi)) = -\begin{vmatrix} a & b \\ c & d \end{vmatrix} \operatorname{area}(\pi). \tag{7}$$

We thus have:

Theorem 2.6. *Let A be a linear transformation with matrix $\begin{pmatrix} a & b \\ c & d \end{pmatrix}$ such that $\begin{vmatrix} a & b \\ c & d \end{vmatrix} \neq 0$. If π is any parallelogram with one vertex at 0, then*

$$\operatorname{area}(A(\pi)) = \left(\text{absolute value of} \begin{vmatrix} a & b \\ c & d \end{vmatrix}\right) \operatorname{area}(\pi).$$

We can derive an interesting consequence from Theorem 2.6. If C is a linear transformation, we write $\det C$ for the determinant of the matrix of C. Now let A, B be two linear transformations. Assume $\det A > 0$, $\det B > 0$. Then A preserves orientation and B preserves orientation. It follows that BA preserves orientation. Let Q be the unit square $Q = \{(x, y) \mid 0 \leqslant x \leqslant 1, 0 \leqslant y \leqslant 1\}$.

$$(BA)(Q) = B(A(Q)).$$

so

$$\operatorname{area}((BA)((Q))) = \operatorname{area}(B(A(Q))) = (\det B)\operatorname{area}(A(Q)),$$

by Theorem 2.6. Hence,

$$\det(BA)\operatorname{area}(Q) = (\det B)(\det A)\operatorname{area}(Q).$$

It follows that

$$\det(BA) = (\det B)(\det A). \tag{8}$$

We have obtained this under the assumption $\det A > 0$ and $\det B > 0$. Recall the earlier result:

$$\begin{vmatrix} ax_1 + by_1 & ax_2 + by_2 \\ cx_1 + dy_1 & cx_2 + dy_2 \end{vmatrix} = \begin{vmatrix} a & b \\ c & d \end{vmatrix} \cdot \begin{vmatrix} x_1 & x_2 \\ y_1 & y_2 \end{vmatrix}.$$

For this formula, $\begin{pmatrix} a & b \\ c & d \end{pmatrix}$ and $\begin{pmatrix} x_1 & x_2 \\ y_1 & y_2 \end{pmatrix}$ are any two matrices, and on the left-hand side we have the matrix of $\begin{pmatrix} a & b \\ c & d \end{pmatrix}\begin{pmatrix} x_1 & x_2 \\ y_1 & y_2 \end{pmatrix}$. So (8) is true without restriction. We have:

Theorem 2.7. *If A, B are two linear transformations, then*

$$\det(BA) = (\det B)(\det A).$$

Exercise 1. Calculate the product of the matrices $\begin{pmatrix} 1 & 2 \\ 3 & 4 \end{pmatrix}$ and $\begin{pmatrix} -1 & 1 \\ 0 & 5 \end{pmatrix}$ and verify Theorem 2.7 when $A = \begin{pmatrix} 1 & 2 \\ 3 & 4 \end{pmatrix}$ and $B = \begin{pmatrix} -1 & 1 \\ 0 & 5 \end{pmatrix}$.

Exercise 2. Let Q be the square of side 1 whose edges are parallel to the coordinate axes and whose lower left-hand corner is at $\begin{pmatrix} 2 \\ 5 \end{pmatrix}$. If A is a linear transformation, define

$$A(Q) = \{A(\mathbf{X}) \mid \text{the vector } \mathbf{X} \text{ is in } Q\}.$$

In each of the following cases, sketch $A(Q)$ and find area($A(Q)$).

(i) Matrix of A is $\begin{pmatrix} 1 & 0 \\ 1 & 1 \end{pmatrix}$.

(ii) Matrix of A is $\begin{pmatrix} 2 & 0 \\ 1 & 3 \end{pmatrix}$.

(iii) Matrix of A is $\begin{pmatrix} a & b \\ c & d \end{pmatrix}$.

Exercise 3. Show that the conclusion of Theorem 2.6 remains valid when π is any parallelogram, not necessarily with one vertex at 0.

Exercise 4. In this exercise Q_1, Q_2, etc., are rectangles with sides parallel to the axes. Q_1 is the square of side 10 with lower left-hand corner at $\begin{pmatrix} 0 \\ 0 \end{pmatrix}$. Q_2 and Q_3 are squares of side 2 with lower left-hand corners at $\begin{pmatrix} 2 \\ 6 \end{pmatrix}$ and $\begin{pmatrix} 6 \\ 6 \end{pmatrix}$, respectively. Q_4 is a square of side 1 with lower left-hand corner at $\begin{pmatrix} 4.5 \\ 4 \end{pmatrix}$. Q_5 is the rectangle of height 1, base 4, with lower left-hand corner at $\begin{pmatrix} 3 \\ 1 \end{pmatrix}$. We denote by W the region obtained by removing from Q_1 the figures Q_2, Q_3, Q_4, Q_5.

(a) Draw W on graph paper.
(b) Let A be the linear transformation having matrix $\begin{pmatrix} 1 & 0 \\ 1 & 1 \end{pmatrix}$.
(c) Draw the image $A(W)$ on graph paper.
(d) What is the area of $A(W)$?

§1. Isometries of the Plane

Let us find all linear transformation T which *preserve length*, i.e., such that for every segment, the length of the image of the segment under T equals the length of the segment, or, in other words, whenever $\mathbf{X}_1$ and $\mathbf{X}_2$ are two vectors, then

$$|T(\mathbf{X}_1) - T(\mathbf{X}_2)| = |\mathbf{X}_1 - \mathbf{X}_2|. \tag{9}$$

Such a transformation is called an *isometry*.

We know that $T(\mathbf{X}_1) - T(\mathbf{X}_2) = T(\mathbf{X}_1 - \mathbf{X}_2)$. So (9) says that

$$|T(\mathbf{X}_1 - \mathbf{X}_2)| = |\mathbf{X}_1 - \mathbf{X}_2|.$$

Hence (9) holds, provided we have

$$|T(\mathbf{X})| = |\mathbf{X}| \qquad \text{for every vector } \mathbf{X}. \tag{10}$$

Conversely, if (9) holds, we get (10) by setting $\mathbf{X}_1 = \mathbf{X}$, $\mathbf{X}_2 = \mathbf{0}$. So (9) and (10) are equivalent conditions.

Let T be a linear transformation satisfying (10) and denote by $\begin{pmatrix} a & b \\ c & d \end{pmatrix}$ the matrix of T. What consequences follow for the entries a, b, c, d from the fact that T preserves length, i.e., that (10) is true?

Set $\mathbf{X} = \begin{pmatrix} x \\ y \end{pmatrix}$. Then

$$T(\mathbf{X}) = \begin{pmatrix} a & b \\ c & d \end{pmatrix}\begin{pmatrix} x \\ y \end{pmatrix} = \begin{pmatrix} ax + by \\ cx + dy \end{pmatrix},$$

and so

$$|T(\mathbf{X})| = \sqrt{(ax + by)^2 + (cx + dy)^2}\ .$$

Since $|T(\mathbf{X})| = |\mathbf{X}| = \sqrt{x^2 + y^2}$, we have

$$\sqrt{(ax + by)^2 + (cx + dy)^2} = \sqrt{x^2 + y^2}\ ,$$

and so, simplifying, we get

$$a^2x^2 + 2abxy + b^2y^2 + c^2x^2 + 2cdxy + d^2y^2 = x^2 + y^2,$$

i.e.,

$$(a^2 + c^2)x^2 + (b^2 + d^2)y^2 + (2ab + 2cd)xy = x^2 + y^2. \tag{11}$$

This holds for every vector $\mathbf{X} = \begin{pmatrix} x \\ y \end{pmatrix}$. Setting $\mathbf{X} = \begin{pmatrix} 1 \\ 0 \end{pmatrix}$, we get

(i) $$a^2 + c^2 = 1$$

and setting $\mathbf{X} = \begin{pmatrix} 0 \\ 1 \end{pmatrix}$, we get

(ii) $$b^2 + d^2 = 1.$$

Inserting this information in (11) and simplifying, we get

$$(2ab + 2cd)xy = 0.$$

Setting $x = 1$, $y = 1$ and dividing by 2, we get

(iii) $$ab + cd = 0.$$

Thus relations (i), (ii), and (iii) are consequences of (10). We can interpret these relations geometrically. Set

$$\mathbf{U} = \binom{a}{c}, \qquad \mathbf{V} = \binom{b}{d}.$$

Then (i), (ii), and (iii) say that:

$$|\mathbf{U}| = 1, \quad |\mathbf{V}| = 1, \quad \text{and} \quad \mathbf{U} \cdot \mathbf{V} = 0.$$

Since $|\mathbf{U}| = 1$, we can write $\binom{a}{c} = \mathbf{U} = \begin{pmatrix} \cos\theta \\ \sin\theta \end{pmatrix}$, where θ is the polar angle of $\mathbf{U}$, so $a = \cos\theta$, $c = \sin\theta$.

Since $\mathbf{U} \cdot \mathbf{V} = 0$ and $|\mathbf{V}| = 1$, $\mathbf{V}$ is obtained from $\mathbf{U}$ either by a positive or a negative rotation by $\pi/2$. In the first case, $\binom{b}{d} = \mathbf{V} = R_{\pi/2}\begin{pmatrix} \cos\theta \\ \sin\theta \end{pmatrix} = \begin{pmatrix} -\sin\theta \\ \cos\theta \end{pmatrix}$, so $b = -\sin\theta$, $d = \cos\theta$. Hence,

$$\begin{pmatrix} a & b \\ c & d \end{pmatrix} = \begin{pmatrix} \cos\theta & -\sin\theta \\ \sin\theta & \cos\theta \end{pmatrix}. \tag{12}$$

In the second case, $\binom{b}{d} = \mathbf{V} = R_{-\pi/2}\begin{pmatrix} \cos\theta \\ \sin\theta \end{pmatrix} = \begin{pmatrix} \sin\theta \\ -\cos\theta \end{pmatrix}$, so $b = \sin\theta$, $d = -\cos\theta$. Hence,

$$\begin{pmatrix} a & b \\ c & d \end{pmatrix} = \begin{pmatrix} \cos\theta & \sin\theta \\ \sin\theta & -\cos\theta \end{pmatrix}. \tag{13}$$

We recognize the matrix (12) as the matrix of the rotation R_θ. (See Fig. 2.41.) Also, we recall that in Chapter 2.3, we saw that $\begin{pmatrix} \cos 2\theta & \sin 2\theta \\ \sin 2\theta & -\cos 2\theta \end{pmatrix}$ is the matrix of the reflection in the line through the origin in which forms an angle θ with the positive x-axis. It follows that the matrix $\begin{pmatrix} \cos\theta & \sin\theta \\ \sin\theta & -\cos\theta \end{pmatrix}$ in (13) is the matrix of reflection through the line forming an angle $\frac{1}{2}\theta$ with the positive x-axis. So we have:

Theorem 2.8. *Let T be a length-preserving linear transformation, i.e., assume that T satisfies* (10). *Then either the matrix of T is $\begin{pmatrix} \cos\theta & -\sin\theta \\ \sin\theta & \cos\theta \end{pmatrix}$ for some number θ and then T is rotation R_θ, or else the matrix of T is $\begin{pmatrix} \cos\theta & \sin\theta \\ \sin\theta & -\cos\theta \end{pmatrix}$, and then T is reflection through the line through the origin which forms an angle of $\theta/2$ with the positive x-axis.*

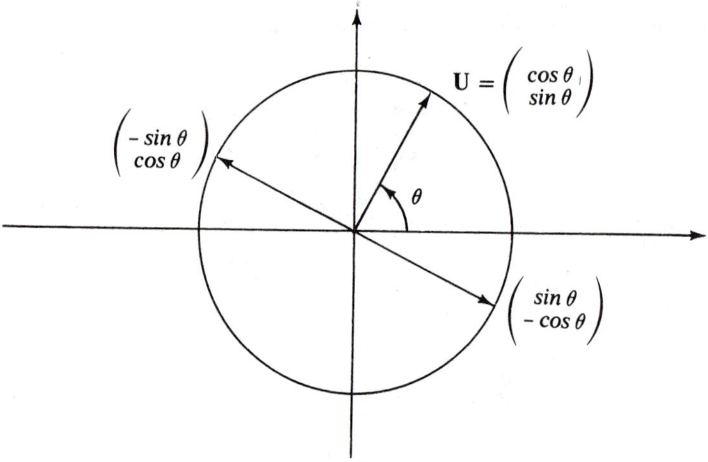

Figure 2.41

Exercise 5. For each of the following matrices, the corresponding transformation is either a rotation or a reflection. Decide which case occurs for each matrix. When it is a rotation, find the angle of rotation, and when it is a reflection, find the line in which it reflects.

(a) $\begin{pmatrix} -1/\sqrt{2} & 1/\sqrt{2} \\ 1/\sqrt{2} & 1/\sqrt{2} \end{pmatrix}$,

(b) $\begin{pmatrix} 3/5 & 4/5 \\ -4/5 & 3/5 \end{pmatrix}$,

(c) $\begin{pmatrix} 0 & 1 \\ 1 & 0 \end{pmatrix}$.

Exercise 6. A transformation T is length preserving and the matrix of T is $\begin{pmatrix} a & b \\ c & d \end{pmatrix}$.

(i) Show that the determinant $\begin{vmatrix} a & b \\ c & d \end{vmatrix}$ is either 1 or -1.

(ii) Show that T is a rotation when the determinant is 1 and a reflection when the determinant is -1.

Exercise 7. Let T_1, T_2 be two length-preserving transformations.

(a) Show that T_1T_2 is length preserving.
(b) Show that if T_1 and T_2 are both reflections, then T_1T_2 is a rotation.
(c) Show that if T_1 is a rotation and T_2 is a reflection, then T_1T_2 is a reflection.

Exercise 8. Let T_1 be reflection through the line along $\begin{pmatrix} 1 \\ 1 \end{pmatrix}$ and let T_2 be reflection through the line along $\begin{pmatrix} 1 \\ 2 \end{pmatrix}$. Write T_1T_2 in the form $T_1T_2 = R_\theta$ and find the number θ.

Having studied the effect of a linear transformation on area and length, we can ask what happens to *angles*. Let L_1, L_2 be two rays beginning at the origin and let θ be the angle from L_1 to L_2, measured counterclockwise. Let

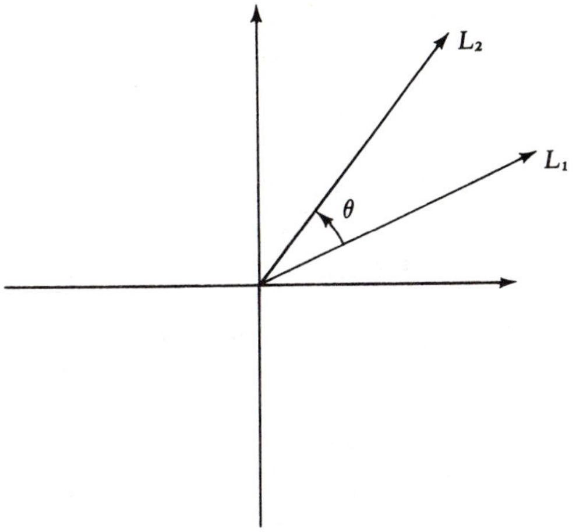

Figure 2.42

A be a linear transformation having an inverse. The images $A(L_1)$ and $A(L_2)$ are two new rays beginning at **0**. Denote by θ' the angle from $A(L_1)$ to $A(L_2)$. (See Figs. 2.42 and 2.43.) If $\theta' = \theta$ for each pair of rays L_1, L_2, then we say that A *preserves angle*.

We devote the next set of exercises to studying those linear transformations which preserve angles.

Exercise 9. Let A and B be two linear transformations which preserve angles. Show that the transformation AB and BA preserve angles.

Exercise 10.

(a) Show that each rotation R_θ preserves angles.
(b) Show that each stretching D_r preserves angles.

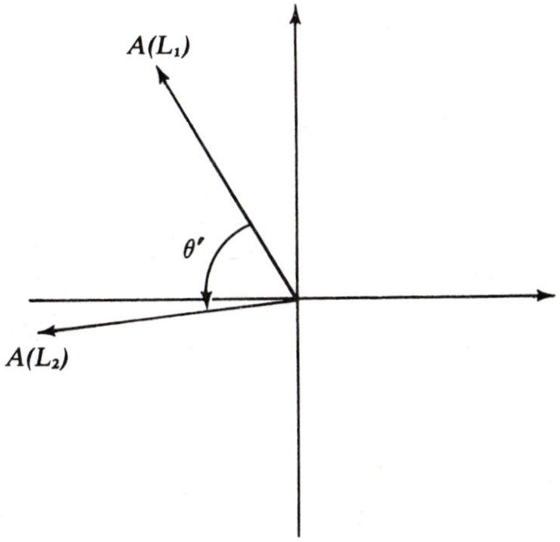

Figure 2.43

Exercise 11. Let R_θ and D_r be rotation by θ and stretching by t, respectively. Then

$$m(R_\theta) = \begin{pmatrix} \cos\theta & -\sin\theta \\ \sin\theta & \cos\theta \end{pmatrix}$$

and

$$m(D_r) = \begin{pmatrix} r & 0 \\ 0 & r \end{pmatrix}.$$

(a) Show that $D_r R_\theta$ preserves angles.
(b) Find the matrix $m(D_r R_\theta)$.
(c) Show there exist numbers a, b, not both 0, such that $m(D_r R_\theta) = \begin{pmatrix} a & b \\ -b & a \end{pmatrix}$.

Exercise 12. Show that if a, b are any two numbers not both 0, then the transformation whose matrix is $\begin{pmatrix} a & b \\ -b & a \end{pmatrix}$ preserves angles.

Exercise 13. Let A have the matrix $\begin{pmatrix} t_1 & 0 \\ 0 & t_2 \end{pmatrix}$. Show that A preserves angles if and only if $t_1 = t_2$.

In the following exercises, A denotes a linear transformation which preserves angles and $\begin{pmatrix} a & b \\ c & d \end{pmatrix} = m(A)$.

Exercise 14. Set $\mathbf{E}_1 = \begin{pmatrix} 1 \\ 0 \end{pmatrix}$, $\mathbf{E}_2 = \begin{pmatrix} 0 \\ 1 \end{pmatrix}$. Then $A(\mathbf{E}_1) = \begin{pmatrix} a \\ c \end{pmatrix}$, $A(\mathbf{E}_2) = \begin{pmatrix} b \\ d \end{pmatrix}$. Show that the angle from $\begin{pmatrix} a \\ c \end{pmatrix}$ to $\begin{pmatrix} b \\ d \end{pmatrix}$ is $\pi/2$.

Exercise 15. Write $\begin{pmatrix} a \\ c \end{pmatrix}$ in the form $\begin{pmatrix} a \\ c \end{pmatrix} = \begin{pmatrix} t_1\cos\theta \\ t_1\sin\theta \end{pmatrix}$ where $t_1 > 0$. Show that $\begin{pmatrix} b \\ d \end{pmatrix}$ can be expressed in the form $\begin{pmatrix} b \\ d \end{pmatrix} = \begin{pmatrix} -t_2\sin\theta \\ t_2\cos\theta \end{pmatrix}$, where $t_2 > 0$.

Exercise 16. By the preceding,

$$\begin{pmatrix} a & b \\ c & d \end{pmatrix} = \begin{pmatrix} t_1\cos\theta & -t_2\sin\theta \\ t_1\sin\theta & t_2\cos\theta \end{pmatrix}.$$

Show that

$$\begin{pmatrix} a & b \\ c & d \end{pmatrix} = \begin{pmatrix} \cos\theta & -\sin\theta \\ \sin\theta & \cos\theta \end{pmatrix}\begin{pmatrix} t_1 & 0 \\ 0 & t_2 \end{pmatrix}.$$

Exercise 17. Denoting by B the transformation with matrix $\begin{pmatrix} t_1 & 0 \\ 0 & t_2 \end{pmatrix}$, by the preceding exercise, we get $A = R_\theta B$.

(a) Show that B preserves angles.
(b) Using Exercise 13, show that $t_1 = t_2$ and deduce that B equals the stretching D_r, where $r = t_1$.

Exercise 18. Use the preceding exercise to prove the following result: if A is a linear transformation which preserves angles, then A is the product of a stretching and a rotation.

Exercise 19. Show that if A is a linear transformation which preserves angles, then the matrix of A has the form $\begin{pmatrix} a & b \\ -b & a \end{pmatrix}$.

§2. Determinants of Shears and Permutations

Recall the elementary matrices

$$m(H_k) = \begin{pmatrix} 1 & 0 \\ k & 1 \end{pmatrix}, \quad m(J_k) = \begin{pmatrix} 1 & k \\ 0 & 1 \end{pmatrix}, \quad m(K) = \begin{pmatrix} 0 & 1 \\ 1 & 0 \end{pmatrix}.$$

Exercise 20. Find the determinants for $m(H_k)$, $m(J_k)$, and $m(K)$.

CHAPTER 2.6

Eigenvalues

EXAMPLE 1. Let L be a line through the origin and let S be the transformation which reflects each vector in L. If $\mathbf{X}$ is on the line L, then $S(\mathbf{X}) = \mathbf{X}$. If $\mathbf{X}$ is on the line L' which goes through the origin and is perpendicular to L, then $S(\mathbf{X}) = -\mathbf{X}$.

Let T be a linear transformation. Fix a scalar t. If there is a vector $\mathbf{X} \neq \mathbf{0}$ such that $T(\mathbf{X}) = t\mathbf{X}$, then we say that t is an *eigenvalue* of T. If t is an eigenvalue of T, then each vector $\mathbf{Y}$ such that $T(\mathbf{Y}) = t\mathbf{Y}$ is called an *eigenvector* corresponding to t.

In the preceding example, $t = 1$ and $t = -1$ are eigenvalues of the reflection S. Every vector $\mathbf{Y}$ on L is an eigenvector of S corresponding to $t = 1$, since $S(\mathbf{Y}) = \mathbf{Y} = 1 \cdot \mathbf{Y}$. Every vector $\mathbf{Y}$ on L' is an eigenvector of S corresponding to $t = -1$, since $S(\mathbf{Y}) = -\mathbf{Y} = (-1) \cdot \mathbf{Y}$ (see Fig. 2.44).

Let T be a linear transformation. A vector $\mathbf{X} \neq \mathbf{0}$ is an eigenvector of T, corresponding to some eigenvalue, if and only if T takes $\mathbf{X}$ into a scalar multiple of itself. In other words, $\mathbf{X}$ is an eigenvector of T if and only if $\mathbf{X}$ and $T(\mathbf{X})$ lie on the same straight line through the origin (see Fig. 2.45).

EXAMPLE 2. Let D_r be stretching by r. Then for every vector $\mathbf{X}$, $D_r(\mathbf{X}) = r\mathbf{X}$. Hence, r is an eigenvalue of D_r. Every vector $\mathbf{X}$ is an eigenvector of D corresponding to the eigenvalue r.

EXAMPLE 3. Let $R_{\pi/2}$ be rotation by $\pi/2$ radians. If $\mathbf{X}$ is any vector $\neq \mathbf{0}$, it is clear that $\mathbf{X}$ and $R_{\pi/2}(\mathbf{X})$ do not lie on the same straight line through the origin. It follows that $R_{\pi/2}$ has *no* eigenvalue (see Fig. 2.46).

Exercise 1. Let L be a straight line through the origin and let P be the transformation which projects each vector $\mathbf{X}$ to L (see Fig. 2.47).

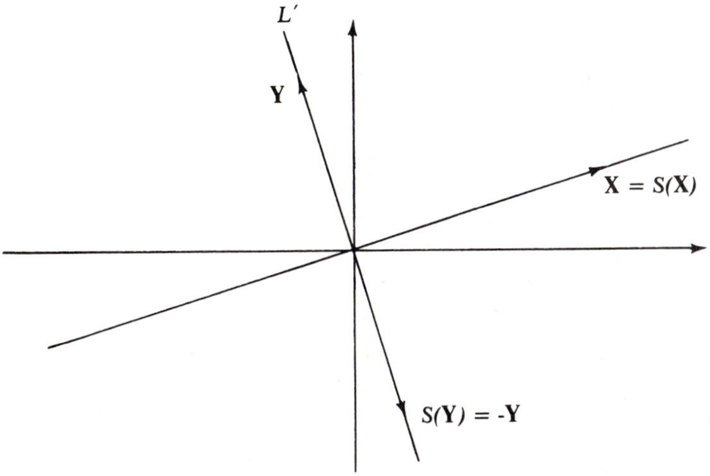

Figure 2.44

(a) Show that 0 and 1 are eigenvalues of P.
(b) Find all the eigenvectors which correspond to each of these eigenvalues.
(c) Show that P has no eigenvalues except for 0 and 1.

Exercise 2. Find all eigenvalues and corresponding eigenvectors for each of the following transformations:

(a) Rotation by π radians,
(b) I,
(c) 0.

Let A be a linear transformation and $\begin{pmatrix} a & b \\ c & d \end{pmatrix}$ its matrix. Assume t is an eigenvalue of A and $\begin{pmatrix} x \\ y \end{pmatrix}$ is a corresponding eigenvector with $\begin{pmatrix} x \\ y \end{pmatrix} \neq \begin{pmatrix} 0 \\ 0 \end{pmatrix}$.

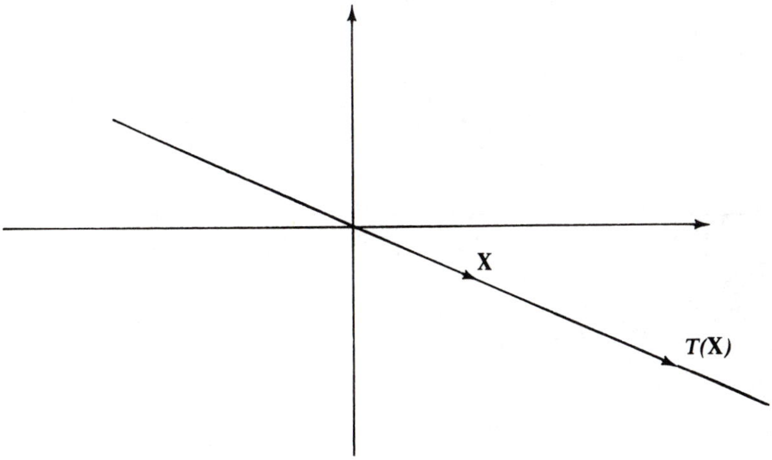

Figure 2.45

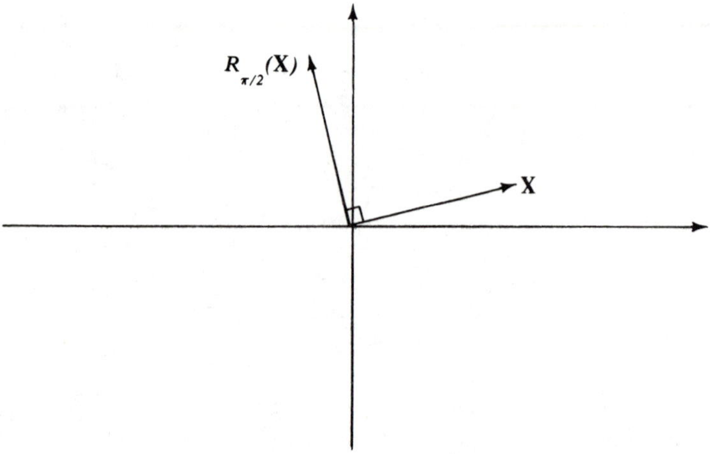

Figure 2.46

Then $A\begin{pmatrix} x \\ y \end{pmatrix} = t\begin{pmatrix} x \\ y \end{pmatrix}$, so

$$\begin{pmatrix} a & b \\ c & d \end{pmatrix}\begin{pmatrix} x \\ y \end{pmatrix} = t\begin{pmatrix} x \\ y \end{pmatrix},$$

or

$$ax + by = tx,$$
$$cx + dy = ty,$$

and so

$$(a - t)x + by = 0,$$
$$cx + (d - t)y = 0.$$

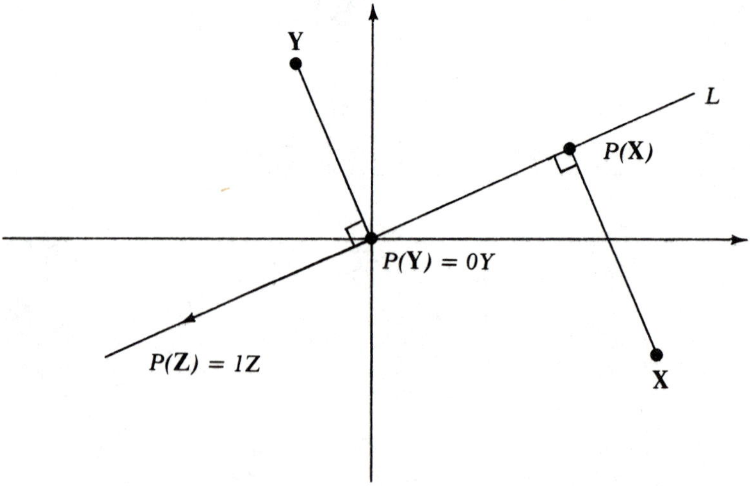

Figure 2.47

But x, y are not both 0. So we can apply Proposition 1, Chapter 2.4, and conclude that

$$(a-t)(d-t)-bc=0. \tag{1}$$

Equation (1) can be expressed in the equivalent forms:

$$\begin{vmatrix} a-t & b \\ c & d-t \end{vmatrix} = 0, \tag{2}$$

and

$$t^2-(a+d)t+(ad-bc)=0. \tag{3}$$

We call Eq. (3) the *characteristic equation* of A. We have seen that if t is an eigenvalue of A, then t is a root of Eq. (3). Furthermore, since t is a real number by definition, t is a real root.

Conversely, suppose that t is a real root of (3). Let us show that t is then an eigenvalue of A. By assumption,

$$\begin{vmatrix} a-t & b \\ c & d-t \end{vmatrix} = 0.$$

By Proposition 1, Chapter 2.4, there exists a pair of scalars x, y with $\begin{pmatrix} x \\ y \end{pmatrix} \neq \begin{pmatrix} 0 \\ 0 \end{pmatrix}$ such that $\begin{pmatrix} a-t & b \\ c & d-t \end{pmatrix}\begin{pmatrix} x \\ y \end{pmatrix} = \begin{pmatrix} 0 \\ 0 \end{pmatrix}$, and so

$$(a-t)x+by=0$$

and

$$cx+(d-t)y=0.$$

This implies that

$$ax+by=tx,$$
$$cx+dy=ty,$$

so

$$\begin{pmatrix} a & b \\ c & d \end{pmatrix}\begin{pmatrix} x \\ y \end{pmatrix} = t\begin{pmatrix} x \\ y \end{pmatrix}.$$

Thus, t is an eigenvalue of A, and $\begin{pmatrix} x \\ y \end{pmatrix}$ is an eigenvector corresponding to t.

In summary, we now know:

Theorem 2.9. *A real number t is an eigenvalue of the linear transformation A if and only if t is a root of the characteristic equation of A.*

Exercise 3. Find the characteristic equation and calculate its roots for each of the following transformations:

(a) stretching D_r;
(b) rotation R_θ;
(c) reflection in the y-axis;
(d) reflection in the line along $\begin{pmatrix} 1 \\ 1 \end{pmatrix}$.

Note. By an eigenvalue (or eigenvector) *of a matrix* we shall mean an eigenvalue (or eigenvector) of the corresponding linear transformation.

EXAMPLE 4. Find all eigenvalues and eigenvectors of the matrix $\begin{pmatrix} 3 & 4 \\ 4 & -3 \end{pmatrix}$.

SOLUTION. The characteristic equation here is

$$\begin{vmatrix} 3-t & 4 \\ 4 & -3-t \end{vmatrix} = 0,$$

or

$$t^2 - 25 = 0.$$

Its roots are $t = 5$ and $t = -5$. So, the values 5 and -5 are the eigenvalues. Let us find the eigenvectors which correspond to $t = 5$. We seek $\begin{pmatrix} x \\ y \end{pmatrix}$ with

$$\begin{pmatrix} 3 & 4 \\ 4 & -3 \end{pmatrix}\begin{pmatrix} x \\ y \end{pmatrix} = 5\begin{pmatrix} x \\ y \end{pmatrix}.$$

Thus,

$$\begin{aligned} 3x + 4y &= 5x, \\ 4x - 3y &= 5y, \end{aligned}$$

so

$$\begin{aligned} -2x + 4y &= 0, \\ 4x - 8y &= 0. \end{aligned}$$

It follows that $x = 2y$. Thus, an eigenvector with eigenvalue 5 has the form

$$\begin{pmatrix} 2y \\ y \end{pmatrix}.$$

Conversely, *every* vector of this form is an eigenvector for

$$\begin{pmatrix} 3 & 4 \\ 4 & -3 \end{pmatrix}\begin{pmatrix} 2y \\ y \end{pmatrix} = \begin{pmatrix} 10y \\ 5y \end{pmatrix} = 5\begin{pmatrix} 2y \\ y \end{pmatrix}.$$

Notice that the eigenvectors we have found fill up a straight line through the origin. Try to find the eigenvectors of $\begin{pmatrix} 3 & 4 \\ 4 & -3 \end{pmatrix}$ which have -5 as their eigenvalue.

We now turn our attention to a class of matrices which occur in many applications of linear algebra, the *symmetric matrices*. A matrix $\begin{pmatrix} a & b \\ c & d \end{pmatrix}$ is called *symmetric* if $b = c$, i.e., if the matrix has the form $\begin{pmatrix} s & t \\ t & u \end{pmatrix}$. The matrix $\begin{pmatrix} 3 & 4 \\ 4 & -3 \end{pmatrix}$, which we studied in Example 4, is symmetric.

Let A be a linear transformation and assume $m(A) = \begin{pmatrix} a & b \\ b & c \end{pmatrix}$, so that $m(A)$ is symmetric. The characteristic equation of A is

$$\begin{vmatrix} a - t & b \\ b & c - t \end{vmatrix} = 0,$$

or

$$(a - t)(c - t) - b^2 = t^2 - (a + c)t + (ac - b^2) = 0. \tag{4}$$

The roots of (4) are

$$t = \frac{a + c \pm \sqrt{(a + c)^2 - 4(ac - b^2)}}{2}.$$

Simplifying, we obtain

$$\begin{aligned}(a + c)^2 - 4(ac - b^2) &= a^2 + 2ac + c^2 - 4ac + 4b^2 \\ &= a^2 - 2ac + c^2 + 4b^2 = (a - c)^2 + 4b^2.\end{aligned}$$

Since $(a - c)^2 + 4b^2 \geqslant 0$, its square root is a real number. We consider two possible cases:

(i) $(a - c)^2 + 4b^2 = 0$;
then $a = c$ and $b = 0$, so $m(A) = \begin{pmatrix} a & b \\ b & c \end{pmatrix} = \begin{pmatrix} a & 0 \\ 0 & a \end{pmatrix}$, and so A is stretching, $A = aI$.

(ii) $(a - c)^2 + 4b^2 > 0$;
then (4) has the two distinct real roots

$$t_1 = \frac{(a + c) + \sqrt{(a - c)^2 + 4b^2}}{2}$$

and

$$t_2 = \frac{(a + c) - \sqrt{(a - c)^2 + 4b^2}}{2}.$$

By Theorem 2.9, t_1 and t_2 are eigenvalues of A. We have proved:

Proposition 1. *Let A be a linear transformation with symmetric matrix $\begin{pmatrix} a & b \\ b & c \end{pmatrix}$. Then either $A = aI$ or A has two distinct eigenvalues t_1, t_2 where*

$$t_1 = \tfrac{1}{2}\left((a + c) + \sqrt{(a - c)^2 + 4b^2}\right),$$

$$t_2 = \tfrac{1}{2}\left((a + c) - \sqrt{(a - c)^2 + 4b^2}\right).$$

Exercise 4. For each of the following matrices, find two eigenvalues of the corresponding linear transformation:

(i) $\begin{pmatrix} 3 & 0 \\ 0 & -5 \end{pmatrix}$,
(ii) $\begin{pmatrix} 1 & 1 \\ 1 & 1 \end{pmatrix}$,
(iii) $\begin{pmatrix} 0 & \sqrt{2} \\ \sqrt{2} & \pi \end{pmatrix}$.

Exercise 5. For each of the linear transformations in the preceding exercise, find one nonzero eigenvector corresponding to each eigenvalue. Show that in each case, if $\mathbf{X}_1, \mathbf{X}_2$ are eigenvectors corresponding to distinct eigenvalues, then $\mathbf{X}_1$ and $\mathbf{X}_2$ are orthogonal.

Exercise 5 suggests that the following theorem may be true.

Theorem 2.10. *Let A be a linear transformation with symmetric matrix $\begin{pmatrix} a & b \\ b & c \end{pmatrix}$ and let t_1, t_2 be distinct eigenvalues of A. Choose nonzero eigenvectors $\mathbf{X}_1, \mathbf{X}_2$ corresponding to t_1 and t_2, respectively. Then $\mathbf{X}_1$ and $\mathbf{X}_2$ are orthogonal.*

PROOF. First assume that $b \neq 0$. Set $\mathbf{X}_1 = \begin{pmatrix} x_1 \\ y_1 \end{pmatrix}$, $\mathbf{X}_2 = \begin{pmatrix} x_2 \\ y_2 \end{pmatrix}$. Then $\begin{pmatrix} a & b \\ b & c \end{pmatrix}\begin{pmatrix} x_1 \\ y_1 \end{pmatrix} = t_1\begin{pmatrix} x_1 \\ y_1 \end{pmatrix}$. So

$$ax_1 + by_1 = t_1x_1$$

or

$$by_1 = (t_1 - a)x_1 .$$

If $x_1 = 0$, then $by_1 = 0$, and since $b \neq 0$ by assumption, then $y_1 = 0$, so $\mathbf{X}_1 = \begin{pmatrix} 0 \\ 0 \end{pmatrix}$, contrary to hypothesis. So, $x_1 \neq 0$ and

$$\frac{y_1}{x_1} = \frac{t_1 - a}{b} .$$

Similarly, $x_2 \neq 0$ and

$$\frac{y_2}{x_2} = \frac{t_2 - a}{b} .$$

Since y_1/x_1 and y_2/x_2 are the slopes of $\mathbf{X}_1$ and $\mathbf{X}_2$, to prove that $\mathbf{X}_1$ and $\mathbf{X}_2$ are orthogonal amounts to showing that

$$\left(\frac{y_1}{x_1}\right)\left(\frac{y_2}{x_2}\right) = -1, \quad \text{i.e.,} \quad \left(\frac{t_1 - a}{b}\right)\left(\frac{t_2 - a}{b}\right) = -1. \quad (5)$$

Try showing that (5) is true, using the values of t_1, t_2 obtained in the last theorem, before reading the rest of the proof.

By Proposition 1, in this chapter,

$$t_1 = \tfrac{1}{2}a + \tfrac{1}{2}c + \tfrac{1}{2}\sqrt{(a-c)^2 + 4b^2}$$

and

$$t_2 = \tfrac{1}{2}a + \tfrac{1}{2}c - \tfrac{1}{2}\sqrt{(a-c)^2 + 4b^2}\ .$$

So,

$$t_2 - a = \tfrac{1}{2}(c-a) - \tfrac{1}{2}\sqrt{(a-c)^2 + 4b^2}$$

and

$$t_1 - a = \tfrac{1}{2}(c-a) + \tfrac{1}{2}\sqrt{(a-c)^2 + 4b^2}\ .$$

Then

$$(t_1 - a)(t_2 - a) = \tfrac{1}{4}(c-a)^2 - \tfrac{1}{4}\left[(a-c)^2 + 4b^2\right] = -b^2.$$

Hence,

$$\left(\frac{t_1 - a}{b}\right)\left(\frac{t_2 - a}{b}\right) = \frac{-b^2}{b^2} = -1,$$

as desired. Thus, $\mathbf{X}_1$ and $\mathbf{X}_2$ are orthogonal.

Now if $b = 0$, then $\begin{pmatrix} a & b \\ b & c \end{pmatrix} = \begin{pmatrix} a & 0 \\ 0 & c \end{pmatrix}$. So $t_1 = a$, $\mathbf{X}_1 = \begin{pmatrix} x_1 \\ 0 \end{pmatrix}$ and $t_2 = c$, $\mathbf{X}_2 = \begin{pmatrix} 0 \\ y_2 \end{pmatrix}$. Since $\begin{pmatrix} x_1 \\ 0 \end{pmatrix}$ and $\begin{pmatrix} 0 \\ y_2 \end{pmatrix}$ are orthogonal, the desired conclusion holds here as well.

An alternative proof of Theorem 2.10 can be obtained from the following exercises.

Exercise 6. Let A be the linear transformation which occurs in Theorem 2.10. Let $\mathbf{X}, \mathbf{Y}$ be any two vectors. Show that

$$A(\mathbf{X}) \cdot \mathbf{Y} = \mathbf{X} \cdot A(\mathbf{Y}).$$

Exercise 7. Let A be as in the preceding exercise and let t_1, t_2 be distinct eigenvalues of A, and $\mathbf{X}_1, \mathbf{X}_2$ the corresponding eigenvectors.

(a) Show $A(\mathbf{X}_1) \cdot \mathbf{X}_2 = t_1(\mathbf{X}_1 \cdot \mathbf{X}_2)$ and $A(\mathbf{X}_2) \cdot \mathbf{X}_1 = t_2(\mathbf{X}_1 \cdot \mathbf{X}_2)$.
(b) Using Exercise 6, deduce from (a) that $t_1(\mathbf{X}_1 \cdot \mathbf{X}_2) = t_2(\mathbf{X}_1 \cdot \mathbf{X}_2)$.
(c) Using the fact that $t_1 \neq t_2$, conclude that $\mathbf{X}_1 \cdot \mathbf{X}_2 = 0$. Thus, Theorem 2.10 holds.

Exercise 8. For each of the following matrices, find all eigenvalues and eigenvectors:

(a) $\begin{pmatrix} 1 & 2 \\ 3 & 4 \end{pmatrix}$,
(b) $\begin{pmatrix} 4 & 3 \\ 2 & 1 \end{pmatrix}$,
(c) $\begin{pmatrix} p & 0 \\ 0 & q \end{pmatrix}$,
(d) $\begin{pmatrix} 0 & 7 \\ 0 & 0 \end{pmatrix}$,
(e) $\begin{pmatrix} 0 & 5 \\ -1 & 1 \end{pmatrix}$,
(f) $\begin{pmatrix} \cos\theta & -\sin\theta \\ \sin\theta & \cos\theta \end{pmatrix}$,
(g) $\begin{pmatrix} \cos\theta & \sin\theta \\ \sin\theta & -\cos\theta \end{pmatrix}$.

Exercise 9. Given numbers a, b, c, d, show that the matrices $\begin{pmatrix} a & b \\ c & d \end{pmatrix}$ and $\begin{pmatrix} d & c \\ b & a \end{pmatrix}$ have the same eigenvalues.

Exercise 10. Denote by N a linear transformation such that $N^2 = 0$. Show that 0 is the only eigenvalue of N.

Exercise 11. Let B be a linear transformation such that B has the eigenvalue 0 and no other eigenvalue. Show that $B^2 = 0$.

Exercise 12. Let E be a linear transformation such that $E^2 = E$. What are the eigenvalues of E?

Exercise 13. Let C be a linear transformation such that C has eigenvalues 0 and 1. Show that $C^2 = C$.

Exercise 14. Let T be a linear transformation with nonzero eigenvectors $\mathbf{X}_1, \mathbf{X}_2$ and corresponding eigenvalues t_1, t_2, where $t_1 \neq t_2$.
Set $S = (T - t_1 I)(T - t_2 I)$.

(a) Show that $S(\mathbf{X}_2) = 0$.
(b) Show that $S = (T - t_2 I)(T - t_1 I)$.
(c) Show that $S(\mathbf{X}_1) = 0$.
(d) Show that $S(c_1\mathbf{X}_1 + c_2\mathbf{X}_2) = 0$, where c_1, c_2 are given constants.
(e) Show that $S = 0$, i.e.,

$$(T - t_1 I)(T - t_2 I) = 0.$$

Exercise 15. Let T be a linear transformation and let $\begin{pmatrix} a & b \\ c & d \end{pmatrix}$ be its matrix. Assume T has eigenvalues t_1, t_2 with $t_1 \neq t_2$.
Using part (e) of Exercise 14, show that

$$T^2 - (a + d)T + (ad - bc)I = 0. \tag{6}$$

Exercise 16. Let T be a linear transformation and let $\begin{pmatrix} a & b \\ c & d \end{pmatrix}$ be its matrix. Do not assume that T has any eigenvalues. Show by direct calculation that (6) is still true.

Exercise 17. Verify formula (6) when T has matrix

(a) $\begin{pmatrix} 1 & 1 \\ 1 & 1 \end{pmatrix}$,

(b) $\begin{pmatrix} 2 & 0 \\ 0 & 3 \end{pmatrix}$,

(c) $\begin{pmatrix} 0 & 5 \\ 0 & 0 \end{pmatrix}$.

CHAPTER 2.7

Classification of Conic Sections

We can use matrix multiplication to keep track of the action of a transformation A on a pair of vectors $\begin{pmatrix} x_1 \\ y_1 \end{pmatrix}$ and $\begin{pmatrix} x_2 \\ y_2 \end{pmatrix}$. Let $m(A)$ be the matrix of A, and consider the matrix $\begin{pmatrix} x_1 & x_2 \\ y_1 & y_2 \end{pmatrix}$ whose columns are $\begin{pmatrix} x_1 \\ y_1 \end{pmatrix}$ and $\begin{pmatrix} x_2 \\ y_2 \end{pmatrix}$. If we set $\begin{pmatrix} x_1' \\ y_1' \end{pmatrix} = A\begin{pmatrix} x_1 \\ y_1 \end{pmatrix}$ and $\begin{pmatrix} x_2' \\ y_2' \end{pmatrix} = A\begin{pmatrix} x_2 \\ y_2 \end{pmatrix}$, then, as we shall prove,

$$m(A)\begin{pmatrix} x_1 & x_2 \\ y_1 & x_1 \end{pmatrix} = \begin{pmatrix} x_1' & x_2' \\ y_1' & y_2' \end{pmatrix}. \tag{1}$$

EXAMPLE 1. $m(A) = \begin{pmatrix} 1 & 2 \\ 3 & 4 \end{pmatrix}$, $\begin{pmatrix} x_1 & x_2 \\ y_1 & y_2 \end{pmatrix} = \begin{pmatrix} 0 & -1 \\ 1 & 0 \end{pmatrix}$. Then $A\begin{pmatrix} 0 \\ 1 \end{pmatrix} = \begin{pmatrix} 2 \\ 4 \end{pmatrix}$, $A\begin{pmatrix} -1 \\ 0 \end{pmatrix} = \begin{pmatrix} -1 \\ -3 \end{pmatrix}$. By (1),

$$\begin{pmatrix} 1 & 2 \\ 3 & 4 \end{pmatrix}\begin{pmatrix} 0 & -1 \\ 1 & 0 \end{pmatrix} = \begin{pmatrix} 2 & -1 \\ 4 & -3 \end{pmatrix}.$$

Direct computation verifies this equation.

To prove (1) in general, we write

$$m(A) = \begin{pmatrix} a & b \\ c & d \end{pmatrix}.$$

Then

$$A\begin{pmatrix} x_1 \\ y_1 \end{pmatrix} = \begin{pmatrix} ax_1 + by_1 \\ cx_1 + dy_1 \end{pmatrix}, \qquad A\begin{pmatrix} x_2 \\ y_2 \end{pmatrix} = \begin{pmatrix} ax_2 + by_2 \\ cx_2 + dy_2 \end{pmatrix}.$$

Formula (1) states that

$$m(A)\begin{pmatrix} x_1 & x_2 \\ y_1 & y_2 \end{pmatrix} = \begin{pmatrix} ax_1 + by_1 & ax_2 + by_2 \\ cx_1 + dy_1 & cx_2 + dy_2 \end{pmatrix},$$

which is true because of the way we have defined multiplication of matrices. So formula (1) holds in general.

Now suppose that

$$A\begin{pmatrix} x_1 \\ y_1 \end{pmatrix} = t_1\begin{pmatrix} x_1 \\ y_1 \end{pmatrix}, \qquad A\begin{pmatrix} x_2 \\ y_2 \end{pmatrix} = t_2\begin{pmatrix} x_2 \\ y_2 \end{pmatrix},$$

or, in other words, that $\begin{pmatrix} x_1 \\ y_1 \end{pmatrix}$ and $\begin{pmatrix} x_2 \\ y_2 \end{pmatrix}$ are eigenvectors of A. Then (1) gives

$$m(A)\begin{pmatrix} x_1 & x_2 \\ y_1 & y_2 \end{pmatrix} = \begin{pmatrix} t_1x_1 & t_2x_2 \\ t_1y_1 & t_2y_2 \end{pmatrix}.$$

The matrix on the right-hand side equals

$$\begin{pmatrix} x_1 & x_2 \\ y_1 & y_2 \end{pmatrix}\begin{pmatrix} t_1 & 0 \\ 0 & t_2 \end{pmatrix},$$

so we have found the following result.

Let the linear transformation A have eigenvectors $\begin{pmatrix} x_1 \\ y_1 \end{pmatrix}$ and $\begin{pmatrix} x_2 \\ y_2 \end{pmatrix}$ corresponding to eigenvalues t_1, t_2. Then

$$m(A)\begin{pmatrix} x_1 & x_2 \\ y_1 & y_2 \end{pmatrix} = \begin{pmatrix} x_1 & x_2 \\ y_1 & y_2 \end{pmatrix}\begin{pmatrix} t_1 & 0 \\ 0 & t_2 \end{pmatrix}. \tag{2}$$

In addition, now suppose that the eigenvectors $\begin{pmatrix} x_1 \\ y_1 \end{pmatrix}$ and $\begin{pmatrix} x_2 \\ y_2 \end{pmatrix}$ are not *linearly dependent*. It follows by (25), Chapter 2.0, that the determinant $\begin{vmatrix} x_1 & x_2 \\ y_1 & y_2 \end{vmatrix}$ is different from 0, and so, by Theorem 2.4, the matrix $\begin{pmatrix} x_1 & x_2 \\ y_1 & y_2 \end{pmatrix}$ possesses an inverse $\begin{pmatrix} x_1 & x_2 \\ y_1 & y_2 \end{pmatrix}^{-1}$. We now multiply both sides of Eq. (2) on the right-hand side by $\begin{pmatrix} x_1 & x_2 \\ y_1 & y_2 \end{pmatrix}^{-1}$. This yields

$$m(A) = m(A)\begin{pmatrix} x_1 & x_2 \\ y_1 & y_2 \end{pmatrix}\begin{pmatrix} x_1 & x_2 \\ y_1 & y_2 \end{pmatrix}^{-1} = \begin{pmatrix} x_1 & x_2 \\ y_1 & y_2 \end{pmatrix}\begin{pmatrix} t_1 & 0 \\ 0 & t_2 \end{pmatrix}\begin{pmatrix} x_1 & x_2 \\ y_1 & y_2 \end{pmatrix}^{-1}.$$

We introduce the linear transformations P and D with $m(P) = \begin{pmatrix} x_1 & x_2 \\ y_1 & y_2 \end{pmatrix}$, $m(D) = \begin{pmatrix} t_1 & 0 \\ 0 & t_2 \end{pmatrix}$. The last equation can now be written

$$m(A) = m(P)m(D)m(P)^{-1}.$$

It follows that $A = PDP^{-1}$. We have proved:

Theorem 2.11. *Let A be a linear transformation with linearly independent eigenvectors $\begin{pmatrix} x_1 \\ y_1 \end{pmatrix}$ and $\begin{pmatrix} x_2 \\ y_2 \end{pmatrix}$, corresponding to the eigenvalues t_1 and t_2. Then*

$$A = PDP^{-1} \tag{3}$$

where $m(P) = \begin{pmatrix} x_1 & x_2 \\ y_1 & y_2 \end{pmatrix}$ *and* $m(D) = \begin{pmatrix} t_1 & 0 \\ 0 & t_2 \end{pmatrix}$.

Note: Assume t_1 and t_2 are eigenvalues of A and $\mathbf{X}_1, \mathbf{X}_2$ are corresponding nonzero eigenvectors. If $t_1 \neq t_2$, then it follows that $\mathbf{X}_1$ and $\mathbf{X}_2$ are linearly independent. To see this, suppose the contrary, i.e., suppose $\mathbf{X}_2 = s\mathbf{X}_1$ for some scalar s with $s \neq 0$. Then

$$A(\mathbf{X}_2) = A(s\mathbf{X}_1) = sA(\mathbf{X}_1) = st_1\mathbf{X}_1.$$

and so $st_1\mathbf{X}_1 = A(\mathbf{X}_2) = t_2\mathbf{X}_2 = t_2 s\mathbf{X}_1$. It follows that $st_1 = t_2 s$, and so $t_1 = t_2$, which contradicts our assumption. So $\mathbf{X}_1$ and $\mathbf{X}_2$ are linearly independent, as claimed, as long as t_1 and t_2 are distinct.

Whenever A is a linear transformation whose characteristic polynomial has distinct real roots, then formula (3) is valid.

Note: Recall that a matrix whose entries are 0 except for those on the diagonal, i.e., whose form is $\begin{pmatrix} s & 0 \\ 0 & t \end{pmatrix}$, is called a *diagonal* matrix. The matrix of the transformation D above is a diagonal matrix.

It is easy to compute the powers of a diagonal matrix.

$$\begin{pmatrix} s & 0 \\ 0 & t \end{pmatrix}^2 = \begin{pmatrix} s & 0 \\ 0 & t \end{pmatrix}\begin{pmatrix} s & 0 \\ 0 & t \end{pmatrix} = \begin{pmatrix} s^2 & 0 \\ 0 & t^2 \end{pmatrix},$$

$$\begin{pmatrix} s & 0 \\ 0 & t \end{pmatrix}^3 = \begin{pmatrix} s & 0 \\ 0 & t \end{pmatrix}^2\begin{pmatrix} s & 0 \\ 0 & t \end{pmatrix} = \begin{pmatrix} s^2 & 0 \\ 0 & t^2 \end{pmatrix}\begin{pmatrix} s & 0 \\ 0 & t \end{pmatrix} = \begin{pmatrix} s^3 & 0 \\ 0 & t^3 \end{pmatrix}.$$

Continuing in this way, we see that the nth power of the diagonal matrix $\begin{pmatrix} s & 0 \\ 0 & t \end{pmatrix}$ is the diagonal matrix $\begin{pmatrix} s^n & 0 \\ 0 & t^n \end{pmatrix}$ whose entries on the diagonal are the nth powers of the original entries.

Exercise 1. For each of the following matrices, find the nth power of the matrix when $n = 2, 3, 7, 100$.

(i) $\begin{pmatrix} -1 & 0 \\ 0 & 1 \end{pmatrix}$,

(ii) $\begin{pmatrix} 2 & 0 \\ 0 & 10 \end{pmatrix}$,

(iii) $\begin{pmatrix} 0 & 0 \\ 0 & 4 \end{pmatrix}$.

Now let A and D be the linear transformations which occur in Theorem 2.11. By (3), $A = PDP^{-1}$. Hence,

$$A^2 = (PDP^{-1})(PDP^{-1}) = PD(P^{-1}P)DP^{-1} = PDIDP^{-1}$$
$$= PDDP^{-1} = PD^2P^{-1},$$
$$A^3 = AA^2 = (PDP^{-1})(PD^2P^{-1}) = PD(P^{-1}P)D^2P^{-1}$$
$$= PDID^2P^{-1} = PD^3P^{-1}.$$

Continuing in this way, we find that

$$A^4 = PD^4P^{-1}, \qquad A^5 = PD^5P^{-1},$$

and in general,

$$A^n = PD^nP^{-1}, \tag{4}$$

where n is a positive integer. It follows that

$$m(A^n) = m(PD^nP^{-1}) = m(P)m(D^n)m(P^{-1}) = m(P)(m(D))^n m(P^{-1}).$$

Since $m(D) = \begin{pmatrix} t_1 & 0 \\ 0 & t_2 \end{pmatrix}$, we know that $(m(D))^n = \begin{pmatrix} t_1^n & 0 \\ 0 & t_2^n \end{pmatrix}$. So we have

$$(m(A))^n = m(A^n) = m(P)\begin{pmatrix} t_1^n & 0 \\ 0 & t_2^n \end{pmatrix} m(P^{-1}). \tag{5}$$

Note: Formula (5) allows us to calculate the nth power of $m(A)$ in a practical way, as the following example and exercises illustrate.

Example 2. Let us find the nth power of the matrix $\begin{pmatrix} 3 & 4 \\ 4 & -3 \end{pmatrix}$ for $n = 1, 2, 3 \ldots$.

In Example 4, Chapter 2.6, we found that the eigenvalues are $t_1 = 5$, $t_2 = -5$. As corresponding eigenvectors, we can take

$$\mathbf{X}_1 = \begin{pmatrix} 2 \\ 1 \end{pmatrix} \quad \text{and} \quad \mathbf{X}_2 = \begin{pmatrix} -1 \\ 2 \end{pmatrix}.$$

The transformation P of Theorem 2.11 then has matrix $m(P) = \begin{pmatrix} 2 & -1 \\ 1 & 2 \end{pmatrix}$. Then

$$m(P^{-1}) = (m(P))^{-1} = \begin{pmatrix} 2/5 & 1/5 \\ -1/5 & 2/5 \end{pmatrix}.$$

By (5),

$$\begin{pmatrix} 3 & 4 \\ 4 & -3 \end{pmatrix}^n = \begin{pmatrix} 2 & -1 \\ 1 & 2 \end{pmatrix}\begin{pmatrix} 5^n & 0 \\ 0 & (-5)^n \end{pmatrix}\begin{pmatrix} 2/5 & 1/5 \\ -1/5 & 2/5 \end{pmatrix}.$$

For instance, taking $n = 3$, we get

$$\begin{pmatrix} 3 & 4 \\ 4 & -3 \end{pmatrix}^3 = \begin{pmatrix} 2 & -1 \\ 1 & 2 \end{pmatrix}\begin{pmatrix} 125 & 0 \\ 0 & -125 \end{pmatrix}\begin{pmatrix} 2/5 & 1/5 \\ -1/5 & 2/5 \end{pmatrix}$$
$$= \begin{pmatrix} 2 & -1 \\ 1 & 2 \end{pmatrix}\begin{pmatrix} 50 & 25 \\ 25 & -50 \end{pmatrix} = \begin{pmatrix} 75 & 100 \\ 100 & -75 \end{pmatrix}.$$

Exercise 2. Let A be the linear transformation whose matrix is $\begin{pmatrix} 1 & 3 \\ 3 & -1 \end{pmatrix}$. Find a linear transformation D with diagonal matrix and find a linear transformation P, using Theorem 2.11 such that $A = PDP^{-1}$.

Exercise 3. Let A be as in the preceding exercise. Calculate the matrix $m(A^{10}) = \begin{pmatrix} 1 & 3 \\ 3 & -1 \end{pmatrix}^{10}$.

Exercise 4. Using Theorem 2.11, calculate $\begin{pmatrix} 1 & 1 \\ 1 & 1 \end{pmatrix}^8$.

Exercise 5. Using Theorem 2.11, calculate $\begin{pmatrix} 3 & 0 \\ 4 & 2 \end{pmatrix}^6$.

Exercise 6. Fix scalars a, b. Show that if n is an even integer, then $\begin{pmatrix} a & b \\ b & -a \end{pmatrix}^n$ is a diagonal matrix.

The second application of eigenvalues which we shall discuss in this chapter concerns quadratic forms. Let a, b, c be given scalars. For every pair of numbers x, y, we define

$$H(x, y) = ax^2 + 2bxy + cy^2.$$

H is called a *quadratic form*, i.e., a polynomial in x and y, each of whose terms is of the second degree.

Associated with H, we consider the curve whose equation is

$$ax^2 + 2bxy + cy^2 = 1. \tag{6}$$

We denote this curve by C_H.

EXAMPLE 3.

(i) $a = 1$, $b = 0$, $c = 1$. Then C_H is the circle: $x^2 + y^2 = 1$.
(ii) $a = 1$, $b = 0$, $c = -1$. Then C_H is the hyperbola: $x^2 - y^2 = 1$.

Question. Given numbers a, b, c, how can we decide what kind of curve C_H is?

Let us introduce new coordinate axes, to be called the u-axis and the v-axis, by rotating the x- and y-axes about the origin (see Fig. 2.48).

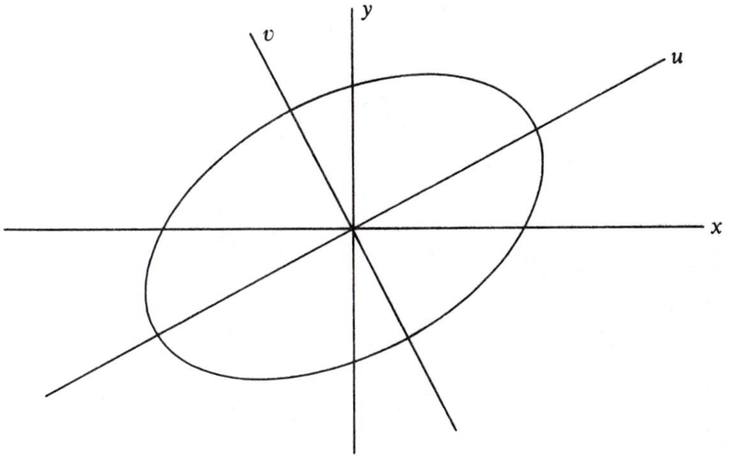

Figure 2.48

Expressed in terms of the new u and v coordinates, the equation of C_H may look more familiar. This will happen if C_H has an axis of symmetry and we manage to choose the u-axis along that axis of symmetry.

EXAMPLE 4. $a = 0$, $2b = 3$, $c = 0$. C_H has the equation: $3xy = 1$ or $xy = \frac{1}{3}$. Evidently the line $x = y$ is an axis of symmetry of C_H. Let us choose the u-axis along this line. Then the v-axis falls on the line $x = -y$.

Suppose a point $\mathbf{X}$ has old coordinates $\binom{x}{y}$ and new coordinates $\binom{u}{v}$. Denote by α the polar angle of $\mathbf{X}$ in the (u, v)-system. Then the polar angle of $\mathbf{X}$ in the (x, y)-system is $\alpha + \pi/4$ (see Fig. 2.49). Hence,

$$\begin{pmatrix} x \\ y \end{pmatrix} = |\mathbf{X}| \begin{pmatrix} \cos(\alpha + \pi/4) \\ \sin(\alpha + \pi/4) \end{pmatrix} = |\mathbf{X}| \begin{pmatrix} (\cos\alpha)\sqrt{2}/2 - (\sin\alpha)\sqrt{2}/2 \\ (\sin\alpha)\sqrt{2}/2 + (\cos\alpha)\sqrt{2}/2 \end{pmatrix}$$

$$= \frac{\sqrt{2}}{2} \begin{pmatrix} |\mathbf{X}|\cos\alpha - |\mathbf{X}|\sin\alpha \\ |\mathbf{X}|\sin\alpha + |\mathbf{X}|\cos\alpha \end{pmatrix}.$$

Also,

$$\binom{u}{v} = |\mathbf{X}| \binom{\cos\alpha}{\sin\alpha},$$

so

$$u = |\mathbf{X}|\cos\alpha,$$
$$v = |\mathbf{X}|\sin\alpha.$$

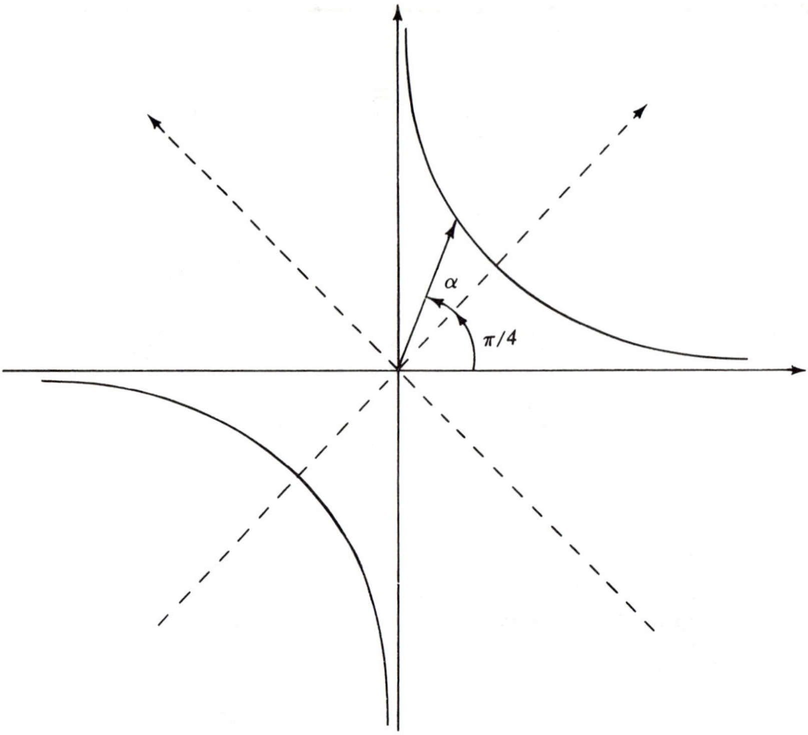

Figure 2.49

Thus,

$$\begin{pmatrix} x \\ y \end{pmatrix} = \frac{\sqrt{2}}{2}\begin{pmatrix} u - v \\ u + v \end{pmatrix},$$

$$\begin{cases} x = \dfrac{\sqrt{2}}{2}u - \dfrac{\sqrt{2}}{2}v, \\[2ex] y = \dfrac{\sqrt{2}}{2}u + \dfrac{\sqrt{2}}{2}v. \end{cases} \tag{7}$$

Now suppose $\mathbf{X}$ is a point on C_H. Then $3xy = 1$. Hence,

$$3\left(\frac{\sqrt{2}}{2}u - \frac{\sqrt{2}}{2}v\right)\left(\frac{\sqrt{2}}{2}u + \frac{\sqrt{2}}{2}v\right) = 1,$$

or

$$3\left(\frac{\sqrt{2}}{2}\right)^2 (u - v)(u + v) = 1,$$

or

$$3 \cdot \tfrac{1}{2}(u^2 - v^2) = 1.$$

We have found: if $\mathbf{X}$ is a point on C_H with new coordinates $\binom{u}{v}$, then

$$\tfrac{3}{2}u^2 - \tfrac{3}{2}v^2 = 1. \tag{8}$$

Equation (8) is an equation for C_H in the (u, v)-system. We recognize (8) as describing a hyperbola with one axis along the u-axis.

Now let a, b, c be given numbers with $b \neq 0$ and let $H(x, y) = ax^2 + 2bxy + cy^2$. We can express $H(x, y)$ as the dot product

$$\begin{pmatrix} ax + by \\ bx + cy \end{pmatrix} \cdot \begin{pmatrix} x \\ y \end{pmatrix} = (ax + by)x + (bx + cy)y = H(x, y).$$

Also,

$$\begin{pmatrix} ax + by \\ bx + cy \end{pmatrix} = \begin{pmatrix} a & b \\ b & c \end{pmatrix} \begin{pmatrix} x \\ y \end{pmatrix}.$$

Thus,

$$H(x, y) = \left(\begin{pmatrix} a & b \\ b & c \end{pmatrix} \begin{pmatrix} x \\ y \end{pmatrix} \right) \cdot \begin{pmatrix} x \\ y \end{pmatrix}. \tag{9}$$

Now, $\begin{pmatrix} a & b \\ b & c \end{pmatrix}$ is a symmetric matrix, so Proposition 1, Chapter 2.6, tells us that $\begin{pmatrix} a & b \\ b & c \end{pmatrix}$ has eigenvalues t_1, t_2. Since $b \neq 0$, $t_1 \neq t_2$. Let $\mathbf{X}_1$ be an eigenvector corresponding to t_1 such that $|\mathbf{X}_1| = 1$.

We choose new coordinate axes as follows: The u-axis passes through $\mathbf{X}_1$, directed so that $\mathbf{X}_1$ points in the positive direction. The v-axis is chosen orthogonal to the u-axis and oriented so that the positive u-direction goes over into the positive v-direction by a counterclockwise rotation of $\pi/2$ radians.

Set $\mathbf{X}_1 = \begin{pmatrix} x_1 \\ y_1 \end{pmatrix}$ and set $\mathbf{X}_2 = \begin{pmatrix} -y_1 \\ x_1 \end{pmatrix}$. Then $\mathbf{X}_2$ lies on the v-axis. We know that each eigenvector corresponding to t_2 is orthogonal to $\mathbf{X}_1$ and, hence, lies on the v-axis. Hence, every vector lying on the v-axis is an eigenvector corresponding to t_2. In particular, $\mathbf{X}_2$ is such an eigenvector (see Fig. 2.50).

Let X be any point, with $\binom{x}{y}$ its old coordinates and $\binom{u}{v}$ its new coordinates. Then

$$\mathbf{X} = u\mathbf{X}_1 + v\mathbf{X}_2.$$

Then

$$\begin{cases} u = \mathbf{X} \cdot \mathbf{X}_1 = \begin{pmatrix} x \\ y \end{pmatrix} \cdot \begin{pmatrix} x_1 \\ y_1 \end{pmatrix} = xx_1 + yy_1, \\ v = \mathbf{X} \cdot \mathbf{X}_2 = \begin{pmatrix} x \\ y \end{pmatrix} \cdot \begin{pmatrix} -y_1 \\ x_1 \end{pmatrix} = -xy_1 + yx_1. \end{cases} \tag{10}$$

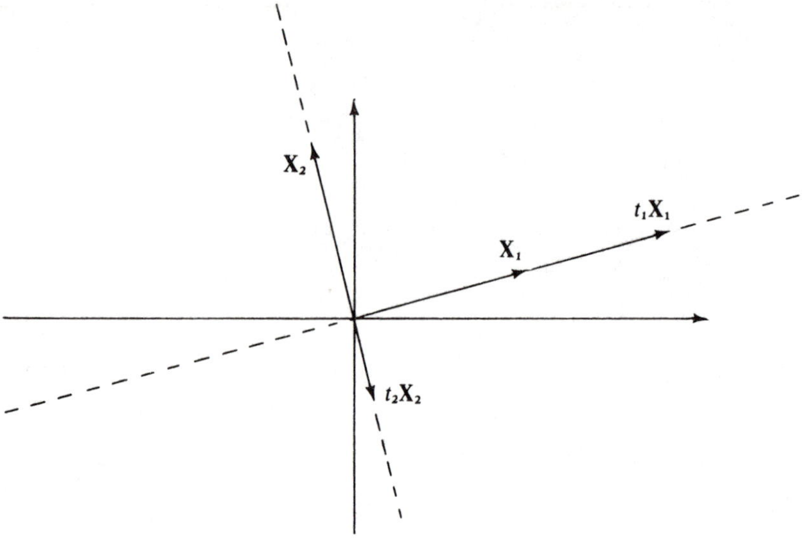

Figure 2.50

Using (9), we have

$$H(x, y) = \begin{pmatrix} a & b \\ b & c \end{pmatrix}(\mathbf{X}) \cdot \mathbf{X} = \left[\begin{pmatrix} a & b \\ b & c \end{pmatrix}(u\mathbf{X}_1 + v\mathbf{X}_2)\right] \cdot [u\mathbf{X}_1 + v\mathbf{X}_2]$$
$$= (ut_1\mathbf{X}_1 + vt_2\mathbf{X}_2) \cdot (u\mathbf{X}_1 + v\mathbf{X}_2)$$
$$= u^2t_1 + v^2t_2.$$

Thus, we have found:

Theorem 2.12. *Let t_1, t_2 be the eigenvalues of the matrix $\begin{pmatrix} a & b \\ b & c \end{pmatrix}$, where $b \neq 0$. Let* **X** *be any point, and let $\begin{pmatrix} x \\ y \end{pmatrix}$ be its old coordinates and $\begin{pmatrix} u \\ v \end{pmatrix}$ be its new coordinates. Then*

$$ax^2 + 2bxy + cy^2 = t_1u^2 + t_2v^2. \tag{11}$$

Now let **X** be a point on the curve C_H whose equation is $ax^2 + 2bxy + cy^2 = 1$. Let $\begin{pmatrix} x \\ y \end{pmatrix}$ and $\begin{pmatrix} u \\ v \end{pmatrix}$ be, respectively, the old coordinates and the new coordinates of **X**. Then using (11), we get

$$t_1u^2 + t_2v^2 = ax^2 + 2bxy + cy^2 = 1,$$

since $\begin{pmatrix} x \\ y \end{pmatrix}$ lies on C_H. So

$$t_1u^2 + t_2v^2 = 1. \tag{12}$$

Equation (12) is valid for every point on C_H and only for such points. This means that (12) is *an equation for* C_H *in the* (u,v)*-system*.

EXAMPLE 5. Describe and sketch the curve C_H: $3x^2 + 8xy - 3y^2 = 1$. Here $a = 3$, $b = 4$, $c = -3$. The matrix $\begin{pmatrix} a & b \\ b & c \end{pmatrix} = \begin{pmatrix} 3 & 4 \\ 4 & -3 \end{pmatrix}$. The eigenvalues are $t_1 = 5$ and $t_2 = -5$.

$$\begin{pmatrix} 3 & 4 \\ 4 & -3 \end{pmatrix}\begin{pmatrix} 2 \\ 1 \end{pmatrix} = 5\begin{pmatrix} 2 \\ 1 \end{pmatrix}.$$

Since we need an eigenvector $\mathbf{X}_1 = \begin{pmatrix} x_1 \\ y_1 \end{pmatrix}$ of length 1, we set $\mathbf{X}_1 = (1/\sqrt{5})\begin{pmatrix} 2 \\ 1 \end{pmatrix} = \begin{bmatrix} 2/\sqrt{5} \\ 1/\sqrt{5} \end{bmatrix}$. So

$$x_1 = \frac{2}{\sqrt{5}}, \qquad y_1 = \frac{1}{\sqrt{5}}.$$

We choose the u-axis to pass through $\mathbf{X}_1$, so it is the line $y = 2x$, and the v-axis is the line $y = -\frac{1}{2}x$. In the new system, the equation of C_H is

$$5u^2 - 5v^2 = 1, \tag{13}$$

where we have used (12) with $t_1 = 5$, $t_2 = -5$.

We can check Eq. (13) by using the relations between u, v and x, y. By (10) we know

$$u = x_1 x + y_1 y = \frac{2}{\sqrt{5}}x + \frac{1}{\sqrt{5}}y,$$

$$v = -y_1 x + x_1 y = -\frac{1}{\sqrt{5}}x + \frac{2}{\sqrt{5}}y.$$

Hence,

$$\begin{aligned} 5u^2 - 5v^2 &= 5\left(\frac{2}{\sqrt{5}}x + \frac{1}{\sqrt{5}}y\right)^2 - 5\left(-\frac{1}{\sqrt{5}}x + \frac{2}{\sqrt{5}}y\right)^2 \\ &= 5\cdot\tfrac{1}{5}(2x+y)^2 - 5\cdot\tfrac{1}{5}(-x+2y)^2 \\ &= 4x^2 + 4xy + y^2 - x^2 + 4xy - 4y^2 \\ &= 3x^2 + 8xy - 3y^2. \end{aligned}$$

Thus, for every point $\mathbf{X}$ in the plane, if $\begin{pmatrix} x \\ y \end{pmatrix}$ are the old and $\begin{pmatrix} u \\ v \end{pmatrix}$ are the new coordinates of $\mathbf{X}$, then we have

$$5u^2 - 5v^2 = 3x^2 + 8xy - 3y^2. \tag{14}$$

By definition of C_H, $\mathbf{X}$ lies on C_H if and only if $3x^2 + 8xy - 3y^2 = 1$ and so if and only if $5u^2 - 5v^2 = 1$. So (13) is verified.

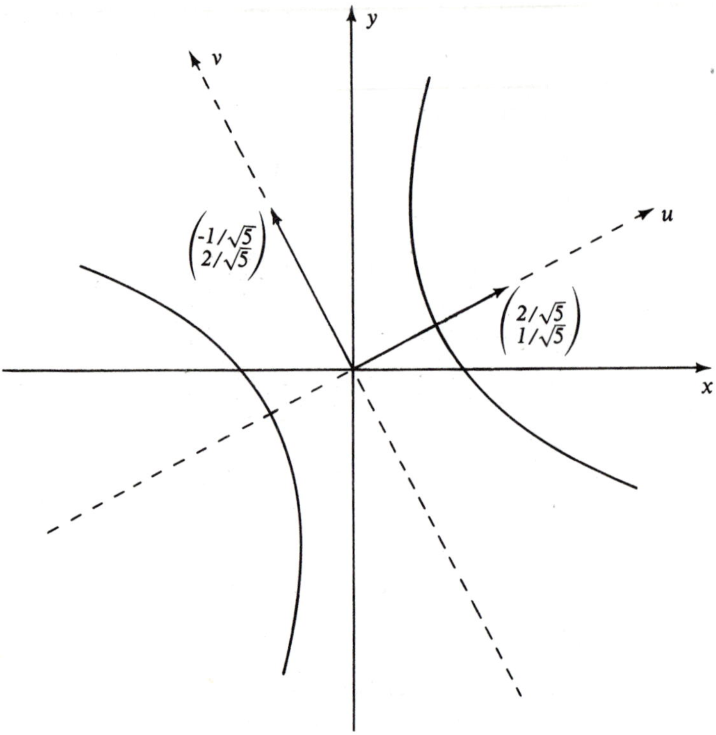

Figure 2.51

From Eq. (13), we find that C_H is a hyperbola with its axes of symmetry along the u- and v-axes (see Fig. 2.51).

Now let $H(x, y) = ax^2 + 2bxy + cy^2$ be a given quadratic form. Assume $b \neq 0$. Let t_1, t_2 denote the eigenvalues of the matrix $\begin{pmatrix} a & b \\ b & c \end{pmatrix}$. We have:

Theorem 2.13. *Let C_H denote the curve $ax^2 + 2bxy + cy^2 = 1$.*

(i) *If t_1, t_2 are both > 0, then C_H is an ellipse.*
(ii) *If $t_1, t_2 < 0$, then C_H is empty.*
(iii) *If t_1, t_2 have opposite signs, then C_H is a hyperbola.*

Exercise 7. Using the fact that in the (u, v)-system, C_H has equation

$$t_1u^2 + t_2v^2 = 1,$$

prove Theorem 2.13.

EXAMPLE 6. Describe the curve C_H,

$$x^2 + 2xy + y^2 = 1. \tag{15}$$

Here $\begin{pmatrix} a & b \\ b & c \end{pmatrix} = \begin{pmatrix} 1 & 1 \\ 1 & 1 \end{pmatrix}$ and t_1, t_2 are the roots of the polynomial

$$\begin{vmatrix} 1-t & 1 \\ 1 & 1-t \end{vmatrix} = t^2 - 2t.$$

So $t_1 = 0$, $t_2 = 2$. In the (u, v)-system, C_H has the equation

$$0u^2 + 2v^2 = 1 \quad \text{or} \quad v^2 = \tfrac{1}{2}.$$

This equation describes a locus consisting of two parallel straight lines $v = 1/\sqrt{2}$ and $v = 1/\sqrt{2}$. Thus, here C_H consists of two straight lines. We can also see this directly by writing (15) in the form

$$(x+y)^2 = 1 \quad \text{or} \quad (x+y)^2 - 1 = 0$$

or, equivalently, $((x+y)+1)((x+y)-1) = 0$. So C_H consists of the two lines $x + y + 1 = 0$ and $x + y - 1 = 0$.

Thus, in addition to the possibilities of an ellipse, hyperbola, and empty locus, noted in Theorem 2.13, C_H may consist of two lines.

Exercise 8. Classify and sketch the curve $2xy - y^2 = 1$.

Exercise 9. Classify and sketch the curve $4x^2 + 2\sqrt{2}\,xy + 3y^2 = 1$.

Exercise 10. Classify and sketch the curve $x^2 - 2xy + y^2 = 1$.

The quadratic form $H(x, y) = ax^2 + 2bxy + cy^2$ is called *positive definite* if $H(x, y) > 0$ whenever $(x, y) \neq (0, 0)$. For instance, $H(x, y) = 2x^2 + 3y^2$ is positive definite and $H(x, y) = x^2 - y^2$ is not positive definite.

Exercise 11. Give conditions on the coefficients a, b, c in order that $H(x, y)$ is positive definite. *Hint*: Make use of formula (11).

CHAPTER 2.8

Differential Systems

EXAMPLE 1. We are given two tanks of capacity 100 gallons, each filled with a mixture of salt and water. The tanks are connected by pipes as shown in Fig. 2.52 and at all times the mixture in each tank is kept uniform by stirring.

The mixture from tank I flows into tank II through a pipe at 10 gal/min, and in the reverse direction, the mixture flows into tank I from tank II through a second pipe at 5 gal/min. Also, the mixture leaves tank II through a third pipe at 5 gal/min, while fresh water flows into tank I through another pipe at 5 gal/min.

Denote by $x(t)$ the amount of salt (in lbs) in tank I at time t, and by $y(t)$ the corresponding amount in tank II. Suppose, at time $t = 0$, there are x_0 lbs of salt in tank I, and 0 lbs of salt in tank II. Find expressions for $x(t)$ and $y(t)$ in terms of t.

Consider the time interval from time t to time $t + \Delta t$. During that time interval, each gallon flowing into tank I from tank II contains $y(t)/100$ lbs of salt, while each gallon flowing from tank I to tank II contains $x(t)/100$ lbs of salt. Hence, the net change of the amount of salt in tank I during the time interval is

$$\Delta x = \frac{5y(t)\Delta t}{100} - \frac{10x(t)\Delta t}{100},$$

while the corresponding change for tank II is

$$\Delta y = \frac{10x(t)\Delta t}{100} - \frac{10y(t)\Delta t}{100}.$$

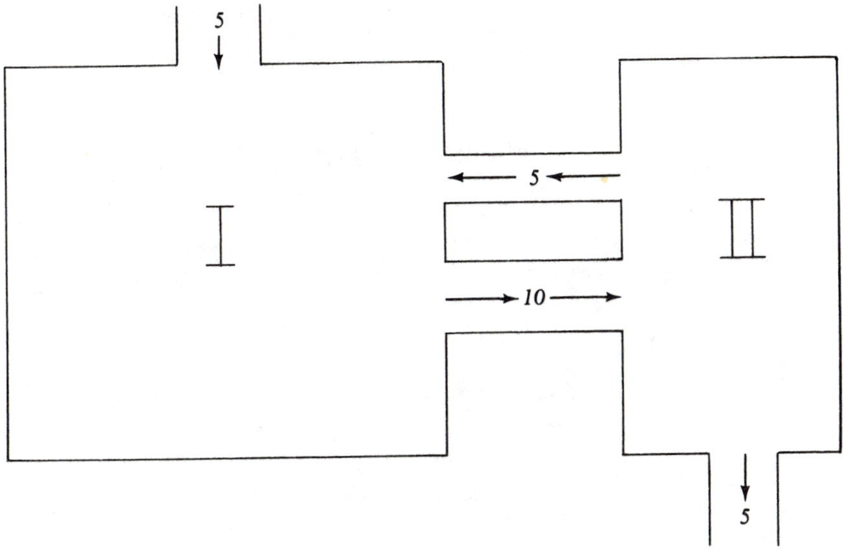

Figure 2.52

Dividing both equations by Δt and letting $\Delta t \to 0$, we get

$$\begin{cases} \dfrac{dx}{dt}(t) = \dfrac{5}{100}\, y(t) - \dfrac{10}{100}\, x(t), \\ \dfrac{dy}{dt}(t) = \dfrac{10}{100}\, x(t) - \dfrac{10}{100}\, y(t). \end{cases} \tag{1}$$

In addition, we know that

$$x(0) = x_0, \qquad y(0) = 0. \tag{2}$$

The functions $t \to x(t)$, $t \to y(t)$ must be determined from conditions (1) and (2).

A system of equations involving two unknown functions x and y which has the form

$$\begin{cases} \dfrac{dx}{dt} = ax + by, \\ \dfrac{dy}{dt} = cx + dy, \end{cases} \tag{3}$$

where a, b, c, d are given constants, is called a *differential system*.

Thus (1) is a differential system with $a = -\frac{10}{100}$, $b = \frac{5}{100}$, $c = \frac{10}{100}$, $d = -\frac{10}{100}$. The condition

$$x(0) = x_0, \qquad y(0) = y_0,$$

where x_0, y_0 are given constants, is called an *initial condition* for the system (3). Thus (2) is an initial condition.

We shall use the notion of a vector-valued function of t. A vector-valued function $\mathbf{X}(t)$ assigns to each number t a vector $\mathbf{X}(t) = \begin{pmatrix} x_1(t) \\ x_2(t) \end{pmatrix}$. Thus,

$$\mathbf{X}(t) = \begin{pmatrix} t^2 \\ t^3 + 1 \end{pmatrix} \quad \text{and} \quad \mathbf{X}(t) = \begin{pmatrix} \cos t \\ \sin t \end{pmatrix}$$

are vector-valued functions. If $\mathbf{X}(t) = \begin{pmatrix} x_1(t) \\ x_2(t) \end{pmatrix}$, then $t \to x_1(t)$ and $t \to x_2(t)$ are scalar-valued functions. We define the *derivative* of the function $t \to \mathbf{X}(t) = \begin{pmatrix} x_1(t) \\ x_2(t) \end{pmatrix}$ by

$$\frac{d\mathbf{X}}{dt} = \begin{pmatrix} dx_1/dt \\ dx_2/dt \end{pmatrix}.$$

Thus, if $\mathbf{X}(t) = \begin{pmatrix} t^2 \\ t^3 + 1 \end{pmatrix}$, then $d\mathbf{X}/dt = \begin{pmatrix} 2t \\ 3t^2 \end{pmatrix}$, while if $\mathbf{X}(t) = \begin{pmatrix} \cos t \\ \sin t \end{pmatrix}$, then $d\mathbf{X}/dt = \begin{pmatrix} -\sin t \\ \cos t \end{pmatrix}$. Note that $d\mathbf{X}/dt$ is again a vector-valued function.

Exercise 1. Fix a vector $\mathbf{Y}$. Define a vector-valued function $t \to \mathbf{Y}(t)$ by setting $\mathbf{Y}(t) = t^n\mathbf{Y}$. Show that

$$\frac{d\mathbf{Y}}{dt} = nt^{n-1}\mathbf{Y}.$$

Now let the scalar-valued functions $t \to x(t)$, $t \to y(t)$ be a solution of the differential system (3). In vector form, we can write

$$\begin{pmatrix} dx/dt \\ dy/dt \end{pmatrix} = \begin{pmatrix} ax + by \\ cx + dy \end{pmatrix}. \tag{4}$$

We define the vector-valued function $t \to \mathbf{X}(t)$ by $\mathbf{X}(t) = \begin{pmatrix} x(t) \\ y(t) \end{pmatrix}$. Then the left-hand side of (4) is $d\mathbf{X}/dt$, and the right-hand side of (4) is

$$\begin{pmatrix} ax + by \\ cx + dy \end{pmatrix} = \begin{pmatrix} a & b \\ c & d \end{pmatrix}\begin{pmatrix} x \\ y \end{pmatrix} = \begin{pmatrix} a & b \\ c & d \end{pmatrix}(\mathbf{X}(t)).$$

Thus (4) may be written in the form

$$\frac{d\mathbf{X}}{dt} = \begin{pmatrix} a & b \\ c & d \end{pmatrix}\mathbf{X}(t). \tag{5}$$

How shall we solve Eq. (5) for $\mathbf{X}(t)$? Recall that letting a matrix $\begin{pmatrix} a & b \\ c & d \end{pmatrix}$ act on a vector $\mathbf{X}$ to give the vector $\begin{pmatrix} a & b \\ c & d \end{pmatrix}\mathbf{X}$ is analogous to multiplying a number x by a scalar a to give the number ax. So Eq. (5) is analogous to the equation

$$\frac{dx}{dt} = ax, \tag{6}$$

where x is now a scalar-valued function of t and a is a given scalar. We know how to solve Eq. (6). The solutions have the form

$$x(t) = Ce^{at},$$

where C is a constant. Setting $t = 0$, we get $x(0) = C$, so

$$x(t) = x(0)e^{at} = e^{ta}(x(0)),$$

where we have changed the order of multiplication with malice aforethought. Let us look for a solution to Eq. (5) by looking for an analogue of $e^{ta}(x(0))$. We take

$$\mathbf{X}(t) = e^{tm}(\mathbf{X}(0)), \qquad \text{with} \quad m = \begin{pmatrix} a & b \\ c & d \end{pmatrix}. \tag{7}$$

First we must define the *exponential of a matrix*. In §1 we shall define, given a matrix m, a matrix to be denoted e^m or $\exp(m)$ and to be called the *exponential of* m.

Applying the matrix e^{tm} to a fixed vector $\mathbf{X}(0)$, we then obtain a vector for each t, and thus we get the vector-valued function $t \to \mathbf{X}(t)$ defined in (7). We shall then show that $\mathbf{X}(t)$ solves (5).

In what follows we shall use the symbol I for *the matrix* $\begin{pmatrix} 1 & 0 \\ 0 & 1 \end{pmatrix}$, which is properly denoted $m(I)$. This simplifies the formulas, and should cause no confusion.

§1. The Exponential of a Matrix

Let $m = \begin{pmatrix} a & b \\ c & d \end{pmatrix}$ be a matrix. Since we have defined addition and multiplication of matrices, we can write expressions such as m^2 or $m^3 - 3m + I$. We interpret $m^3 - 3m + I$ as the result of applying the polynomial $P(x) = x^3 - 3x + 1$ to the matrix m:

$$P(m) = m^3 - 3m + I.$$

More generally, if $Q(x)$ is the polynomial

$$Q(x) = c_n x^n + c_{n-1}x^{n-1} + \cdots + c_1 x + c_0,$$

where $c_n, c_{n-1}, \ldots, c_1, c_0$ are scalars, we set

$$Q(m) = c_n m^n + c_{n-1}m^{n-1} + \cdots + c_1 m + c_0 I,$$

and we regard $Q(m)$ as the matrix obtained by *applying the polynomial* Q *to the matrix* m.

We now replace the polynomial Q by the *exponential function* $\exp(x)$. We know that $\exp(x)$ is given by an infinite series

$$\exp(x) = 1 + x + \frac{x^2}{2!} + \cdots + \frac{x^n}{n!} + \cdots, \tag{8}$$

where the series converges for every number x. We wish to apply the function $\exp(x)$ to the matrix m. We *define*

$$\exp(m) = I + m + \frac{m^2}{2!} + \cdots + \frac{m^n}{n!} + \cdots. \tag{9}$$

An infinite series is understood as a *limit*. Thus, Eq. (8) means that the sequence of numbers

$$1, 1 + x, 1 + x + \frac{x^2}{2!}, \ldots, 1 + x + \frac{x^2}{2!} + \cdots + \frac{x^n}{n!}, \ldots$$

converges to the limit $\exp(x)$ as $n \to \infty$. Similarly, we interpret Eq. (9) to say that $\exp(m)$ is defined as the *limit of the sequence of matrices*

$$I, I + m, I + m + \frac{m^2}{2!}, \ldots, I + m + \frac{m^2}{2!} + \cdots + \frac{m^n}{n!}, \ldots.$$

Of course, $\exp(m)$ is then itself a matrix.

EXAMPLE 2. Let m be the diagonal matrix

$$m = \begin{pmatrix} s & 0 \\ 0 & t \end{pmatrix},$$

where s, t are scalar. What is $\exp(m)$? Recall the formula for $\begin{pmatrix} s & 0 \\ 0 & t \end{pmatrix}^n$ found in Chapter 2.7.

$$\begin{aligned}
1 + m + \frac{m^2}{2!} + \cdots + \frac{m^n}{n!}
&= \begin{pmatrix} 1 & 0 \\ 0 & 1 \end{pmatrix} + \begin{pmatrix} s & 0 \\ 0 & t \end{pmatrix} + \frac{1}{2!}\begin{pmatrix} s^2 & 0 \\ 0 & t^2 \end{pmatrix} + \frac{1}{3!}\begin{pmatrix} s^3 & 0 \\ 0 & t^3 \end{pmatrix} + \cdots + \frac{1}{n!}\begin{pmatrix} s^n & 0 \\ 0 & t^n \end{pmatrix} \\
&= \begin{pmatrix} 1 & 0 \\ 0 & 1 \end{pmatrix} + \begin{pmatrix} s & 0 \\ 0 & t \end{pmatrix} + \begin{pmatrix} s^2/2! & 0 \\ 0 & t^2/2! \end{pmatrix} \\
&\quad + \begin{pmatrix} s^3/3! & 0 \\ 0 & t^3/3! \end{pmatrix} + \cdots + \begin{pmatrix} s^n/n! & 0 \\ 0 & t^n/n! \end{pmatrix} \\
&= \begin{pmatrix} 1 + s + s^2/2! + s^3/3! + \cdots + s^n/n! & 0 \\ 0 & 1 + t + t^2/2! + t^3/3! + \cdots + t^n/n! \end{pmatrix}.
\end{aligned}$$

Letting $n \to \infty$, we find that

$$\begin{aligned}
\exp(m) &= \lim_{n\to\infty} \left(I + m + \frac{m^2}{2!} + \cdots + \frac{m^n}{n!} \right) \\
&= \begin{bmatrix} \lim_{n\to\infty} (1 + s + \cdots + s^n/n!) & 0 \\ 0 & \lim_{n\to\infty} (1 + t + \cdots + t^n/n!) \end{bmatrix} = \begin{bmatrix} e^s & 0 \\ 0 & e^t \end{bmatrix}.
\end{aligned}$$

Thus,

$$\exp\left[\begin{pmatrix} s & 0 \\ 0 & t \end{pmatrix}\right] = \begin{pmatrix} e^s & 0 \\ 0 & e^t \end{pmatrix}.$$

Example 3. Find $\exp\left[\begin{pmatrix} 0 & 1 \\ 0 & 0 \end{pmatrix}\right]$.

$$\begin{pmatrix} 0 & 1 \\ 0 & 0 \end{pmatrix}^2 = \begin{pmatrix} 0 & 1 \\ 0 & 0 \end{pmatrix}\begin{pmatrix} 0 & 1 \\ 0 & 0 \end{pmatrix} = \begin{pmatrix} 0 & 0 \\ 0 & 0 \end{pmatrix}.$$

Hence, $\begin{pmatrix} 0 & 1 \\ 0 & 0 \end{pmatrix}^n = 0$ for $n = 2, 3, \ldots$. So

$$\exp\left[\begin{pmatrix} 0 & 1 \\ 0 & 0 \end{pmatrix}\right] = \begin{pmatrix} 1 & 0 \\ 0 & 1 \end{pmatrix} + \begin{pmatrix} 0 & 1 \\ 0 & 0 \end{pmatrix} = \begin{pmatrix} 1 & 1 \\ 0 & 1 \end{pmatrix}.$$

Let A be a linear transformation. We define exp(A) as the linear transformation whose matrix is $\exp[m(A)]$.

Example 4. Let $R_{\pi/2}$ be rotation by $\pi/2$. Find $\exp(R_{\pi/2})$.

Set $m = m(R_{\pi/2}) = \begin{pmatrix} 0 & -1 \\ 1 & 0 \end{pmatrix}$.

$$m^2 = \begin{pmatrix} 0 & -1 \\ 1 & 0 \end{pmatrix}\begin{pmatrix} 0 & -1 \\ 1 & 0 \end{pmatrix} = \begin{pmatrix} -1 & 0 \\ 0 & -1 \end{pmatrix} = (-1)I.$$

Hence, for every positive integer k,

$$m^{2k} = ((-1)I)^k = (-1)^k I^k = (-1)^k I,$$

and so

$$m^{2k+1} = m^{2k}m = ((-1)^k I)m = (-1)^k m.$$

So

$$m^3 = (-1)m, \qquad m^4 = I, \qquad m^5 = m, \qquad m^6 = (-1)I, \qquad m^7 = (-1)m,$$

and so on. Hence,

$$\begin{aligned}\exp(m) &= I + m + \frac{1}{2!}(-1)I + \frac{1}{3!}(-1)m + \frac{1}{4!}I \\ &\quad + \frac{1}{5!}m + \frac{1}{6!}(-1)I + \frac{1}{7!}(-1)m + \cdots \\ &= \left(1 - \frac{1}{2!} + \frac{1}{4!} - \frac{1}{6!} + \cdots\right)I + \left(1 - \frac{1}{3!} + \frac{1}{5!} - \frac{1}{7!} + \cdots\right)m.\end{aligned}$$

We can simplify this formula by recalling that

$$\cos x = 1 - \frac{x^2}{2!} + \frac{x^4}{4!} - \frac{x^6}{6!} + \cdots,$$

and

$$\sin x = x - \frac{x^3}{3!} + \frac{x^5}{5!} - \frac{x^7}{7!} + \cdots,$$

so

$$\cos 1 = 1 - \frac{1}{2!} + \frac{1}{4!} - \frac{1}{6!} + \cdots,$$

and

$$\sin 1 = 1 - \frac{1}{3!} + \frac{1}{5!} - \frac{1}{7!} + \cdots.$$

So

$$\exp(m) = (\cos 1)I + (\sin 1)m = (\cos 1)\begin{pmatrix} 1 & 0 \\ 0 & 1 \end{pmatrix} + (\sin 1)\begin{pmatrix} 0 & -1 \\ 1 & 0 \end{pmatrix}$$

$$= \begin{pmatrix} \cos 1 & -\sin 1 \\ \sin 1 & \cos 1 \end{pmatrix}.$$

So $\exp(R_{\pi/2})$ is the linear transformation whose matrix is

$$\begin{pmatrix} \cos 1 & -\sin 1 \\ \sin 1 & \cos 1 \end{pmatrix}.$$

Exercise 2. Fix a scalar t and consider the matrix $\begin{pmatrix} 0 & -t \\ t & 0 \end{pmatrix}$. Show that

$$\exp\left[\begin{pmatrix} 0 & -t \\ t & 0 \end{pmatrix}\right] = \begin{pmatrix} \cos t & -\sin t \\ \sin t & \cos t \end{pmatrix}. \tag{10}$$

Exercise 3. Set $m = \begin{pmatrix} 1 & 1 \\ 1 & 1 \end{pmatrix}$.

(i) Calculate m^k for $k = 2, 3, 4, \ldots$.
(ii) Calculate $\exp(m)$ and simplify.

Exercise 4. Set $m = \begin{pmatrix} 3 & 1 \\ 0 & 3 \end{pmatrix}$.

(i) Calculate m^k for $k = 2, 3, 4, \ldots$.
(ii) Calculate $\exp(m)$ and simplify.

In Chapter 2.7, we considered a linear transformation A having eigenvalues t_1, t_2 with $t_1 \neq t_2$ and eigenvectors $\mathbf{X}_1 = \begin{pmatrix} x_1 \\ y_1 \end{pmatrix}$ and $\mathbf{X}_2 = \begin{pmatrix} x_2 \\ y_2 \end{pmatrix}$. We defined linear transformations P and D with

$$m(P) = \begin{pmatrix} x_1 & x_2 \\ y_1 & y_2 \end{pmatrix}, \qquad m(D) = \begin{pmatrix} t_1 & 0 \\ 0 & t_2 \end{pmatrix},$$

and we showed, in formula (5) of Chapter 2.7, that

$$(m(A))^n = m(P)\begin{pmatrix} t_1^n & 0 \\ 0 & t_2^n \end{pmatrix} m(P^{-1}), \qquad n = 1, 2, 3, \ldots.$$

It follows that

$$\exp(m(A)) = I + m(A) + \frac{1}{2!}(m(A))^2 + \cdots$$
$$= I + m(P)\begin{pmatrix} t_1 & 0 \\ 0 & t_2 \end{pmatrix} m(P^{-1}) + \frac{m(P)}{2!}\begin{pmatrix} t_1^2 & 0 \\ 0 & t_2^2 \end{pmatrix} m(P^{-1}) + \cdots$$
$$= m(P)\left[I + \begin{pmatrix} t_1 & 0 \\ 0 & t_2 \end{pmatrix} + \frac{1}{2!}\begin{pmatrix} t_1^2 & 0 \\ 0 & t_2^2 \end{pmatrix} + \cdots \right] m(P^{-1})$$

(where we have used that $m(P) \cdot m(P^{-1}) = I$)

$$= m(P)\begin{bmatrix} 1 + t_1 + (1/2!)t_1^2 + \cdots & 0 \\ 0 & 1 + t_2 + (1/2!)t_2^2 + \cdots \end{bmatrix} m(P^{-1})$$
$$= m(P)\begin{pmatrix} e^{t_1} & 0 \\ 0 & e^{t_2} \end{pmatrix} m(P^{-1}).$$

Thus, we have shown:

Theorem 2.14.

$$\exp(m(A)) = m(P)\begin{pmatrix} e^{t_1} & 0 \\ 0 & e^{t_2} \end{pmatrix} m(P^{-1}). \tag{11}$$

EXAMPLE 5. Calculate $\exp\left[\begin{pmatrix} 3 & 4 \\ 4 & -3 \end{pmatrix}\right]$.

Here

$$t_1 = 5, \qquad t_2 = -5, \qquad \mathbf{X}_1 = \begin{pmatrix} 2 \\ 1 \end{pmatrix}, \qquad \mathbf{X}_2 = \begin{pmatrix} -1 \\ 2 \end{pmatrix}.$$

So

$$m(P) = \begin{pmatrix} 2 & -1 \\ 1 & 2 \end{pmatrix}, \qquad m(P^{-1}) = \begin{pmatrix} 2/5 & 1/5 \\ -1/5 & 2/5 \end{pmatrix}.$$

By (11), we have

$$\exp\left[\begin{pmatrix} 3 & 4 \\ 4 & -3 \end{pmatrix}\right] = \begin{pmatrix} 2 & -1 \\ 1 & 2 \end{pmatrix}\begin{pmatrix} e^5 & 0 \\ 0 & e^{-5} \end{pmatrix}\begin{pmatrix} 2/5 & 1/5 \\ -1/5 & 2/5 \end{pmatrix}$$
$$= \begin{pmatrix} 2 & -1 \\ 1 & 2 \end{pmatrix}\begin{pmatrix} (2/5)e^5 & (1/5)e^5 \\ -(1/5)e^{-5} & (2/5)e^{-5} \end{pmatrix}$$
$$= \begin{pmatrix} (4/5)e^5 + (1/5)e^{-5} & (2/5)e^5 - (2/5)e^{-5} \\ (2/5)e^5 - (2/5)e^{-5} & (1/5)e^5 + (4/5)e^{-5} \end{pmatrix}.$$

EXAMPLE 6. Calculate $\exp\left[\begin{pmatrix} 1 & 0 \\ 0 & 0 \end{pmatrix}\right]$.

Since $\begin{vmatrix} 1-t & 0 \\ 0 & -t \end{vmatrix} = t^2 - t = t(t-1)$, the eigenvalues are $t_1 = 1$, $t_2 = 0$. The corresponding eigenvectors are $\mathbf{X}_1 = \begin{pmatrix} 1 \\ 0 \end{pmatrix}$, $\mathbf{X}_2 = \begin{pmatrix} 0 \\ 1 \end{pmatrix}$. So $m(P) = \begin{pmatrix} 1 & 0 \\ 0 & 1 \end{pmatrix} = I$, and then $m(P^{-1}) = I$. Hence, by (11),

$$\exp\left[\begin{pmatrix} 1 & 0 \\ 0 & 0 \end{pmatrix}\right] = I\begin{pmatrix} e & 0 \\ 0 & 1 \end{pmatrix}I = \begin{pmatrix} e & 0 \\ 0 & 1 \end{pmatrix}.$$

Exercise 5.

(a) Compute $\begin{pmatrix} 1 & 0 \\ 0 & 0 \end{pmatrix}^n$ for $n = 1, 2, 3, \ldots$.

(b) Compute $\exp\left[\begin{pmatrix} 1 & 0 \\ 0 & 0 \end{pmatrix}\right]$ directly from the definition and compare your answer with the result of Example 6.

Exercise 6. Using Theorem 2.14, calculate $\exp\left[\begin{pmatrix} 1 & 3 \\ 3 & -1 \end{pmatrix}\right]$.

Exercise 7. Calculate $\exp\left[\begin{pmatrix} 3 & 0 \\ 4 & 2 \end{pmatrix}\right]$.

Exercise 8. Calculate $\exp\left[\begin{pmatrix} 1 & 2 \\ 1 & 1 \end{pmatrix}\right]$.

Recall Eq. (5): $d\mathbf{X}/dt = m(\mathbf{X}(t))$, where $m = \begin{pmatrix} a & b \\ c & d \end{pmatrix}$. We fix a vector $\mathbf{X}$ and define $\mathbf{X}(t) = \exp(tm)(\mathbf{X}_0)$. In §2, we shall show that $\mathbf{X}(t)$ solves (5) and satisfies the initial condition $\mathbf{X}(0) = \mathbf{X}_0$, and we shall study examples and applications.

§2. Solutions of Differential Systems

We fix a matrix m and a vector $\mathbf{X}_0$.

$$\exp(tm) = I + tm + \frac{t^2}{2!}m^2 + \frac{t^3m^3}{3!} + \cdots,$$

so

$$(\exp(tm))(\mathbf{X}_0) = \mathbf{X}_0 + tm(\mathbf{X}_0) + \frac{t^2}{2!}m^2(\mathbf{X}_0) + \frac{t^3}{3!}m^3(\mathbf{X}_0) + \cdots.$$

Both sides of the last equation are vector-valued functions of t. It can be shown that the derivative of the sum of the infinite series is obtained by differentiating the series term by term. In other words,

$$\frac{d}{dt}\{(\exp(tm))(\mathbf{X}_0)\} = \frac{d}{dt}(tm(\mathbf{X}_0)) + \frac{d}{dt}\left(\frac{t^2}{2!}m^2(\mathbf{X}_0)\right) + \cdots. \quad (12)$$

The right-hand side of (12) is equal to

$$m(\mathbf{X}_0) + \frac{2t}{2!}m^2(\mathbf{X}_0) + \frac{3t^2}{3!}m^3(\mathbf{X}_0) + \frac{4t^3}{4!}m^4(\mathbf{X}_0) + \cdots$$
$$= m(\mathbf{X}_0) + tm^2(\mathbf{X}_0) + \frac{t^2}{2!}m^3(\mathbf{X}_0) + \frac{t^3}{3!}m^4(\mathbf{X}_0) + \cdots$$
$$= m(\mathbf{X}_0) + m(tm(\mathbf{X}_0)) + m\left(\frac{t^2}{2!}m^2(\mathbf{X}_0)\right) + m\left(\frac{t^3}{3!}m^3(\mathbf{X}_0)\right) + \cdots$$
$$= m\left\{\mathbf{X}_0 + tm(\mathbf{X}_0) + \frac{t^2}{2!}m^2(\mathbf{X}_0) + \frac{t^3}{3!}m^3(\mathbf{X}_0) + \cdots\right\}$$
$$= m\{(\exp(tm))(\mathbf{X}_0)\}.$$

So (12) gives us

$$\frac{d}{dt}\{(\exp(tm))(\mathbf{X}_0)\} = m\{(\exp(tm))(\mathbf{X}_0)\}. \tag{13}$$

We define $\mathbf{X}(t) = (\exp(tm))(\mathbf{X}_0))$. Then (13) states that

$$\frac{d\mathbf{X}}{dt}(t) = m(\mathbf{X}(t)). \tag{14}$$

In other words, we have shown that $\mathbf{X}(t)$ solves our original equation (5). Also, setting $t = 0$ in the definition of $\mathbf{X}(t)$, we find that

$$\mathbf{X}(0) = I(\mathbf{X}_0) = \mathbf{X}_0, \tag{15}$$

since $\exp(0) = I + 0 + 0 + \cdots = I$. So we have proved:

Theorem 2.15. *Let m be a matrix. Fix a vector $\mathbf{X}_0$. Set $\mathbf{X}(t) = (\exp(tm))(\mathbf{X}_0)$ for all t. Then,*

$$\frac{d\mathbf{X}}{dt} = m\mathbf{X}(t), \tag{16}$$

and

$$\mathbf{X}(0) = \mathbf{X}_0. \tag{17}$$

Example 7. Solve the differential system

$$\begin{cases} \dfrac{dx}{dt} = -y, \\ \dfrac{dy}{dt} = x, \end{cases} \tag{18}$$

with the initial condition: $x(0) = 1$, $y(0) = 0$.

In vector form, with $\mathbf{X}(t) = \begin{pmatrix} x(t) \\ y(t) \end{pmatrix}$, we have

$$\frac{d\mathbf{X}}{dt} = \begin{pmatrix} 0 & -1 \\ 1 & 0 \end{pmatrix}(\mathbf{X})$$

with initial condition $\mathbf{X}(0) = \begin{pmatrix} 1 \\ 0 \end{pmatrix}$. Set $m = \begin{pmatrix} 0 & -1 \\ 1 & 0 \end{pmatrix}$, $\mathbf{X}_0 = \begin{pmatrix} 1 \\ 0 \end{pmatrix}$ and set

$$\mathbf{X}(t) = \exp(tm)(\mathbf{X}_0).$$

By Exercise 2 in this chapter,

$$\exp(tm) = \exp\left[\begin{pmatrix} 0 & -t \\ t & 0 \end{pmatrix}\right] = \begin{pmatrix} \cos t & -\sin t \\ \sin t & \cos t \end{pmatrix}.$$

So

$$\mathbf{X}(t) = \begin{pmatrix} \cos t & -\sin t \\ \sin t & \cos t \end{pmatrix}\begin{pmatrix} 1 \\ 0 \end{pmatrix} = \begin{pmatrix} \cos t \\ \sin t \end{pmatrix}.$$

Since $\mathbf{X}(t) = \begin{pmatrix} x(t) \\ y(t) \end{pmatrix}$, we obtain $x(t) = \cos t$, $y(t) = \sin t$. Inserting these functions in (18), we see that it checks. Also, $x(0) = 1$, $y(0) = 0$, so the initial condition checks also.

Example 8. Solve the differential system (18) with initial condition $x(0) = x_0, y(0) = y_0$.

We take $\mathbf{X}_0 = \begin{pmatrix} x_0 \\ y_0 \end{pmatrix}$ and set

$$\mathbf{X}(t) = \left(\exp\begin{bmatrix} 0 & -t \\ t & 0 \end{bmatrix}\right)(\mathbf{X}_0) = \begin{pmatrix} \cos t & -\sin t \\ \sin t & \cos t \end{pmatrix}\begin{pmatrix} x_0 \\ y_0 \end{pmatrix},$$

so

$$\mathbf{X}(t) = \begin{pmatrix} (\cos t)x_0 - (\sin t)y_0 \\ (\sin t)x_0 + (\cos t)y_0 \end{pmatrix}.$$

So

$$x(t) = (\cos t)x_0 - (\sin t)y_0, \qquad y(t) = (\sin t)x_0 + (\cos t)y_0.$$

We check that these functions satisfy (18) and that $x(0) = x_0, y(0) = y_0$.

Exercise 9. Calculate $\exp\left[\begin{pmatrix} 3t & 4t \\ 4t & -3t \end{pmatrix}\right]$, where t is a given number.

Exercise 10. Using the result of Exercise 9, solve the system

$$\begin{cases} \dfrac{dx}{dt} = 3x + 4y \\ \dfrac{dy}{dt} = 4x - 3y \end{cases} \quad \text{with} \quad \begin{Bmatrix} x(0) = 1 \\ y(0) = 0 \end{Bmatrix}, \tag{19}$$

by using Theorem 2.15 with $m = \begin{pmatrix} 3 & 4 \\ 4 & -3 \end{pmatrix}$ and $\mathbf{X}_0 = \begin{pmatrix} 1 \\ 0 \end{pmatrix}$.

Exercise 11. Solve the system (19) with $x(0) = s_1, y(0) = s_2$.

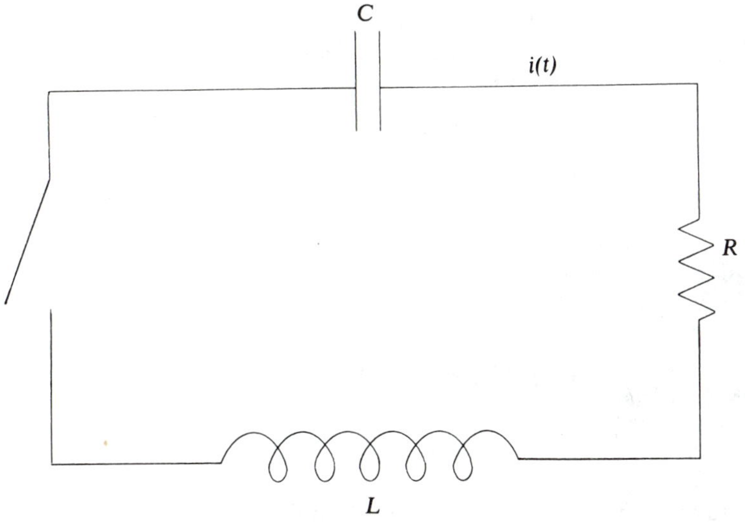

Figure 2.53

Exercise 12. Solve the system

$$\begin{cases} \dfrac{dx}{dt} = 2x + 4y \\ \dfrac{dy}{dt} = 4x + 6y \end{cases} \quad \text{with} \quad \begin{Bmatrix} x(0) = 1 \\ y(0) = 0 \end{Bmatrix}. \tag{20}$$

Exercise 13. Solve the system (1) (at the beginning of this chapter) with initial condition (2).

EXAMPLE 9. Consider an electric circuit consisting of a condenser of capacitance C connected to a resistance of R ohms and an inductance of L henries. A switch is inserted in the circuit (see Fig. 2.53). The condenser is charged with a charge of Q_0 coulombs, with the switch open. At time $t = 0$, the switch is closed and the condenser begins to discharge, causing a current to flow in the circuit. Denote by $i(t)$ the current flowing at time t and by $Q(t)$ the charge on the condenser at time t. The laws of electricity tell us the following: the voltage drop at time t equals $(1/C)Q(t)$ across the condenser, while the voltage drop across the resistance is $Ri(t)$ and the voltage drop across the inductance is $L(di/dt)$. The sum of all the voltage drops equals 0 at every time $t > 0$, since the circuit is closed. Thus, we have

$$\frac{1}{C}\,Q(t) + Ri(t) + L\frac{di}{dt} = 0$$

or

$$\frac{di}{dt} = -\frac{1}{LC}\,Q(t) - \frac{R}{L}\,i(t).$$

Also, the current at time t equals the negative of dQ/dt or $i(t) = -dQ/dt$. So the two functions: $t \to i(t)$ and $t \to Q(t)$ satisfy

$$\begin{cases} \dfrac{di}{dt} = ai + bQ, \\ \dfrac{dQ}{dt} = -i, \end{cases} \tag{21}$$

where $a = -R/L$, $b = -1/LC$. So to calculate the current flowing in the circuit at any time t, we must solve the differential system (21) with initial condition $Q(0) = Q_0$, $i(0) = 0$.

EXAMPLE 10. Let c_1, c_2 be two scalars. We wish to solve the *second-order differential equation*

$$\frac{d^2x}{dt^2} + c_1\frac{dx}{dt} + c_2x = 0 \tag{22}$$

by a function $t \to x(t)$ defined for all t, and we want to satisfy the initial conditions

$$x(0) = x_0, \qquad \frac{dx}{dt}(0) = y_0. \tag{23}$$

We shall reduce the problem (22) to a *first-order differential system* of the form (3). To this end we define $y(t) = (dx/dt)(t)$. Then (22) can be written: $dy/dt + c_1y + c_2x = 0$ or

$$\frac{dy}{dt} = -c_2x - c_1y.$$

So x and y satisfy the differential system

$$\begin{cases} \dfrac{dx}{dt} = y, \\ \dfrac{dy}{dt} = -c_2x - c_1y. \end{cases} \tag{24}$$

EXAMPLE 11. We study the equation

$$\frac{d^2x}{dt^2} + x = 0, \qquad x(0) = x_0, \qquad \frac{dx}{dt}(0) = y_0. \tag{25}$$

Setting $y = dx/dt$, (25) turns into

$$\begin{cases} \dfrac{dx}{dt} = y, \\ \dfrac{dy}{dt} = -x, \end{cases} \qquad x(0) = x_0, \quad y(0) = y_0. \tag{26}$$

Exercise 14. Fix a scalar t. Show that

$$\exp\left[t\begin{pmatrix} 0 & 1 \\ -1 & 0 \end{pmatrix}\right] = \begin{pmatrix} \cos t & \sin t \\ -\sin t & \cos t \end{pmatrix}.$$

Exercise 15. Using the result of Exercise 14, solve first the equations (26) and then Eq. (25).

In Theorem 2.15, we showed that the problem $d\mathbf{X}/dt = m\mathbf{X}(t)$, $\mathbf{X}(0) = \mathbf{X}_0$ has $\mathbf{X}(t) = (\exp(tm)(\mathbf{X}_0))$ as a solution for all t. We shall now show that this is the *only* solution, or, in other words, we shall prove uniqueness of solutions.

Suppose $\mathbf{X}$, $\mathbf{Y}$ are two solutions. Then $d\mathbf{X}/dt = m\mathbf{X}(t)$, $\mathbf{X}(0) = \mathbf{X}_0$ and $d\mathbf{Y}/dt = m\mathbf{Y}(t)$, $\mathbf{Y}(0) = \mathbf{X}_0$. Set $\mathbf{Z}(t) = \mathbf{X}(t) - \mathbf{Y}(t)$. Our aim is to prove that $\mathbf{Z}(t) = 0$ for all t. We have

$$\begin{aligned}\frac{d\mathbf{Z}}{dt} = \frac{d\mathbf{X}}{dt} - \frac{d\mathbf{Y}}{dt} &= m\mathbf{X}(t) - m\mathbf{Y}(t)\\ &= m(\mathbf{X}(t) - \mathbf{Y}(t)) = m\mathbf{Z}(t). \qquad (27)\end{aligned}$$

Also

$$\mathbf{Z}(0) = \mathbf{X}(0) - \mathbf{Y}(0) = \mathbf{X}_0 - \mathbf{X}_0 = 0. \qquad (28)$$

We now shall use (27) and (28) to show that $\mathbf{Z}(t) = \mathbf{0}$ for all t. We denote by $f(t)$ the squared length of $\mathbf{Z}(t)$, i.e.,

$$f(t) = |\mathbf{Z}(t)|^2.$$

$t \to f(t)$ is a scalar-valued function. It satisfies

$$f(t) \geqslant 0 \qquad \text{for all } t \quad \text{and} \quad f(0) = 0.$$

Exercise 16. If $\mathbf{A}(t)$, $\mathbf{B}(t)$ are two vector-valued functions, then

$$\frac{d}{dt}(\mathbf{A}(t) \cdot \mathbf{B}(t)) = \mathbf{A}(t) \cdot \frac{d\mathbf{B}}{dt} + \mathbf{B}(t) \cdot \frac{d\mathbf{A}}{dt}.$$

It follows from Exercise 16 that

$$\frac{df}{dt} = \frac{d}{dt}(\mathbf{Z}(t) \cdot \mathbf{Z}(t)) = \mathbf{Z}(t) \cdot \frac{d\mathbf{Z}}{dt} + \mathbf{Z}(t) \cdot \frac{d\mathbf{Z}}{dt} = 2\mathbf{Z}(t) \cdot \frac{d\mathbf{Z}}{dt}.$$

Using (27), this gives

$$\frac{df}{dt}(t) = 2\mathbf{Z}(t) \cdot m\mathbf{Z}(t). \qquad (29)$$

We set $m = \begin{pmatrix} a & b \\ c & d \end{pmatrix}$. Fix t and set $\mathbf{Z}(t) = \mathbf{Z} = \begin{pmatrix} z_1 \\ z_2 \end{pmatrix}$. Then

$$\begin{aligned}2\mathbf{Z}(t) \cdot m\mathbf{Z}(t) &= 2\begin{pmatrix} z_1 \\ z_2 \end{pmatrix} \cdot \left\{\begin{pmatrix} a & b \\ c & d \end{pmatrix}\begin{pmatrix} z_1 \\ z_2 \end{pmatrix}\right\}\\ &= 2\begin{pmatrix} z_1 \\ z_2 \end{pmatrix} \cdot \begin{pmatrix} az_1 + bz_2 \\ cz_1 + dz_2 \end{pmatrix}\\ &= 2\left(az_1^2 + bz_1z_2 + cz_2z_1 + dz_2^2\right).\end{aligned}$$

Let K be a constant greater than $|a|$, $|b|$, $|c|$, $|d|$. Then

$$|2\mathbf{Z}(t)\cdot m\mathbf{Z}(t)| \leqslant 2\big(|a|z_1^2 + |b|\,|z_1|\,|z_2| + |c|\,|z_2|\,|z_1| + |d|\,|z_2|^2\big)$$
$$\leqslant 2K\big(|z_1|^2 + 2|z_1|\,|z_2| + |z_2|^2\big).$$

Also,

$$2|z_1|\,|z_2| \leqslant |z_1|^2 + |z_2|^2.$$

So

$$|2\mathbf{Z}(t)\cdot m\mathbf{Z}(t)| \leqslant 2K\big(2|z_1|^2 + 2|z_2|^2\big) = 4K\big(|z_1|^2 + |z_2|^2\big)$$
$$= 4K|\mathbf{Z}|^2 = 4Kf(t).$$

By (29), setting $M = 4K$, this gives

$$\frac{df}{dt}(t) \leqslant Mf(t). \tag{30}$$

Consider the derivative

$$\frac{d}{dt}\left(\frac{f(t)}{e^{Mt}}\right) = \frac{e^{Mt}(df/dt) - f(t)Me^{M(t)}}{e^{2Mt}} = \frac{(df/dt) - Mf(t)}{e^{Mt}}.$$

By (30), the numerator of the right-hand term $\leqslant 0$ for all t. So

$$\frac{d}{dt}\left(\frac{f(t)}{e^{Mt}}\right) \leqslant 0,$$

so $f(t)/e^{Mt}$ is a decreasing function of t. Also, $f(t)/e^{Mt} \geqslant 0$ and $= 0$ at $t = 0$. But a decreasing function of t, defined on $t \geqslant 0$ which is $\geqslant 0$ for all t and $= 0$ at $t = 0$, is identically 0.

So $f(t)/e^{Mt} = 0$ for all t. Thus $|\mathbf{Z}(t)|^2 = f(t) = 0$, and so $\mathbf{Z}(t) = 0$, and so $\mathbf{X}(t) = \mathbf{Y}(t)$ for all t.

We have proved:

Uniqueness Property. *The only solution of the problem considered in Theorem* 2.15 *is* $\mathbf{X}(t) = (\exp(tm))(\mathbf{X}_0)$.

CHAPTER 3.0

Vector Geometry in 3-Space

Just as in the plane, we may use vectors to express the analytic geometry of 3-dimensional space.

We define a *vector* in 3-space as a triplet of numbers $\begin{bmatrix} x_1 \\ x_2 \\ x_3 \end{bmatrix}$ written in column form, with x_1, x_2, and x_3 as the *first*, *second*, and *third coordinates*. We designate this vector by a single capital letter **X**, i.e., we write $\mathbf{X} = \begin{bmatrix} x_1 \\ x_2 \\ x_3 \end{bmatrix}$. We can picture the vector **X** as an arrow or directed segment, starting at the origin and ending at the point $\begin{bmatrix} x_1 \\ x_2 \\ x_3 \end{bmatrix}$. We denote by $\mathbb{R}^3$ the set of all vectors in 3-space, and we denote by $\mathbb{R}^2$ the set of all vectors in the plane.

We *add* two vectors by adding their components, so if $\mathbf{X} = \begin{bmatrix} x_1 \\ x_2 \\ x_3 \end{bmatrix}$ and $\mathbf{U} = \begin{bmatrix} u_1 \\ u_2 \\ u_3 \end{bmatrix}$, then

$$\mathbf{X} + \mathbf{U} = \begin{bmatrix} x_1 + u_1 \\ x_2 + u_2 \\ x_3 + u_3 \end{bmatrix}.$$

We *multiply* a vector by a scalar c by multiplying each of the coordinates by c, so

$$c\mathbf{X} = c\begin{bmatrix} x_1 \\ x_2 \\ x_3 \end{bmatrix} = \begin{bmatrix} cx_1 \\ cx_2 \\ cx_3 \end{bmatrix}.$$

We set $\mathbf{E}_1 = \begin{pmatrix}1\\0\\0\end{pmatrix}$, $\mathbf{E}_2 = \begin{pmatrix}0\\1\\0\end{pmatrix}$, and $\mathbf{E}_3 = \begin{pmatrix}0\\0\\1\end{pmatrix}$, and we call these the *basis vectors* of 3-space. The first *coordinate axis* is then obtained by taking all multiples $x_1\mathbf{E}_1 = x_1\begin{pmatrix}1\\0\\0\end{pmatrix} = \begin{pmatrix}x_1\\0\\0\end{pmatrix}$, of $\mathbf{E}_1$, and the second and third coordinate axes are defined similarly. Any vector $\mathbf{X}$ may be expressed uniquely as a sum of vectors on the three coordinate axes:

$$\mathbf{X} = \begin{pmatrix}x_1\\x_2\\x_3\end{pmatrix} = \begin{pmatrix}x_1\\0\\0\end{pmatrix} + \begin{pmatrix}0\\x_2\\0\end{pmatrix} + \begin{pmatrix}0\\0\\x_3\end{pmatrix} = x_1\mathbf{E}_1 + x_2\mathbf{E}_2 + x_3\mathbf{E}_3 .$$

Geometrically, we may think of $\mathbf{X}$ as a diagonal segment in a rectangular prism with edges parallel to the coordinate axes (see Fig. 3.1).

Let $\mathbf{X} = \begin{pmatrix}x_1\\x_2\\x_3\end{pmatrix}$ and $\mathbf{U} = \begin{pmatrix}u_1\\u_2\\u_3\end{pmatrix}$ be two vectors. What is the geometric description of the vector $\mathbf{X} + \mathbf{U}$? (See Fig. 3.2.) By analogy with the situation in Section 2.0, we expect to obtain $\mathbf{X} + \mathbf{U}$ by moving the segment $\mathbf{U}$ parallel to itself so that its starting point lands at $\mathbf{X}$, and then taking its endpoint. To see that this expectation is correct, we can reason as follows: If we move the segment $\mathbf{U}$ first in the x-direction by x_1 units, then in the

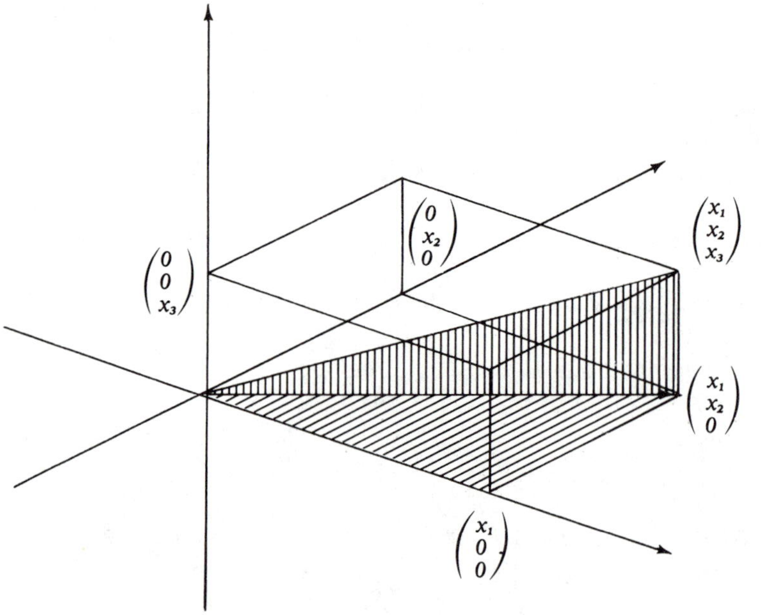

Figure 3.1

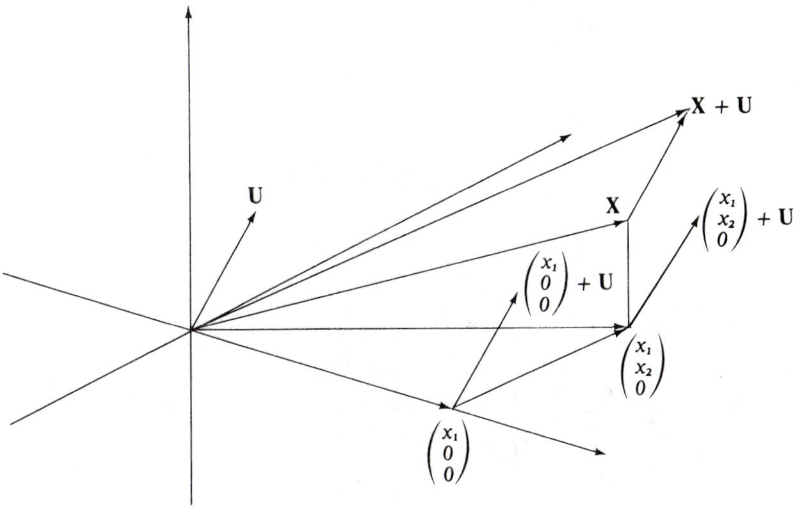

Figure 3.2

y-direction by x_2 units, and finally in the z-direction by x_3 units, it will at all times remain parallel to its original position and in its final position it will start at **X**. If we move the segment **U** by x_1 units in the x-direction, its new endpoint is $\begin{bmatrix} u_1 + x_1 \\ u_2 \\ u_3 \end{bmatrix}$. Taking the corresponding steps in the y- and z-directions, we get $\begin{bmatrix} u_1 + x_1 \\ u_2 + x_2 \\ u_3 + x_3 \end{bmatrix} = \mathbf{U} + \mathbf{X}$ for the final position of the endpoint.

By the *difference* of two vectors **X** and **U**, we mean the vector $\mathbf{X} + (-\mathbf{U})$, which we add to **U** to get **X** (see Fig. 3.3). We may think of $\mathbf{X} + (-\mathbf{U})$ as the vector from the origin which is parallel to the segment from the endpoint of **U** to the endpoint of **X** and has the same length and direction. We often write $\mathbf{X} - \mathbf{U}$ for the sum $\mathbf{X} + (-\mathbf{U})$.

As in the case of the plane, we can establish the following properties of vector addition and scalar multiplication in 3-space: For all vectors **X**, **U**, **A**, and all scalars r, s, we have

(i) $\mathbf{X} + \mathbf{U} = \mathbf{U} + \mathbf{X}$;
(ii) $(\mathbf{X} + \mathbf{U}) + \mathbf{A} = \mathbf{X} + (\mathbf{U} + \mathbf{A})$;
(iii) there is a vector $\mathbf{0}$ such that $\mathbf{X} + \mathbf{0} = \mathbf{X} = \mathbf{0} + \mathbf{X}$ for all **X**;
(iv) For any **X** there is a vector $-\mathbf{X}$ such that $\mathbf{X} + (-\mathbf{X}) = \mathbf{0}$;
(v) $r(\mathbf{X} + \mathbf{U}) = r\mathbf{X} + r\mathbf{U}$;
(vi) $(r + s)(\mathbf{X}) = r\mathbf{X} + s\mathbf{X}$;
(vii) $r(s\mathbf{X}) = (rs)\mathbf{X}$;
(viii) $1 \cdot \mathbf{X} = \mathbf{X}$ for each **X**.

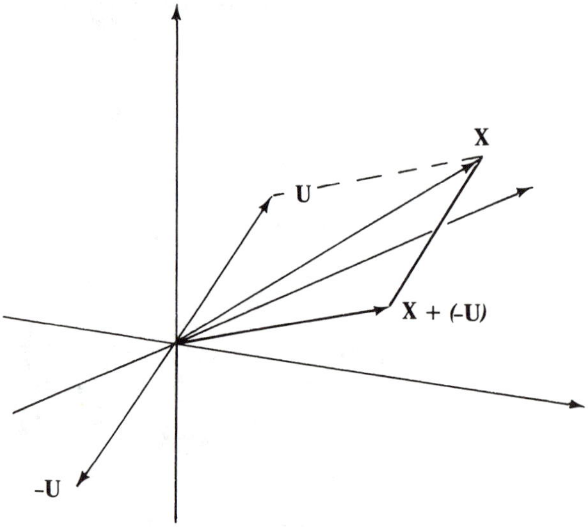

Figure 3.3

Each of these properties can be established by referring to the coordinatewise definitions of addition and scalar multiplication.

If **A** and **B** are vectors in $\mathbb{R}^3$ and s and t are scalars, the expression

$$s\mathbf{A} + t\mathbf{B}$$

is called a *linear combination* of **A** and **B**.

As in two dimensions, the vectors **A** and **B** in $\mathbb{R}^3$ are said to be *linearly dependent* if one is a scalar multiple of the other. A collection of three vectors, **A**, **B**, **C** is said to be *linearly dependent* if one of the vectors is a linear combination of the other two vectors.

EXAMPLE 1. If $\mathbf{A} = \begin{pmatrix} 2 \\ 1 \\ 3 \end{pmatrix}$, $\mathbf{B} = \begin{pmatrix} -1 \\ 0 \\ -2 \end{pmatrix}$, and $\mathbf{C} = \begin{pmatrix} 5 \\ 2 \\ 8 \end{pmatrix}$, then the triplet **A, B, C** is linearly dependent because

$$\mathbf{C} = 2\mathbf{A} + (-1)\mathbf{B}.$$

If a collection of vectors is not linearly dependent, it is said to be *linearly independent*.

EXAMPLE 2. The vectors $\begin{pmatrix} 1 \\ 0 \\ 0 \end{pmatrix}$, $\begin{pmatrix} 0 \\ 1 \\ 0 \end{pmatrix}$, and $\begin{pmatrix} 0 \\ 0 \\ 1 \end{pmatrix}$ are linearly independent since it is impossible to write $\begin{pmatrix} 0 \\ 0 \\ 1 \end{pmatrix}$ as $s\begin{pmatrix} 1 \\ 0 \\ 0 \end{pmatrix} + t\begin{pmatrix} 0 \\ 1 \\ 0 \end{pmatrix}$, and, similarly, for $\begin{pmatrix} 0 \\ 1 \\ 0 \end{pmatrix}$ and $\begin{pmatrix} 0 \\ 0 \\ 1 \end{pmatrix}$.

Exercise 1. Prove that the set of three vectors **A, B, C** is linearly dependent if and only if it is possible to find scalars r, s, t *not all zero*, such that

$$r\mathbf{A} + s\mathbf{B} + t\mathbf{C} = \mathbf{0}.$$

Show that under these conditions we can write one of the vectors as a linear combination of the other two and show the converse.

An extremely useful notion which helps to express many of the ideas of the geometry of 3-space is the *dot product*. We define

$$\mathbf{X} \cdot \mathbf{U} = \begin{pmatrix} x_1 \\ x_2 \\ x_3 \end{pmatrix} \cdot \begin{pmatrix} u_1 \\ u_2 \\ u_3 \end{pmatrix} = x_1u_1 + x_2u_2 + x_3u_3 .$$

The dot product of two vectors is a *scalar*, e.g.,

$$\begin{pmatrix} 2 \\ 1 \\ 3 \end{pmatrix} \cdot \begin{pmatrix} 2 \\ 4 \\ -6 \end{pmatrix} = 2 \cdot 2 + 1 \cdot 4 + 3 \cdot (-6) = -10.$$

We have $\mathbf{E}_1 \cdot \mathbf{E}_2 = \begin{pmatrix} 1 \\ 0 \\ 0 \end{pmatrix} \cdot \begin{pmatrix} 0 \\ 1 \\ 0 \end{pmatrix} = 0$ and $\mathbf{E}_1 \cdot \mathbf{E}_1 = \begin{pmatrix} 1 \\ 0 \\ 0 \end{pmatrix} \cdot \begin{pmatrix} 1 \\ 0 \\ 0 \end{pmatrix} = 1$. Similarly, $\mathbf{E}_2 \cdot \mathbf{E}_3 = 0 = \mathbf{E}_3 \cdot \mathbf{E}_1$, while $\mathbf{E}_2 \cdot \mathbf{E}_2 = 1 = \mathbf{E}_3 \cdot \mathbf{E}_3$.

As in the plane, the dot product behaves somewhat like the ordinary product of numbers. We have the distributive property $(\mathbf{X} + \mathbf{Y}) \cdot \mathbf{U} = \mathbf{X} \cdot \mathbf{U} + \mathbf{Y} \cdot \mathbf{U}$ and the commutative property $\mathbf{X} \cdot \mathbf{U} = \mathbf{U} \cdot \mathbf{X}$. Moreover, for scalar multiplication, we have $(t\mathbf{X}) \cdot \mathbf{U} = t(\mathbf{X} \cdot \mathbf{U})$. To prove this last statement, note that

$$(t\mathbf{X}) \cdot \mathbf{U} = \begin{pmatrix} tx_1 \\ tx_2 \\ tx_3 \end{pmatrix} \cdot \begin{pmatrix} u_1 \\ u_2 \\ u_3 \end{pmatrix} = (tx_1)u_1 + (tx_2)u_2 + (tx_3)u_3$$
$$= t(x_1u_1 + x_2u_2 + x_3u_3) = t(\mathbf{X} \cdot \mathbf{U}).$$

The other properties also have straightforward proofs in terms of coordinates.

By the Pythagorean Theorem in 3-space, the distance from a point $\mathbf{X} = \begin{pmatrix} x_1 \\ x_2 \\ x_3 \end{pmatrix}$ to the origin is $\sqrt{x_1^2 + x_2^2 + x_3^2}$, and we define this number to be the *length* of the vector $\mathbf{X}$, written $|\mathbf{X}|$. For example, $\left|\begin{pmatrix} 3 \\ 4 \\ 12 \end{pmatrix}\right| = \sqrt{169} = 13$, while $|\mathbf{E}_i| = 1$ for each i and $|\mathbf{0}| = 0$.

In general, $\mathbf{X} \cdot \mathbf{X} = \begin{pmatrix} x_1 \\ x_2 \\ x_3 \end{pmatrix} \cdot \begin{pmatrix} x_1 \\ x_2 \\ x_3 \end{pmatrix} = x_1^2 + x_2^2 + x_3^2$, and so this is the square of the length $|\mathbf{X}|$ of the vector $\mathbf{X}$. Note that for any scalar c, we have

$|c\mathbf{X}| = \sqrt{c\mathbf{X} \cdot c\mathbf{X}} = \sqrt{c^2\mathbf{X} \cdot \mathbf{X}} = |c|\sqrt{\mathbf{X} \cdot \mathbf{X}} = |c|\,|\mathbf{X}|$, where $|c| = \sqrt{c^2}$ is the absolute value of the scalar c. We have $|\mathbf{X}| \geqslant 0$ with $|\mathbf{X}| = 0$ if and only if $\mathbf{X} = \mathbf{0}$.

In terms of the notion of dot product, we shall now treat seven basic geometric problems:

(i) To decide when two given vectors $\mathbf{X}$ and $\mathbf{U}$ are perpendicular.
(ii) To calculate the angle between two vectors.
(iii) To find the projection of a given vector on a given line through the origin.
(iv) To find the projection of a given vector on a given plane through the origin.
(v) To compute the distance from a given point to a given plane through the origin.
(vi) To compute the distance from a given point to a given line through the origin.
(vii) To compute the area of the parallelogram formed by two vectors in 3-space.

(i) As in the case of the plane, we may use the law of cosines to give an interpretation of the dot product in terms of the lengths $|\mathbf{X}|$ and $|\mathbf{U}|$ of the vectors $\mathbf{X}$ and $\mathbf{U}$ and the angle θ between them. The law of cosines states

$$|\mathbf{X} - \mathbf{U}|^2 = |\mathbf{X}|^2 + |\mathbf{U}|^2 - 2|\mathbf{X}|\,|\mathbf{U}|\cos\theta.$$

But

$$\begin{aligned}|\mathbf{X} - \mathbf{U}|^2 &= (\mathbf{X} - \mathbf{U}) \cdot (\mathbf{X} - \mathbf{U}) = \mathbf{X} \cdot \mathbf{X} - 2\mathbf{X} \cdot \mathbf{U} + \mathbf{U} \cdot \mathbf{U} \\ &= |\mathbf{X}|^2 + |\mathbf{U}|^2 - 2\mathbf{X} \cdot \mathbf{U}.\end{aligned}$$

Thus

$$\mathbf{X} \cdot \mathbf{U} = |\mathbf{X}|\,|\mathbf{U}|\cos\theta.$$

The vectors $\mathbf{X}$ and $\mathbf{U}$ will be perpendicular if and only if $\cos\theta = 0$, so if and only if $\mathbf{X} \cdot \mathbf{U} = 0$.

(ii) If $\mathbf{X}$ and $\mathbf{U}$ are not perpendicular, we may use the dot product to compute the cosine of the angle between $\mathbf{X}$ and $\mathbf{U}$. For example, if $\mathbf{X} = \begin{pmatrix} 1 \\ 2 \\ 1 \end{pmatrix}$ and $\mathbf{U} = \begin{pmatrix} -1 \\ 1 \\ 3 \end{pmatrix}$, then $\mathbf{X} \cdot \mathbf{U} = (1)(-1) + (2)(1) + (1)(3) = 4$, $|\mathbf{X}| = \sqrt{6}$, $|\mathbf{U}| = \sqrt{11}$, so $\cos\theta = 4/\sqrt{6}\,\sqrt{11}$.

(iii) The projection $P(\mathbf{X})$ of a given vector $\mathbf{X}$ to the line of multiples of a given vector $\mathbf{U}$ is the vector $t\mathbf{U}$ such that $\mathbf{X} - t\mathbf{U}$ is perpendicular to $\mathbf{U}$ (see Fig. 3.4). This condition enables us to compute t since $0 = (\mathbf{X} - t\mathbf{U}) \cdot \mathbf{U} = \mathbf{X} \cdot \mathbf{U} - (t\mathbf{U}) \cdot \mathbf{U} = \mathbf{X} \cdot \mathbf{U} - t(\mathbf{U} \cdot \mathbf{U})$, so $t = \mathbf{X} \cdot \mathbf{U}/\mathbf{U} \cdot \mathbf{U}$ and we have a

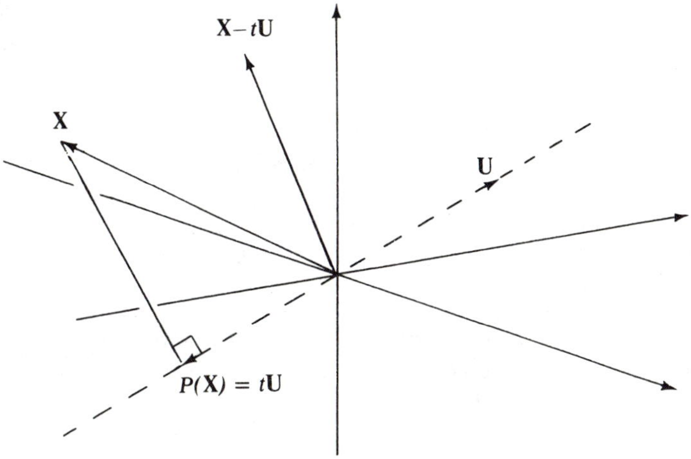

Figure 3.4

formula for the projection:

$$P(\mathbf{X}) = \left(\frac{\mathbf{X}\cdot\mathbf{U}}{\mathbf{U}\cdot\mathbf{U}}\right)\mathbf{U}. \tag{1}$$

Note that this is the same as the formula we obtained in the 2-dimensional case.

Note also that

$$P(\mathbf{X})\cdot\mathbf{U} = \left(\left(\frac{\mathbf{X}\cdot\mathbf{U}}{\mathbf{U}\cdot\mathbf{U}}\right)\mathbf{U}\right)\cdot\mathbf{U} = \left(\frac{\mathbf{X}\cdot\mathbf{U}}{\mathbf{U}\cdot\mathbf{U}}\right)(\mathbf{U}\cdot\mathbf{U}) = \mathbf{X}\cdot\mathbf{U}. \tag{2}$$

(iv) Fix a nonzero vector $\mathbf{U} = \begin{bmatrix} u_1 \\ u_2 \\ u_3 \end{bmatrix}$ and denote by Π the plane through the origin which is orthogonal to $\mathbf{U}$. Since $\mathbf{Y} = \begin{bmatrix} y_1 \\ y_2 \\ y_3 \end{bmatrix}$ lies in Π if and only if $\mathbf{Y}\cdot\mathbf{U} = 0$, an equation of Π is

$$u_1 y_1 + u_2 y_2 + u_3 y_3 = 0. \tag{3}$$

Let $\mathbf{X}$ be a vector. We denote by $Q(\mathbf{X})$ the *projection of* $\mathbf{X}$ *on* Π, i.e., the foot of the perpendicular dropped from $\mathbf{X}$ to Π. Then the segment joining $\mathbf{X}$ and $Q(\mathbf{X})$ is parallel to $\mathbf{U}$, so for some scalar t,

$$\mathbf{X} - Q(\mathbf{X}) = t\mathbf{U}.$$

Then $\mathbf{X} - t\mathbf{U} = Q(\mathbf{X})$. Since $Q(\mathbf{X})$ is in Π, $\mathbf{X} - t\mathbf{U}$ is perpendicular to $\mathbf{U}$. Then, by the discussion of (iii), $\mathbf{X} - Q(\mathbf{X}) = P(\mathbf{X})$ or

$$Q(\mathbf{X}) = \mathbf{X} - P(\mathbf{X}).$$

(v) The *distance* from the point $\mathbf{X}$ to the plane through the origin perpendicular to $\mathbf{U}$ is precisely the length of the projection vector $P(\mathbf{X})$, i.e.,

$$|P(\mathbf{X})| = \left|\frac{\mathbf{X}\cdot\mathbf{U}}{\mathbf{U}\cdot\mathbf{U}}\right||\mathbf{U}| = \frac{|\mathbf{X}\cdot\mathbf{U}|}{|\mathbf{U}|}.$$

It follows that the distance d from the point $\mathbf{X}$ to the plane through the origin perpendicular to $\mathbf{U}$ with the equation $x_1u_1 + x_2u_2 + x_3u_3 = 0$ is given by

$$d = \frac{|x_1u_1 + x_2u_2 + x_3u_3|}{\sqrt{u_1^2 + u_2^2 + u_3^2}}. \tag{4}$$

(vi) *The distance from the point* $\mathbf{X}$ *to the line along* $\mathbf{U}$ with $\mathbf{U} \neq \mathbf{0}$ is the length of the difference vector $|\mathbf{X} - P(\mathbf{X})|$. Since $\mathbf{X} - P(\mathbf{X})$ is perpendicular to $P(\mathbf{X})$, we get $|\mathbf{X}|^2 = |P(\mathbf{X})|^2 + |\mathbf{X} - P(\mathbf{X})|^2$, so

$$|\mathbf{X} - P(\mathbf{X})|^2 = |\mathbf{X}|^2 - |P(\mathbf{X})|^2$$

and hence

$$|\mathbf{X} - P(\mathbf{X})|^2 = \mathbf{X}\cdot\mathbf{X} - \frac{|\mathbf{X}\cdot\mathbf{U}|^2}{|\mathbf{U}|^2},$$

(see Fig. 3.5) and so the distance is given by

$$\sqrt{\mathbf{X}\cdot\mathbf{X} - \frac{(\mathbf{X}\cdot\mathbf{U})^2}{(\mathbf{U}\cdot\mathbf{U})}} = \frac{\sqrt{(\mathbf{X}\cdot\mathbf{X})(\mathbf{U}\cdot\mathbf{U}) - (\mathbf{X}\cdot\mathbf{U})^2}}{|\mathbf{U}|}. \tag{5}$$

(vii) Let $\mathbf{A}$, $\mathbf{B}$ be two vectors and let Π denote the parallelogram with two sides along $\mathbf{A}$ and $\mathbf{B}$ (see Fig. 3.6). The area of Π is the product of the base

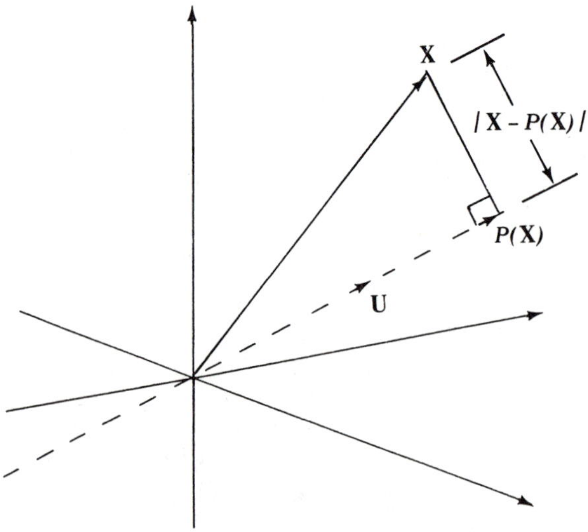

Figure 3.5

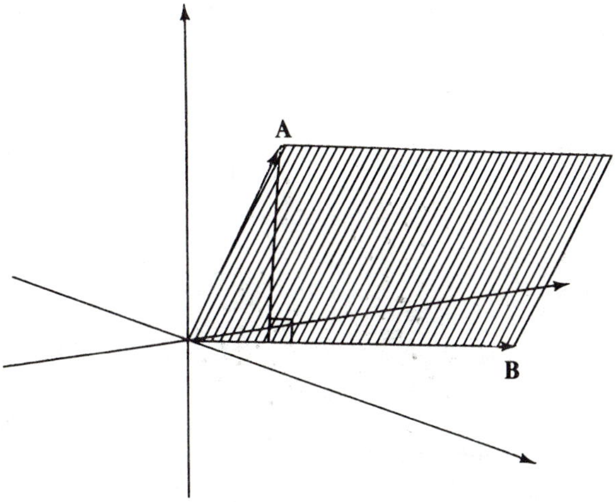

Figure 3.6

$|\mathbf{B}|$, and the altitude on that base which is the distance from **A** to **B**. By (5), this distance $= (1/|\mathbf{B}|)\sqrt{(\mathbf{A}\cdot\mathbf{A})(\mathbf{B}\cdot\mathbf{B})-(\mathbf{A}\cdot\mathbf{B})^2}$, so

$$\text{area}\,\Pi = \sqrt{(\mathbf{A}\cdot\mathbf{A})(\mathbf{B}\cdot\mathbf{B})-(\mathbf{A}\cdot\mathbf{B})^2}\,. \tag{6}$$

§1. The Cross Product and Systems of Equations

Consider a system of two equations in three unknowns:

$$\begin{aligned} a_1x_1 + a_2x_2 + a_3x_3 &= 0,\\ b_1x_1 + b_2x_2 + b_3x_3 &= 0. \end{aligned} \tag{7}$$

We set $\mathbf{A} = \begin{bmatrix} a_1\\ a_2\\ a_3 \end{bmatrix}$ and $\mathbf{B} = \begin{bmatrix} b_1\\ b_2\\ b_3 \end{bmatrix}$. A solution vector $\mathbf{X} = \begin{bmatrix} x_1\\ x_2\\ x_3 \end{bmatrix}$ for (7) satisfies

$$\mathbf{A}\cdot\mathbf{X} = 0, \qquad \mathbf{B}\cdot\mathbf{X} = 0.$$

We may find such an **X** by multiplying the first equation by b_1 and the second by a_1 and subtracting

$$\begin{aligned} a_1b_1x_1 + a_2b_1x_2 + a_3b_1x_3 &= 0,\\ a_1b_1x_1 + a_1b_2x_2 + a_1b_3x_3 &= 0,\\ (a_2b_1 - a_1b_2)x_2 + (a_3b_1 - a_1b_3)x_3 &= 0. \end{aligned} \tag{8a}$$

Similarly, we may multiply the first equation by b_2 and the second by a_2 and subtract to get

$$(a_1b_2 - a_2b_1)x_1 + (a_3b_2 - a_2b_3)x_3 = 0. \tag{8b}$$

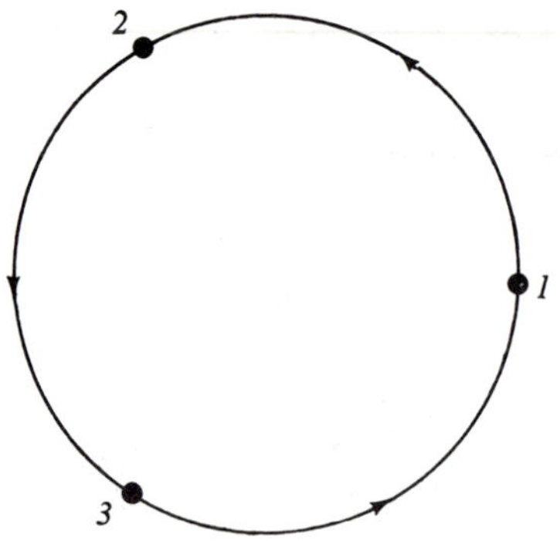

Figure 3.7

We can obtain a solution to the system (8a), (8b) by choosing

$$x_1 = (a_2b_3 - a_3b_2), \qquad x_2 = (a_3b_1 - a_1b_3), \qquad x_3 = (a_1b_2 - a_2b_1). \quad (9)$$

Note that if we think of the subscripts 1, 2, 3 on a wheel, with 1 followed by 2, followed by 3, and three followed by 1 (see Fig. 3.7), then in (9), x_2 is obtained from x_1, and x_3 is obtained from x_2 by following this succession of subscripts of a_i, b_i. We define the *cross product* $\mathbf{A} \times \mathbf{B}$ of $\mathbf{A}$ and $\mathbf{B}$ to be the vector

$$\mathbf{A} \times \mathbf{B} = \begin{bmatrix} a_2b_3 - a_3b_2 \\ a_3b_1 - a_1b_3 \\ a_1b_2 - a_2b_1 \end{bmatrix}. \quad (10)$$

The vector $\mathbf{X} = \mathbf{A} \times \mathbf{B}$ indeed satisfies the conditions

$$\mathbf{A} \cdot \mathbf{X} = 0, \qquad \mathbf{B} \cdot \mathbf{X} = 0,$$

which we set out to satisfy, and this is so since in the expression

$$\mathbf{A} \cdot \mathbf{X} = a_1(a_2b_3 - a_3b_2) + a_2(a_3b_1 - a_1b_3) + a_3(a_1b_2 - a_2b_1)$$

all terms cancel, leaving 0. The same happens for $\mathbf{B} \cdot \mathbf{X}$.

We shall see that the cross product is very useful in solving geometric problems in 3 dimensions.

We may easily verify that the cross product has the following properties:

(i) $\mathbf{A} \times \mathbf{A} = \mathbf{0}$, for every vector $\mathbf{A}$.
(ii) $\mathbf{B} \times \mathbf{A} = -\mathbf{A} \times \mathbf{B}$, for all $\mathbf{A}$, $\mathbf{B}$.
(iii) $\mathbf{A} \times (\mathbf{B} + \mathbf{C}) = \mathbf{A} \times \mathbf{B} + \mathbf{A} \times \mathbf{C}$, for all $\mathbf{A}$, $\mathbf{B}$, $\mathbf{C}$.
(iv) $(t\mathbf{A}) \times \mathbf{B} = t(\mathbf{A} \times \mathbf{B})$ if t is a scalar.

Exercise 2. Show that $\mathbf{E}_1 \times \mathbf{E}_2 = \mathbf{E}_3$, $\mathbf{E}_2 \times \mathbf{E}_3 = \mathbf{E}_1$, and $\mathbf{E}_1 \times \mathbf{E}_3 = -\mathbf{E}_2$.

Exercise 3. Show that $\mathbf{A} \times \mathbf{B} = -\mathbf{B} \times \mathbf{A}$.

Exercise 4. Show that $\mathbf{A} \times (\mathbf{B} + \mathbf{C}) = (\mathbf{A} \times \mathbf{B}) + (\mathbf{A} \times \mathbf{C})$.

Exercise 5. Show that $t\mathbf{A} \times \mathbf{B} = t(\mathbf{A} \times \mathbf{B})$.

Exercise 6. Show that $\mathbf{A} \cdot (\mathbf{B} \times \mathbf{C}) = \mathbf{B} \cdot (\mathbf{C} \times \mathbf{A}) = \mathbf{C} \cdot (\mathbf{A} \times \mathbf{B})$.

We now prove some propositions about the cross product.

(v). *If* **A** *and* **B** *are linearly dependent, then* $\mathbf{A} \times \mathbf{B} = \mathbf{0}$.

PROOF. If $\mathbf{B} = \mathbf{0}$, then $\mathbf{A} \times \mathbf{B} = \mathbf{0}$ automatically. If $\mathbf{B} \neq \mathbf{0}$ and $\mathbf{A} = t\mathbf{B}$, then $\mathbf{A} \times \mathbf{B} = (t\mathbf{B}) \times \mathbf{B} = t(\mathbf{B} \times \mathbf{B}) = \mathbf{0}$.

(vi). *Conversely, if* $\mathbf{A} \times \mathbf{B} = \mathbf{0}$, *then* **A** *and* **B** *are linearly dependent.*

PROOF. If $\mathbf{A} \times \mathbf{B} = \mathbf{0}$, then either $\mathbf{B} = \mathbf{0}$ or at least one of the components of **B** is nonzero. Assume $b_3 \neq 0$. Then $a_3b_2 - a_2b_3 = 0$, so $a_2 = (a_3/b_3)b_2$ and $-a_3b_1 + a_1b_3 = 0$, so $a_1 = (a_3/b_3)b_1$. It follows that

$$\mathbf{A} = \begin{bmatrix} a_1 \\ a_2 \\ a_3 \end{bmatrix} = \begin{bmatrix} (a_3/b_3)b_1 \\ (a_3/b_3)b_2 \\ (a_3/b_3)b_3 \end{bmatrix} = (a_3/b_3)\begin{bmatrix} b_1 \\ b_2 \\ b_3 \end{bmatrix} = (a_3/b_3)\mathbf{B}.$$

Therefore, **A** is a scalar multiple of **B**, so **A** and **B** are linearly dependent. We reason similarly if $b_2 \neq 0$ or $b_1 \neq 0$.

Exercise 7. If **A**, **B**, **C** are linearly dependent, show that $\mathbf{A} \cdot (\mathbf{B} \times \mathbf{C}) = \mathbf{B} \cdot (\mathbf{C} \times \mathbf{A}) = \mathbf{C} \cdot (\mathbf{A} \times \mathbf{B}) = 0$.

(vii). *Let* Π *denote the parallelogram in* $\mathbb{R}^3$ *with sides along* **A** *and* **B**, *where* $\mathbf{B} \neq \mathbf{0}$ *(see Fig. 3.8). Then*

$$\text{area}\,\Pi = |\mathbf{A} \times \mathbf{B}|. \tag{11}$$

PROOF. We have already found formula (6) for the area Π:

$$\text{area}\,\Pi = \sqrt{(\mathbf{A} \cdot \mathbf{A})(\mathbf{B} \cdot \mathbf{B}) - (\mathbf{A} \cdot \mathbf{B})^2}\,.$$

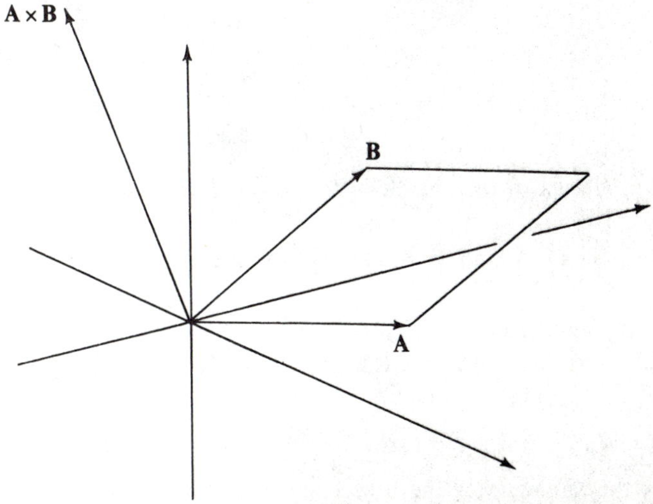

Figure 3.8

In coordinates, this gives

$$
\begin{aligned}
(\text{area}\,\Pi)^2 &= (a_1^2 + a_2^2 + a_3^2)(b_1^2 + b_2^2 + b_3^2) - (a_1b_1 + a_2b_2 + a_3b_3)^2 \\
&= a_1^2b_1^2 + a_1^2b_2^2 + a_1^2b_3^2 - (a_1b_1)^2 - 2a_1b_1a_2b_2 \\
&\quad + a_2^2b_1^2 + a_2^2b_2^2 + a_2^2b_3^2 - (a_2b_2)^2 - 2a_1b_1a_3b_3 \\
&\quad + a_3^2b_1^2 + a_3^2b_2^2 + a_3^2b_3^2 - (a_3b_3)^2 - 2a_2b_2a_3b_3 \\
&= (a_1b_2 - a_2b_1)^2 + (a_3b_1 - a_1b_3)^2 + (a_2b_3 - a_3b_2)^2 \\
&= |\mathbf{A} \times \mathbf{B}|^2.
\end{aligned}
$$

This establishes formula (11).

Observe that if **A** and **B** are linearly dependent, then the parallelogram Π will be contained in a line so its area will be 0, agreeing with the fact that $\mathbf{A} \times \mathbf{B} = \mathbf{0}$ in this case.

Let us now summarize what we have found.

If **A** and **B** are not linearly dependent, then we may describe the vector $\mathbf{A} \times \mathbf{B}$ by saying that it is perpendicular to **A** and **B** and it has length equal to the area of the parallelogram determined by **A** and **B**. Note that this description applies equally well to *two* vectors, $\mathbf{A} \times \mathbf{B}$ and $-(\mathbf{A} \times \mathbf{B})$ lying on opposite sides of the plane containing **A** and **B**.

In Chapter 3.5 we will go more deeply into the significance of the sign of $\mathbf{A} \times \mathbf{B}$.

Next, we shall give a generalization of the Pythagorean Theorem. If we project **A** and **B** into the x_1x_2 plane, we get the vectors $\begin{bmatrix} a_1 \\ a_2 \\ 0 \end{bmatrix}$ and $\begin{bmatrix} b_1 \\ b_2 \\ 0 \end{bmatrix}$. Note that the area of the parallelogram Π_{12} determined by these two vectors is given by

$$\left|\begin{bmatrix} a_1 \\ a_2 \\ 0 \end{bmatrix} \times \begin{bmatrix} b_1 \\ b_2 \\ 0 \end{bmatrix}\right| = \left|\begin{bmatrix} 0 \\ 0 \\ a_1b_2 - a_2b_1 \end{bmatrix}\right| = |a_1b_2 - a_2b_1|.$$

Similarly, the area of the parallelogram Π_{23} determined by the projections $\begin{bmatrix} 0 \\ a_2 \\ a_3 \end{bmatrix}$ and $\begin{bmatrix} 0 \\ b_2 \\ b_3 \end{bmatrix}$ of **A** and **B** to the x_2x_3 plane is given by area $(\Pi_{23}) = |a_3b_2 - a_2b_3|$, and finally area $\Pi_{13} = |a_3b_1 - a_1b_3|$ (see Fig. 3.9).

Formula (vii) thus yields the following striking result which is a generalization of the Pythagorean Theorem:

$$\text{(viii)} \qquad (\text{area}\,\Pi)^2 = (\text{area}\,\Pi_{12})^2 + (\text{area}\,\Pi_{23})^2 + (\text{area}\,\Pi_{13})^2.$$

This is the analogue of the theorem that the square of the length of a vector is the sum of the squares of its projections to the three coordinate

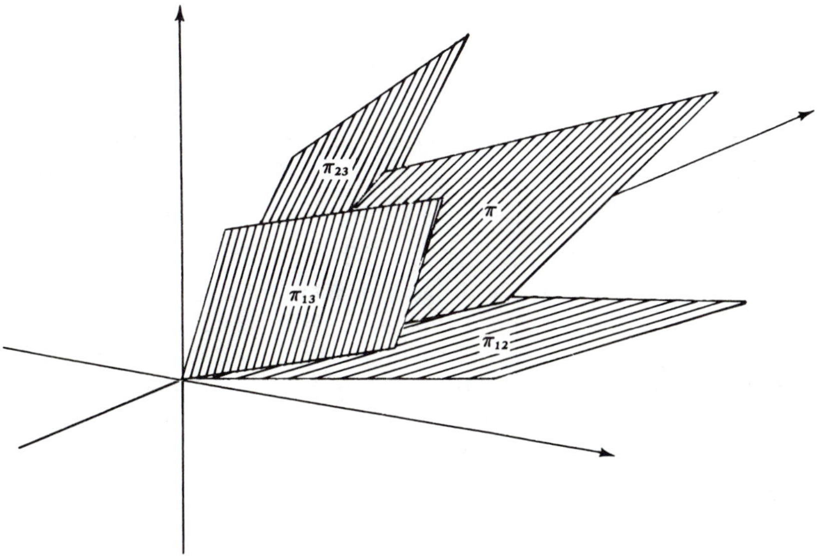

Figure 3.9

axes (see Fig. 3.10):

(ix) Given three vectors **A**, **B**, **C** in 3-space, we shall next obtain a formula for the volume of the parallelepiped determined by these three vectors.

If **A** and **B** are linearly independent, then the distance from the vector **C** to the plane determined by **A** and **B** equals the length of the projection of **C** to the line along $\mathbf{A} \times \mathbf{B}$, since the vector $\mathbf{A} \times \mathbf{B}$ is orthogonal to that plane.

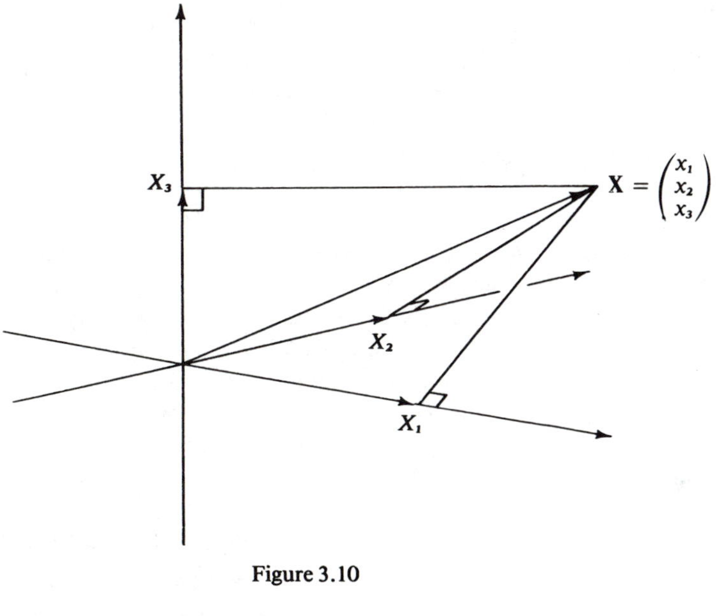

Figure 3.10

So the distance is given by

$$\frac{|\mathbf{C}\cdot(\mathbf{A}\times\mathbf{B})|}{|\mathbf{A}\times\mathbf{B}|}\,. \tag{12}$$

(x) It follows that the volume of the parallelepiped Π with sides along **A**, **B**, and **C** is given by the area of the base $|\mathbf{A}\times\mathbf{B}|$ multiplied by the height $|\mathbf{C}\cdot(\mathbf{A}\times\mathbf{B})|/|\mathbf{A}\times\mathbf{B}|$, i.e.,

$$\text{volume}\,\Pi = |\mathbf{C}\cdot(\mathbf{A}\times\mathbf{B})|. \tag{13}$$

If **A**, **B**, and **C** are linearly dependent, then the parallelepiped is contained in a plane so its volume will be zero, which agrees with the fact that $\mathbf{C}\cdot(\mathbf{A}\times\mathbf{B}) = 0$ in such a case, as we saw in Exercise 7.

We shall next see how the cross product helps us to study systems of linear equations.

We consider a system of three equations in the three unknowns x_1, x_2, x_3:

$$\begin{aligned} a_1x_1 + a_2x_2 + a_3x_3 &= 0,\\ b_1x_1 + b_2x_2 + b_3x_3 &= 0,\\ c_1x_1 + c_2x_2 + c_3x_3 &= 0. \end{aligned} \tag{14}$$

We set $\mathbf{A} = \begin{pmatrix} a_1\\ a_2\\ a_3\end{pmatrix}$, $\mathbf{B} = \begin{pmatrix} b_1\\ b_2\\ b_3\end{pmatrix}$, $\mathbf{C} = \begin{pmatrix} c_1\\ c_2\\ c_3\end{pmatrix}$. The solutions $\mathbf{X} = \begin{pmatrix} x_1\\ x_2\\ x_3\end{pmatrix}$ of the system (14) are the vectors **X** such that **X** is perpendicular to the three vectors **A**, **B**, **C**. We shall show the following:

(xi). *There exists a nonzero solution vector* **X** *of* (14) *if and only if* $\mathbf{C}\cdot(\mathbf{A}\times\mathbf{B}) = 0$.

PROOF. Assume that $\mathbf{C}\cdot(\mathbf{A}\times\mathbf{B}) = 0$. If **A** and **B** are linearly dependent, there is some plane Π through the origin which contains **A**, **B**, and **C**. We choose a nonzero vector **X** perpendicular to Π. Then $\mathbf{X}\cdot\mathbf{A} = 0$, $\mathbf{X}\cdot\mathbf{B} = 0$, $\mathbf{X}\cdot\mathbf{C} = 0$, so **X** is a solution of (14).

If **A** and **B** are linearly independent, then $\mathbf{A}\times\mathbf{B} \neq 0$. By assumption, $(\mathbf{A}\times\mathbf{B})\cdot\mathbf{C} = 0$. Also $(\mathbf{A}\times\mathbf{B})\cdot\mathbf{A} = 0$ and $(\mathbf{A}\times\mathbf{B})\cdot\mathbf{B} = 0$. So $\mathbf{A}\times\mathbf{B}$ is a nonzero solution of (14).

Conversely, assume that $\mathbf{C}\cdot(\mathbf{A}\times\mathbf{B}) \neq 0$. Then **A** and **B** are linearly independent, since otherwise $\mathbf{A}\times\mathbf{B} = \mathbf{0}$ and so $\mathbf{C}\cdot(\mathbf{A}\times\mathbf{B}) = 0$, contrary to assumption. Let Π be the plane through the origin containing **A** and **B**. Then $\mathbf{A}\times\mathbf{B}$ is perpendicular to Π, and every vector perpendicular to Π is a scalar multiple of $\mathbf{A}\times\mathbf{B}$. If **X** is a solution of (14), it follows that $\mathbf{X} = t(\mathbf{A}\times\mathbf{B})$ for some scalar t. Since $\mathbf{X}\cdot\mathbf{C} = 0$, $(t(\mathbf{A}\times\mathbf{B}))\cdot C = t((\mathbf{A}\times\mathbf{B})\cdot\mathbf{C}) = 0$, and so $t = 0$. Hence, $\mathbf{X} = \mathbf{0}$ as claimed.

The following is a fundamental property of the geometry of $\mathbb{R}^3$.

Proposition 1. *If* **A**, **B**, **C** *are three linearly independent vectors in* $\mathbb{R}^3$, *then every vector* **Y** *in* $\mathbb{R}^3$ *can be uniquely expressed as a linear combination of* **A**, **B**, *and* **C**.

PROOF. Since **C** is not a linear combination of **A** and **B**, the line $\{\mathbf{Y} - t\mathbf{C} \mid t \text{ real}\}$ which goes through **Y** and is parallel to **C** will not be parallel to the plane determined by **A** and **B**. Thus, for some t, $\mathbf{Y} - t\mathbf{C}$ lies in that plane and so

$$\mathbf{Y} - t\mathbf{C} = r\mathbf{A} + s\mathbf{B}$$

for suitable scalars r, s. Therefore,

$$\mathbf{Y} = r\mathbf{A} + s\mathbf{B} + t\mathbf{C}. \tag{15}$$

Expression (15) is unique, for if

$$\mathbf{Y} = r'\mathbf{A} + s'\mathbf{B} + t'\mathbf{C},$$

then $r\mathbf{A} + s\mathbf{B} + t\mathbf{C} = r'\mathbf{A} + s'\mathbf{B} + t'\mathbf{C}$, so

$$(r - r')\mathbf{A} + (s - s')\mathbf{B} + (t - t')\mathbf{C} = \mathbf{0}.$$

Since **A**, **B**, **C** are linearly independent, it follows that $r - r' = 0$, $s - s' = 0$, $t - t' = 0$. So the expression (15) is unique, as claimed.

The vectors of length 1 are called *unit vectors*, and their endpoints form the *unit sphere* in 3-space.

Exercise 8. Show that for any choice of angles θ, ϕ, the vector $\begin{bmatrix} \cos\phi\cos\theta \\ \cos\phi\sin\theta \\ \sin\phi \end{bmatrix}$ is a unit vector. Conversely, show that if $1 = u_1^2 + u_2^2 + u_3^2$ and if $u_3 \neq 0$, then it is possible to find an angle ϕ between $-\pi/2$ and $\pi/2$ so that $u_3 = \sin\phi$. The vector $\begin{pmatrix} u_1 \\ u_2 \\ 0 \end{pmatrix}$ then has length $\sqrt{u_1^2 + u_2^2} = \sqrt{1 - u_3^2} = \sqrt{1 - \sin^2\phi} = \cos\phi$, so we may write

$$u_1 = \cos\phi\cos\theta, \qquad u_2 = \cos\phi\sin\theta$$

for some θ between 0 and 2π. Hence $\begin{pmatrix} u_1 \\ u_2 \\ u_3 \end{pmatrix} = \begin{bmatrix} \cos\phi\cos\theta \\ \cos\phi\sin\theta \\ \sin\phi \end{bmatrix}$.

It follows that any nonzero vector **X** in $\mathbb{R}^3$ may be written:

$$\mathbf{X} = |\mathbf{X}| \begin{bmatrix} \cos\phi\cos\theta \\ \cos\phi\sin\theta \\ \sin\phi \end{bmatrix},$$

for some choice of angles ϕ, θ.

CHAPTER 3.1

Transformations of 3-Space

As in the planar case, we define a transformation of 3-space to be a rule T which assigns to every vector $\mathbf{X}$ of 3-space some vector $T(\mathbf{X})$ of 3-space. The vector $T(\mathbf{X})$ is called the *image* of $\mathbf{X}$ under T, and the collection of all vectors which are images of vectors under the transformation T is called the *range* of T. We denote transformations by capital letters, such as A, B, R, S, T, etc.

EXAMPLE 1. Let P denote the transformation which assigns to each vector $\mathbf{X} = \begin{bmatrix} x_1 \\ x_2 \\ x_3 \end{bmatrix}$ the projection to the line along $\mathbf{U} = \begin{bmatrix} 1 \\ 1 \\ 1 \end{bmatrix}$. By formula (1) of Chapter 3.0, we have

$$P(\mathbf{X}) = \left(\frac{\mathbf{X} \cdot \mathbf{U}}{\mathbf{U} \cdot \mathbf{U}}\right)\mathbf{U} = \left(\frac{x_1 + x_2 + x_3}{3}\right)\begin{bmatrix} 1 \\ 1 \\ 1 \end{bmatrix} = \begin{bmatrix} \frac{1}{3}x_1 + \frac{1}{3}x_2 + \frac{1}{3}x_3 \\ \frac{1}{3}x_1 + \frac{1}{3}x_2 + \frac{1}{3}x_3 \\ \frac{1}{3}x_1 + \frac{1}{3}x_2 + \frac{1}{3}x_3 \end{bmatrix}.$$

EXAMPLE 2. Let S denote the transformation which assigns to each vector $\mathbf{X}$ the *reflection of* $\mathbf{X}$ *through the line* along $\mathbf{U} = \begin{bmatrix} 1 \\ 1 \\ 1 \end{bmatrix}$. As in the planar case, $S(\mathbf{X})$ is defined by the condition that the midpoint of the segment between $\mathbf{X}$ and $S(\mathbf{X})$ is the projection of $\mathbf{X}$ to the line along $\mathbf{U}$. Thus

$$S(\mathbf{X}) = 2P(\mathbf{X}) - \mathbf{X}$$

or

$$S\begin{pmatrix} x_1 \\ x_2 \\ x_2 \end{pmatrix} = 2\begin{pmatrix} \frac{1}{3}x_1 + \frac{1}{3}x_2 + \frac{1}{3}x_3 \\ \frac{1}{3}x_1 + \frac{1}{3}x_2 + \frac{1}{3}x_3 \\ \frac{1}{3}x_1 + \frac{1}{3}x_2 + \frac{1}{3}x_3 \end{pmatrix} - \begin{pmatrix} x_1 \\ x_2 \\ x_3 \end{pmatrix} = \begin{pmatrix} -\frac{1}{3}x_1 + \frac{2}{3}x_2 + \frac{2}{3}x_3 \\ \frac{2}{3}x_1 - \frac{1}{3}x_2 + \frac{2}{3}x_3 \\ \frac{2}{3}x_1 + \frac{2}{3}x_2 - \frac{1}{3}x_3 \end{pmatrix}.$$

Exercise 1. In each of the following problems, let P denote projection to the line along $\mathbf{U}$. Find a formula for the coordinates of the image $P\begin{pmatrix} x_1 \\ x_2 \\ x_3 \end{pmatrix}$.

(a) $\mathbf{U} = \begin{pmatrix} 1 \\ 0 \\ 0 \end{pmatrix}$ (so we have projection to the first coordinate axis);

(b) $\mathbf{U} = \begin{pmatrix} 1 \\ 1 \\ 0 \end{pmatrix}$;

(c) $\mathbf{U} = \begin{pmatrix} 1 \\ 1 \\ -1 \end{pmatrix}$;

(d) $\mathbf{U} = \begin{pmatrix} 1 \\ 0 \\ 3 \end{pmatrix}$.

Exercise 2. For each of the vectors in the preceding exercise, find a formula for the reflection $S\begin{pmatrix} x_1 \\ x_2 \\ x_3 \end{pmatrix}$ of the vector $\begin{pmatrix} x_1 \\ x_2 \\ x_2 \end{pmatrix}$ through the line along $\mathbf{U}$.

Example 3. Let $\overline{Q}$ denote projection to the x_2x_3-plane and let $\overline{P}$ denote projection to the x_1-axis. Then

$$\overline{Q}\begin{pmatrix} x_1 \\ x_2 \\ x_3 \end{pmatrix} = \begin{pmatrix} 0 \\ x_2 \\ x_3 \end{pmatrix} \quad \text{and} \quad \overline{P}\begin{pmatrix} x_1 \\ x_2 \\ x_3 \end{pmatrix} = \begin{pmatrix} x_1 \\ 0 \\ 0 \end{pmatrix}.$$

Note that $\overline{Q}(\mathbf{X}) + \overline{P}(\mathbf{X}) = \mathbf{X}$ for each $\mathbf{X}$.

Example 4. Let Q denote projection to the plane through 0 perpendicular to $\mathbf{U} = \begin{pmatrix} 1 \\ 1 \\ 1 \end{pmatrix}$ and let P denote projection to the line along $\mathbf{U} = \begin{pmatrix} 1 \\ 1 \\ 1 \end{pmatrix}$. Then, as in Example 3, we have $Q(\mathbf{X}) + P(\mathbf{X}) = \mathbf{X}$, so by the formula in Example 1, we have

$$Q\begin{pmatrix} x_1 \\ x_2 \\ x_3 \end{pmatrix} = \begin{pmatrix} x_1 \\ x_2 \\ x_3 \end{pmatrix} - \begin{pmatrix} \frac{1}{3}x_1 + \frac{1}{3}x_2 + \frac{1}{3}x_3 \\ \frac{1}{3}x_1 + \frac{1}{3}x_2 + \frac{1}{3}x_3 \\ \frac{1}{3}x_1 + \frac{1}{3}x_2 + \frac{1}{3}x_3 \end{pmatrix} = \begin{pmatrix} \frac{2}{3}x_1 - \frac{1}{3}x_2 - \frac{1}{3}x_3 \\ -\frac{1}{3}x_1 + \frac{2}{3}x_2 - \frac{1}{3}x_3 \\ -\frac{1}{3}x_1 - \frac{1}{3}x_2 + \frac{2}{3}x_3 \end{pmatrix}.$$

Exercise 3. For each of the vectors $\mathbf{U}$ of Exercise 1, let Q denote projection to the plane perpendicular to $\mathbf{U}$. Find the formula for $Q\begin{pmatrix} x_1 \\ x_2 \\ x_3 \end{pmatrix}$ in terms of the coordinates of $\mathbf{X} = \begin{pmatrix} x_1 \\ x_2 \\ x_3 \end{pmatrix}$.

EXAMPLE 5. Let Π denote the plane through the origin perpendicular to $\mathbf{U} = \begin{bmatrix} 1 \\ 1 \\ 1 \end{bmatrix}$. Let R denote reflection through the plane Π. For any vector $\mathbf{X}$, the midpoint of the segment joining $\mathbf{X}$ to $R(\mathbf{X})$ is the projection $Q(\mathbf{X})$ of $\mathbf{X}$ to the plane Π. Therefore,

$$R(\mathbf{X}) = 2Q(\mathbf{X}) - \mathbf{X}$$

where Q is the projection in Example 4. Therefore,

$$R\begin{bmatrix} x_1 \\ x_2 \\ x_3 \end{bmatrix} = 2\begin{bmatrix} \frac{2}{3}x_1 - \frac{1}{3}x_2 - \frac{1}{3}x_3 \\ -\frac{1}{3}x_1 + \frac{2}{3}x_2 - \frac{1}{3}x_3 \\ -\frac{1}{3}x_1 - \frac{1}{3}x_2 + \frac{2}{3}x_3 \end{bmatrix} - \begin{bmatrix} x_1 \\ x_2 \\ x_3 \end{bmatrix} = \begin{bmatrix} \frac{1}{3}x_1 - \frac{2}{3}x_2 - \frac{2}{3}x_3 \\ -\frac{2}{3}x_1 + \frac{1}{3}x_2 - \frac{2}{3}x_3 \\ -\frac{2}{3}x_1 - \frac{2}{3}x_2 + \frac{1}{3}x_3 \end{bmatrix}.$$

Exercise 4. For each of the vectors $\mathbf{U}$ in Exercise 1, let R denote reflection through the plane through the origin perpendicular to $\mathbf{U}$. Find the formula for $R\begin{pmatrix} x_1 \\ x_2 \\ x_3 \end{pmatrix}$ in terms of the coordinates of $\mathbf{X} = \begin{pmatrix} x_1 \\ x_2 \\ x_3 \end{pmatrix}$.

EXAMPLE 6. Let D_t denote the transformation which sends any vector into t times itself, where t is some fixed scalar number. Then

$$D_t(\mathbf{X}) = t\mathbf{X},$$

so

$$D_t\begin{bmatrix} x_1 \\ x_2 \\ x_3 \end{bmatrix} = t\begin{bmatrix} x_1 \\ x_2 \\ x_3 \end{bmatrix} = \begin{bmatrix} tx_1 \\ tx_2 \\ tx_3 \end{bmatrix}.$$

As in the planar case, we call D_t *the stretching by* t.

If $t = 0$, then $D_0(\mathbf{X}) = 0 \cdot \mathbf{X} = \mathbf{0}$ for all $\mathbf{X}$ so D_0 is the *zero transformation*, denoted by 0. If $t = 1$, then $D_1(\mathbf{X}) = 1 \cdot \mathbf{X} = \mathbf{X}$ for all $\mathbf{X}$, so D_1 is the identity transformation denoted by I.

EXAMPLE 7. For a fixed scalar θ with $0 \leqslant \theta < 2\pi$, we define a rotation R_θ^1 of θ radians about the x_1-axis. This rotation leaves the x_1 component fixed and rotates in the x_2x_3-plane according to the rule for rotating in 2-dimensional space, i.e.,

$$R_\theta^1\begin{bmatrix}x_1\\x_2\\x_3\end{bmatrix}=\begin{bmatrix}x_1\\(\cos\theta)x_2-(\sin\theta)x_3\\(\sin\theta)x_2+(\cos\theta)x_3\end{bmatrix}.$$

For example,

$$R_{\pi/2}^1\begin{bmatrix}x_1\\x_2\\x_3\end{bmatrix}=\begin{bmatrix}x_1\\-x_3\\x_2\end{bmatrix}=R_{-3\pi/2}^1\begin{bmatrix}x_1\\x_2\\x_3\end{bmatrix}$$

and

$$R_{-\pi/2}^1\begin{bmatrix}x_1\\x_2\\x_3\end{bmatrix}=\begin{bmatrix}x_1\\x_3\\-x_2\end{bmatrix}=R_{3\pi/2}^1\begin{bmatrix}x_1\\x_2\\x_3\end{bmatrix}.$$

Exercise 5. In terms of the coordinates of $\mathbf{X}=\begin{pmatrix}x_1\\x_2\\x_3\end{pmatrix}$, calculate the images of

(a) $R_\pi^1(\mathbf{X})$,
(b) $R_{\pi/4}^1(\mathbf{X})$, and
(c) $R_{-\pi/2}^1(\mathbf{X})$.

EXAMPLE 8. In a similar way, we may define rotations R_θ^2 and R_θ^3 by θ radians about the x_2-axis and the x_3-axis. We have the formulas

$$R_\theta^2\begin{bmatrix}x_1\\x_2\\x_3\end{bmatrix}=\begin{bmatrix}(\cos\theta)x_1+(\sin\theta)x_3\\x_2\\(-\sin\theta)x_1+(\cos\theta)x_3\end{bmatrix}$$

and

$$R_\theta^3\begin{bmatrix}x_1\\x_2\\x_3\end{bmatrix}=\begin{bmatrix}(\cos\theta)x_1-(\sin\theta)x_2\\(\sin\theta)x_1+(\cos\theta)x_2\\x_3\end{bmatrix}.$$

Note that the algebraic signs for R_θ^2 are different from those of R_θ^1 and R_θ^3.

Exercise 6. Calculate the images

(a) $R_\pi^2(\mathbf{X})$,
(b) $R_{\pi/4}^3(\mathbf{X})$,
(c) $R_{\pi/2}^1(R_{\pi/2}^2(\mathbf{X}))$,
(d) $R_{\pi/2}^2(R_{\pi/2}^1(\mathbf{X}))$.

CHAPTER 3.2

Linear Transformations and Matrices

In Chapter 3.1 we examined a number of transformations T of 3-space, all of which have the property that, in terms of the coordinates of $\mathbf{X} = \begin{bmatrix} x_1 \\ x_2 \\ x_3 \end{bmatrix}$, the coordinates of $T(\mathbf{X})$ are given by *linear functions* of these coordinates. In each case the formulae are of the following type:

$$T\begin{bmatrix} x_1 \\ x_2 \\ x_3 \end{bmatrix} = \begin{bmatrix} a_1x_1 + a_2x_2 + a_3x_3 \\ b_1x_1 + b_2x_2 + b_3x_3 \\ c_1x_1 + c_2x_2 + c_3x_3 \end{bmatrix}.$$

Any transformation of this form is called a *linear transformation of 3-space*. The expression

$$\begin{bmatrix} a_1 & a_2 & a_3 \\ b_1 & b_2 & b_3 \\ c_1 & c_2 & c_3 \end{bmatrix}$$

is called the *matrix* of the transformation T and is denoted by $m(T)$.

We can now list the matrices of the linear transformations in Examples 1–8 in Chapter 3.1.

$$m(P) = \begin{bmatrix} \frac{1}{3} & \frac{1}{3} & \frac{1}{3} \\ \frac{1}{3} & \frac{1}{3} & \frac{1}{3} \\ \frac{1}{3} & \frac{1}{3} & \frac{1}{3} \end{bmatrix}, \qquad (1)$$

$$m(S) = \begin{bmatrix} -\frac{1}{3} & \frac{2}{3} & \frac{2}{3} \\ \frac{2}{3} & -\frac{1}{3} & \frac{2}{3} \\ \frac{2}{3} & \frac{2}{3} & -\frac{1}{3} \end{bmatrix}, \qquad (2)$$

$$m(\bar{Q}) = \begin{bmatrix} 0 & 0 & 0 \\ 0 & 1 & 0 \\ 0 & 0 & 1 \end{bmatrix}, \tag{3}$$

$$m(Q) = \begin{bmatrix} \frac{2}{3} & -\frac{1}{3} & -\frac{1}{3} \\ -\frac{1}{3} & \frac{2}{3} & -\frac{1}{3} \\ -\frac{1}{3} & -\frac{1}{3} & \frac{2}{3} \end{bmatrix}, \tag{4}$$

$$m(R) = \begin{bmatrix} \frac{1}{3} & -\frac{2}{3} & -\frac{2}{3} \\ -\frac{2}{3} & \frac{1}{3} & -\frac{2}{3} \\ -\frac{2}{3} & -\frac{2}{3} & \frac{1}{3} \end{bmatrix}, \tag{5}$$

$$m(D_t) = \begin{bmatrix} t & 0 & 0 \\ 0 & t & 0 \\ 0 & 0 & t \end{bmatrix}, \tag{6}$$

$$m(R_\theta^1) = \begin{bmatrix} 1 & 0 & 0 \\ 0 & \cos\theta & -\sin\theta \\ 0 & \sin\theta & \cos\theta \end{bmatrix}, \tag{7}$$

$$m(R_\theta^2) = \begin{bmatrix} \cos\theta & 0 & \sin\theta \\ & 1 & \\ -\sin\theta & 0 & \cos\theta \end{bmatrix}, \tag{8}$$

$$m(R_\theta^3) = \begin{bmatrix} \cos\theta & -\sin\theta & 0 \\ \sin\theta & \cos\theta & 0 \\ 0 & 0 & 1 \end{bmatrix}.$$

We will denote by id, (read *identity*), the matrix $m(I) = \begin{bmatrix} 1 & 0 & 0 \\ 0 & 1 & 0 \\ 0 & 0 & 1 \end{bmatrix}$.

As in the plane, if T is a linear transformation with matrix $m(T) = \begin{bmatrix} a_1 & a_2 & a_3 \\ b_1 & b_2 & b_3 \\ c_1 & c_2 & c_3 \end{bmatrix}$, then we write

$$\begin{bmatrix} a_1 & a_2 & a_3 \\ b_1 & b_2 & b_3 \\ c_1 & c_2 & c_3 \end{bmatrix} \begin{bmatrix} x_1 \\ x_2 \\ x_3 \end{bmatrix} = \begin{bmatrix} a_1x_1 + a_2x_2 + a_3x_3 \\ b_1x_1 + b_2x_2 + b_3x_3 \\ c_1x_1 + c_2x_2 + c_3x_3 \end{bmatrix}, \tag{9}$$

and we say that the matrix $m(T)$ *acts on the vector* $\begin{bmatrix} x_1 \\ x_2 \\ x_3 \end{bmatrix}$ to yield the vector

$\begin{bmatrix} a_1x_1 + a_2x_2 + a_3x_3 \\ b_1x_1 + b_2x_2 + b_3x_3 \\ c_1x_1 + c_2x_2 + c_3x_3 \end{bmatrix}$.

Example 1.

$$\begin{bmatrix}1 & 2 & 3\\ 4 & 5 & 6\\ 7 & 8 & 9\end{bmatrix}\begin{bmatrix}1\\ 1\\ 1\end{bmatrix} = \begin{bmatrix}1+2+3\\ 4+5+6\\ 7+8+9\end{bmatrix} = \begin{bmatrix}6\\ 15\\ 24\end{bmatrix},$$

$$\begin{bmatrix}1 & 2 & 3\\ 4 & 5 & 6\\ 7 & 8 & 9\end{bmatrix}\begin{bmatrix}0\\ -1\\ 1\end{bmatrix} = \begin{bmatrix}0-2+3\\ 0-5+6\\ 0-8+9\end{bmatrix} = \begin{bmatrix}1\\ 1\\ 1\end{bmatrix},$$

$$\begin{bmatrix}1 & 2 & 3\\ 4 & 5 & 6\\ 7 & 8 & 9\end{bmatrix}\begin{bmatrix}x_1\\ x_2\\ x_3\end{bmatrix} = \begin{bmatrix}x_1+2x_2+3x_3\\ 4x_1+5x_2+6x_3\\ 7x_1+8x_2+9x_3\end{bmatrix}.$$

Example 2.

$$\begin{bmatrix}0 & 1 & 0\\ 1 & 0 & 0\\ 0 & 0 & 1\end{bmatrix}\begin{bmatrix}x_1\\ x_2\\ x_3\end{bmatrix} = \begin{bmatrix}x_2\\ x_1\\ x_3\end{bmatrix}.$$

We now prove two crucial properties of linear transformations which show how they act on sums and scalar products of vectors. If $\mathbf{X} = \begin{bmatrix}x_1\\ x_2\\ x_3\end{bmatrix}$ and $\mathbf{Y} = \begin{bmatrix}y_1\\ y_2\\ y_3\end{bmatrix}$, then

$$\begin{bmatrix}a_1 & a_2 & a_3\\ b_1 & b_2 & b_3\\ c_1 & c_2 & c_3\end{bmatrix}\left(\begin{bmatrix}x_1\\ x_2\\ x_3\end{bmatrix} + \begin{bmatrix}y_1\\ y_2\\ y_3\end{bmatrix}\right) = \begin{bmatrix}a_1 & a_2 & a_3\\ b_1 & b_2 & b_3\\ c_1 & c_2 & c_3\end{bmatrix}\begin{bmatrix}x_1+y_1\\ x_2+y_2\\ x_3+y_3\end{bmatrix}$$

$$= \begin{bmatrix}a_1(x_1+y_1)+a_2(x_2+y_2)+a_3(x_3+y_3)\\ b_1(x_1+y_1)+b_2(x_2+y_2)+b_3(x_3+y_3)\\ c_1(x_1+y_1)+c_2(x_2+y_2)+c_3(x_3+y_3)\end{bmatrix}$$

$$= \begin{bmatrix}a_1x_1+a_2x_2+a_3x_3\\ b_1x_1+b_2x_2+b_3x_3\\ c_1x_1+c_2x_2+c_3x_3\end{bmatrix} + \begin{bmatrix}a_1y_1+a_2y_2+a_3y_3\\ b_1y_1+b_2y_2+b_3y_3\\ c_1y_1+c_2y_2+c_3y_3\end{bmatrix}$$

$$= \begin{bmatrix}a_1 & a_2 & a_3\\ b_1 & b_2 & b_3\\ c_1 & c_2 & c_3\end{bmatrix}\begin{bmatrix}x_1\\ x_2\\ x_3\end{bmatrix} + \begin{bmatrix}a_1 & a_2 & a_3\\ b_1 & b_2 & b_3\\ c_1 & c_2 & c_3\end{bmatrix}\begin{bmatrix}y_1\\ y_2\\ y_3\end{bmatrix}.$$

Thus, for the associated transformation T with $m(T) = \begin{bmatrix}a_1 & a_2 & a_3\\ b_1 & b_2 & b_3\\ c_1 & c_2 & c_3\end{bmatrix}$, we have

$$T(\mathbf{X}+\mathbf{Y}) + T(\mathbf{X}) + T(\mathbf{Y}). \tag{10a}$$

Similarly, we may show that

$$T(r\mathbf{X}) = rT(\mathbf{X}) \tag{10b}$$

for any scalar r.

Exercise 1. Prove the statement (10b).

Conversely, if T is a transformation such that $T(\mathbf{X}+\mathbf{Y}) = T(\mathbf{X}) + T(\mathbf{Y})$ and $T(r\mathbf{X}) = rT(\mathbf{X})$ for all vectors $\mathbf{X}$, $\mathbf{Y}$ and scalars r, then we may show that the coordinates of $T\begin{bmatrix}x_1\\x_2\\x_3\end{bmatrix}$ are given by a set of linear equations in the coordinates of $\mathbf{X} = \begin{bmatrix}x_1\\x_2\\x_3\end{bmatrix}$. Specifically,

$$\begin{aligned} T\begin{bmatrix}x_1\\x_2\\x_3\end{bmatrix} &= T\left(\begin{bmatrix}x_1\\0\\0\end{bmatrix} + \begin{bmatrix}0\\x_2\\0\end{bmatrix} + \begin{bmatrix}0\\0\\x_3\end{bmatrix}\right) \\ &= T\left(x_1\begin{bmatrix}1\\0\\0\end{bmatrix} + x_2\begin{bmatrix}0\\1\\0\end{bmatrix} + x_3\begin{bmatrix}0\\0\\1\end{bmatrix}\right) \\ &= x_1T\begin{bmatrix}1\\0\\0\end{bmatrix} + x_2T\begin{bmatrix}0\\1\\0\end{bmatrix} + x_3T\begin{bmatrix}0\\0\\1\end{bmatrix}. \end{aligned}$$

Let $T\begin{bmatrix}1\\0\\0\end{bmatrix} = \begin{bmatrix}a_1\\b_1\\c_1\end{bmatrix}$, $T\begin{bmatrix}0\\1\\0\end{bmatrix} = \begin{bmatrix}a_2\\b_2\\c_2\end{bmatrix}$, $T\begin{bmatrix}0\\0\\1\end{bmatrix} = \begin{bmatrix}a_3\\b_3\\c_3\end{bmatrix}$. Then

$$\begin{aligned} T\begin{bmatrix}x_1\\x_2\\x_3\end{bmatrix} &= x_1\begin{bmatrix}a_1\\b_1\\c_1\end{bmatrix} + x_2\begin{bmatrix}a_2\\b_2\\c_2\end{bmatrix} + x_3\begin{bmatrix}a_3\\b_3\\c_3\end{bmatrix} \\ &= \begin{bmatrix}a_1x_1\\b_1x_1\\c_1x_1\end{bmatrix} + \begin{bmatrix}a_2x_2\\b_2x_2\\c_2x_2\end{bmatrix} + \begin{bmatrix}a_3x_3\\b_3x_3\\c_3x_3\end{bmatrix} \\ &= \begin{bmatrix}a_1x_1 + a_2x_2 + a_3x_3\\b_1x_1 + b_2x_2 + b_3x_3\\c_1x_1 + c_2x_2 + c_3x_3\end{bmatrix} \end{aligned}$$

as predicted. In summary, we have:

Theorem 3.1. *Let T be a transformation of 3-space. Then T is a linear transformation if and only if T satisfies the conditions*

$$T(\mathbf{X}+\mathbf{Y}) = T(\mathbf{X}) + T(\mathbf{Y})$$

and

$$T(r\mathbf{X}) = rT(\mathbf{X})$$

for all vectors $\mathbf{X}$ *and* $\mathbf{Y}$ *and all scalars* r.

Let T be a linear transformation of $\mathbb{R}^3$. Let L be a straight line in $\mathbb{R}^3$. By the image of L under T, we mean the collection of vectors $T(\mathbf{X})$ for all vectors $\mathbf{X}$ with endpoint lying on L. If Π is a plane, we define the image of Π under T in a similar way.

Theorem 3.2. *Let T be a linear transformation of $\mathbb{R}^3$. The image of a straight line L under T is either a straight line or a point. The image of a plane Π under T is either a plane, a straight line, or a point. The image of $\mathbb{R}^3$ under T is either $\mathbb{R}^3$, a plane, a straight line, or a point.*

The proof proceeds in exact analogy with the proof in dimension 2 (proof of Theorem 2.2 in Chapter 2.2).

Exercise 2. Describe the images of a line $L = \{\mathbf{A} + t\mathbf{U} \mid t \text{ real}\}$ under a linear transformation T.

Exercise 3. Let T be a linear transformation. Let Π be the plane $\{\mathbf{C} + t\mathbf{U} + s\mathbf{V} \mid t, s \text{ real}\}$, where $\mathbf{U}$ and $\mathbf{V}$ are linearly independent. Show that the image of Π under T is the collection of vectors $T(\mathbf{C}) + tT(\mathbf{U}) + sT(\mathbf{V})$. Under what conditions will the image be a single point? When will the image be a line?

Exercise 4. The image under T of $\mathbb{R}^3 = \{x_1\mathbf{E}_1 + x_2\mathbf{E}_2 + x_3\mathbf{E}_3 \mid x_1, x_2, x_3 \text{ real}\}$ is $\{x_1T(\mathbf{E}_1) + x_2T(\mathbf{E}_2) + x_3T(\mathbf{E}_3)\}$. Under what conditions on $T(\mathbf{E}_1)$, $T(\mathbf{E}_2)$, and $T(\mathbf{E}_3)$ is this all of $\mathbb{R}^3$?

CHAPTER 3.3

Sums and Products of Linear Transformations

If T and S are linear transformations, then we may define a new transformation $T+S$ by the condition

$$(T+S)(\mathbf{X}) = T(\mathbf{X}) + S(\mathbf{X}) \qquad \text{for every vector } \mathbf{X}.$$

Then by definition, $(T+S)(\mathbf{X}+\mathbf{Y}) = T(\mathbf{X}+\mathbf{Y}) + S(\mathbf{X}+\mathbf{Y})$, and since T and S are linear transformations, this equals $T(\mathbf{X}) + T(\mathbf{Y}) + S(\mathbf{X}) + S(\mathbf{Y})$ $= T(\mathbf{X}) + S(\mathbf{X}) + T(\mathbf{Y}) + S(\mathbf{Y}) = (T+S)(\mathbf{X}) + (T+S)(\mathbf{Y})$. Thus for every pair $\mathbf{X}$, $\mathbf{Y}$, we have

$$(T+S)(\mathbf{X}+\mathbf{Y}) = (T+S)(\mathbf{X}) + (T+S)(\mathbf{Y}).$$

Similarly, we may show

$$(T+S)(t\mathbf{X}) = t(T+S)(\mathbf{X}).$$

Therefore, by Theorem 3.1, $T+S$ is a linear transformation. It is called the *sum* of the transformations T and S.

If the matrix of T is $m(T) = \begin{pmatrix} a_{11} & a_{12} & a_{13} \\ a_{21} & a_{22} & a_{23} \\ a_{31} & a_{32} & a_{33} \end{pmatrix}$ and the matrix of S is

$m(S) = \begin{pmatrix} b_{11} & b_{12} & b_{13} \\ b_{21} & b_{22} & b_{23} \\ b_{31} & b_{32} & b_{33} \end{pmatrix}$, then we may calculate the matrix $m(T+S)$ of

$T+S$ as follows:

$$(T+S)\begin{bmatrix}x_1\\x_2\\x_3\end{bmatrix} = T\begin{bmatrix}x_1\\x_2\\x_3\end{bmatrix} + S\begin{bmatrix}x_1\\x_2\\x_3\end{bmatrix}$$

$$= \begin{bmatrix}a_{11} & a_{12} & a_{13}\\a_{21} & a_{22} & a_{23}\\a_{31} & a_{32} & a_{33}\end{bmatrix}\begin{bmatrix}x_1\\x_2\\x_3\end{bmatrix} + \begin{bmatrix}b_{11} & b_{12} & b_{13}\\b_{21} & b_{22} & b_{23}\\b_{31} & b_{32} & b_{33}\end{bmatrix}\begin{bmatrix}x_1\\x_2\\x_3\end{bmatrix}$$

$$= \begin{bmatrix}a_{11}x_1 + a_{12}x_2 + a_{13}x_3\\a_{21}x_1 + a_{22}x_2 + a_{23}x_3\\a_{31}x_1 + a_{32}x_2 + a_{33}x_3\end{bmatrix} + \begin{bmatrix}b_{11}x_1 + b_{12}x_2 + b_{13}x_3\\b_{21}x_1 + b_{22}x_2 + b_{23}x_3\\b_{31}x_1 + b_{32}x_2 + b_{33}x_3\end{bmatrix}$$

$$= \begin{bmatrix}(a_{11}+b_{11})x_1 + (a_{12}+b_{12})x_2 + (a_{13}+b_{13})x_3\\(a_{21}+b_{21})x_1 + (a_{22}+b_{22})x_2 + (a_{23}+b_{23})x_3\\(a_{31}+b_{31})x_1 + (a_{32}+b_{32})x_2 + (a_{33}+b_{33})x_3\end{bmatrix}$$

$$= \begin{bmatrix}a_{11}+b_{11} & a_{12}+b_{12} & a_{13}+b_{13}\\a_{21}+b_{21} & a_{22}+b_{22} & a_{23}+b_{23}\\a_{31}+b_{31} & a_{32}+b_{32} & a_{33}+b_{33}\end{bmatrix}\begin{bmatrix}x_1\\x_2\\x_3\end{bmatrix}.$$

Thus the matrix for the sum of two linear transformations is just the matrix formed by adding the corresponding entries in the matrices of the two linear transformations.

We define matrix addition componentwise by the formula:

$$\begin{bmatrix}a_{11} & a_{12} & a_{13}\\a_{21} & a_{22} & a_{23}\\a_{31} & a_{32} & a_{33}\end{bmatrix} + \begin{bmatrix}b_{11} & b_{12} & b_{13}\\b_{21} & b_{22} & b_{23}\\b_{31} & b_{32} & b_{33}\end{bmatrix} = \begin{bmatrix}a_{11}+b_{11} & a_{12}+b_{12} & a_{13}+b_{13}\\a_{21}+b_{21} & a_{22}+b_{22} & a_{23}+b_{23}\\a_{31}+b_{31} & a_{32}+b_{32} & a_{33}+b_{33}\end{bmatrix}. \tag{1}$$

Therefore we may write the matrix of $T+S$ as the sum of the matrices of T and S:

$$m(T+S) = m(T) + m(S). \tag{2}$$

Exercise 1. For any linear transformation T and any scalar t, we define a new transformation tT by the condition: $(tT)(\mathbf{X}) = tT(\mathbf{X})$ for all $\mathbf{X}$. Show that tT is a linear transformation by showing that $(tT)(\mathbf{X}+\mathbf{Y}) = (tT)(\mathbf{X}) + (tT)(\mathbf{Y})$ and $(tT)(s\mathbf{X}) = s(tT)(\mathbf{X})$ for any vectors $\mathbf{X}$, $\mathbf{Y}$ and any scalar s. For any matrix, we define the product of the matrix by the scalar t to be the matrix whose entries are all multiplied by t, i.e.,

$$t\begin{pmatrix} a_{11} & a_{12} & a_{13} \\ a_{21} & a_{22} & a_{23} \\ a_{31} & a_{32} & a_{33} \end{pmatrix} = \begin{pmatrix} ta_{11} & ta_{12} & ta_{13} \\ ta_{21} & ta_{22} & ta_{23} \\ ta_{31} & ta_{32} & ta_{33} \end{pmatrix}. \tag{3}$$

Show that $m(tT) = tm(T)$.

EXAMPLE 1. We may use the above ideas to calculate the matrices of some of the transformations we used in Chapter 3.2, e.g., if S is reflection in the x_3-axis and T is projection to the x_3-axis, then $S = 2T - I$, so

$$m(S) = m(2T - I) = m(2T) - m(I) = 2m(T) - m(I)$$

$$= 2\begin{pmatrix} 0 & 0 & 0 \\ 0 & 0 & 0 \\ 0 & 0 & 1 \end{pmatrix} - \begin{pmatrix} 1 & 0 & 0 \\ 0 & 1 & 0 \\ 0 & 0 & 1 \end{pmatrix} = \begin{pmatrix} -1 & 0 & 0 \\ 0 & -1 & 0 \\ 0 & 0 & 1 \end{pmatrix}.$$

As in the case of the plane, we may define the product of two linear transformations T and S to be the transformation R given by $R(\mathbf{X}) = S(T(\mathbf{X}))$.

We write $R = ST$, and we call R *the product of S and T*.

EXAMPLE 2. Let K be reflection in the x_1x_2-plane and J be reflection in the x_2x_3-plane. Then

$$J\begin{pmatrix} x_1 \\ x_2 \\ x_3 \end{pmatrix} = \begin{pmatrix} -x_1 \\ x_2 \\ x_3 \end{pmatrix} \quad \text{and} \quad K(J(\mathbf{X})) = K\begin{pmatrix} -x_1 \\ x_2 \\ x_3 \end{pmatrix} = \begin{pmatrix} -x_1 \\ x_2 \\ -x_3 \end{pmatrix}.$$

Also, $J(K(\mathbf{X})) = J\begin{pmatrix} x_1 \\ x_2 \\ -x_3 \end{pmatrix} = \begin{pmatrix} -x_1 \\ x_2 \\ -x_3 \end{pmatrix}$, so $KJ = JK$.

Exercise 2. Show that if S and T are any linear transformations, then ST and TS are linear transformation, using Theorem 3.1.

EXAMPLE 3. Let P be projection to the x_1x_2-plane and let Q be projection to the x_2x_3-plane. Describe PQ and QP. We find $PQ\begin{pmatrix} x_1 \\ x_2 \\ x_3 \end{pmatrix} =$

$P\left(Q\begin{pmatrix} x_1 \\ x_2 \\ x_3 \end{pmatrix}\right) = P\begin{pmatrix} 0 \\ x_2 \\ x_3 \end{pmatrix} = \begin{pmatrix} 0 \\ x_2 \\ 0 \end{pmatrix}$. Also, $QP\begin{pmatrix} x_1 \\ x_2 \\ x_3 \end{pmatrix} = \begin{pmatrix} 0 \\ x_2 \\ 0 \end{pmatrix}$.

Exercise 3. Let R denote projection to the x_3-axis and let P denote projection to the x_1x_2-plane. Find RP and PR.

EXAMPLE 4. Find PP. Since $PP\begin{pmatrix}x_1\\x_2\\x_3\end{pmatrix} = P\left(P\begin{pmatrix}x_1\\x_2\\x_3\end{pmatrix}\right) = P\begin{pmatrix}x_1\\x_2\\0\end{pmatrix} = \begin{pmatrix}x_1\\x_2\\0\end{pmatrix}$. Therefore, $PP(\mathbf{X}) = P(\mathbf{X})$ for all $\mathbf{X}$, so we have $PP = P$.

Exercise 4. Show that $RR = R$ and $QQ = Q$, where R and Q are as above.

EXAMPLE 5. Find KQ and QK, when K and Q are the transformations defined in Examples 2 and 3. We have

$$KQ\begin{pmatrix}x_1\\x_2\\x_3\end{pmatrix} = K\begin{pmatrix}0\\x_2\\x_3\end{pmatrix} = \begin{pmatrix}0\\x_2\\-x_3\end{pmatrix}, \qquad QK\begin{pmatrix}x_1\\x_2\\x_3\end{pmatrix} = Q\begin{pmatrix}x_1\\x_2\\-x_3\end{pmatrix} = \begin{pmatrix}0\\x_2\\-x_3\end{pmatrix}.$$

Thus $KQ = QK$.

Exercise 5. Find KP and PK, and RK and KR, where K, P, and R are the transformations in Examples 2, 3, and 4.

EXAMPLE 6. Let S be a linear transformation and let I denote the identity transformation. Find SI and IS. For each $\mathbf{X}$, we have $SI(\mathbf{X}) = S(\mathbf{X})$ and $IS(\mathbf{X}) = I(S(\mathbf{X})) = S(\mathbf{X})$. Thus $SI = S = IS$.

If T and S are linear transformations with matrices

$$m(T) = \begin{pmatrix}a_{11} & a_{12} & a_{13}\\ a_{21} & a_{22} & a_{23}\\ a_{31} & a_{32} & a_{33}\end{pmatrix} \quad \text{and} \quad m(S) = \begin{pmatrix}b_{11} & b_{12} & b_{13}\\ b_{21} & b_{22} & b_{23}\\ b_{31} & b_{32} & b_{33}\end{pmatrix},$$

then we know that TS is also a linear transformation, so we may calculate its matrix as follows. To find the first column of $m(TS)$, we must find the image of $\mathbf{E}_1$, i.e., $TS(\mathbf{E}_1) = T(S(\mathbf{E}_1))$, where $S(\mathbf{E}_1) = \begin{pmatrix}b_{11}\\b_{21}\\b_{31}\end{pmatrix}$. Thus

$$T(S(\mathbf{E}_1)) = \begin{pmatrix}a_{11} & a_{12} & a_{13}\\ a_{21} & a_{22} & a_{23}\\ a_{31} & a_{32} & a_{33}\end{pmatrix}\begin{pmatrix}b_{11}\\b_{21}\\b_{31}\end{pmatrix} = \begin{pmatrix}a_{11}b_{11} + a_{12}b_{21} + a_{13}b_{31}\\ a_{21}b_{11} + a_{22}b_{21} + a_{23}b_{31}\\ a_{31}b_{11} + a_{32}b_{21} + a_{33}b_{31}\end{pmatrix}.$$

Similarly, we may find the second column of $m(TS)$ by computing $T(S(\mathbf{E}_2))$ and the third column of $m(TS)$ by computing $T(S(\mathbf{E}_3))$.

We define the product of the two matrices $m(T)$ and $m(S)$ to be the matrix $m(TS)$ of the product transformation TS. Thus $m(T)m(S) =$

$m(TS)$ or

$$\begin{bmatrix} a_{11} & a_{12} & a_{13} \\ a_{21} & a_{22} & a_{23} \\ a_{31} & a_{32} & a_{33} \end{bmatrix}\begin{bmatrix} b_{11} & b_{12} & b_{13} \\ b_{21} & b_{22} & b_{23} \\ b_{31} & b_{32} & b_{33} \end{bmatrix} = m(TS)$$

$$= \begin{bmatrix} a_{11}b_{11} + a_{12}b_{21} + a_{13}b_{31} & a_{11}b_{12} + a_{12}b_{22} + a_{13}b_{32} & a_{11}b_{13} + a_{12}b_{23} + a_{13}b_{33} \\ a_{21}b_{11} + a_{22}b_{21} + a_{23}b_{31} & a_{21}b_{12} + a_{22}b_{22} + a_{23}b_{32} & a_{21}b_{13} + a_{22}b_{23} + a_{23}b_{33} \\ a_{31}b_{11} + a_{32}b_{21} + a_{33}b_{31} & a_{31}b_{12} + a_{32}b_{22} + a_{33}b_{32} & a_{31}b_{13} + a_{32}b_{23} + a_{33}b_{33} \end{bmatrix}. \tag{4}$$

EXAMPLE 7. If $m(T) = \begin{bmatrix} 2 & 1 & 2 \\ 1 & 3 & 1 \\ 3 & 1 & 4 \end{bmatrix}$ and $m(S) = \begin{bmatrix} 1 & 1 & 2 \\ 2 & 3 & 1 \\ 1 & 3 & 1 \end{bmatrix}$, then

$$m(T)m(S)$$

$$= \begin{bmatrix} 2\cdot 1 + 1\cdot 2 + 2\cdot 1 & 2\cdot 1 + 1\cdot 3 + 2\cdot 3 & 2\cdot 2 + 1\cdot 1 + 2\cdot 1 \\ 1\cdot 1 + 3\cdot 2 + 1\cdot 1 & 1\cdot 1 + 3\cdot 3 + 1\cdot 3 & \boxed{1\cdot 2 + 3\cdot 1 + 1\cdot 1} \\ 3\cdot 1 + 1\cdot 2 + 4\cdot 1 & 3\cdot 1 + 1\cdot 3 + 4\cdot 3 & 3\cdot 2 + 1\cdot 1 + 4\cdot 1 \end{bmatrix}$$

$$= \begin{bmatrix} 6 & 11 & 7 \\ 8 & 13 & 6 \\ 9 & 18 & 11 \end{bmatrix}$$

and

$$m(S)m(T) = \begin{bmatrix} 1 & 1 & 2 \\ 2 & 3 & 1 \\ 1 & 3 & 1 \end{bmatrix}\begin{bmatrix} 2 & 1 & 2 \\ 1 & 3 & 1 \\ 3 & 1 & 4 \end{bmatrix} = \begin{bmatrix} 9 & 6 & 11 \\ 10 & 12 & \boxed{11} \\ 8 & 11 & 9 \end{bmatrix}.$$

Note that $m(TS) \neq m(ST)$ in this case.

When computing the product of matrices in the 3-dimensional case, we find that each entry is the dot product of a *row* of the first matrix with a *column* of the second. The entry in the second row, third column, of $m(T)m(S)$ is given by the dot product of the second row of $m(T)$ with the third column of $m(S)$. We indicate this entry in the examples above.

Exercise 6. Calculate the products of the following matrices:

$$\begin{pmatrix} 2 & 1 & 0 \\ 0 & 3 & 3 \\ 0 & 2 & 1 \end{pmatrix}\begin{pmatrix} 7 & 0 & 1 \\ 1 & 1 & 1 \\ 2 & 1 & 2 \end{pmatrix}, \quad \begin{pmatrix} 3 & 0 & 0 \\ 0 & 1 & 0 \\ 0 & 0 & 3 \end{pmatrix}\begin{pmatrix} 2 & 0 & 6 \\ 1 & 1 & 1 \\ 1 & 3 & 2 \end{pmatrix}, \quad \begin{pmatrix} 0 & 1 & 0 \\ 1 & 0 & 0 \\ 0 & 0 & 1 \end{pmatrix}\begin{pmatrix} 1 & 2 & 3 \\ 4 & 5 & 6 \\ 7 & 8 & 9 \end{pmatrix}.$$

Exercise 7. Each of the transformations J, K, P, Q, R of Examples 1, 2, and 3 has the property that the transformation equals its own square, i.e., $JJ = J$. Verify that

$m(J) \cdot m(J) = m(J)$, and verify that each of the matrices of the other linear transformations equals its own square.

Exercise 8. Since $PR = 0$ as linear transformations, it follows that $m(PR) = m(0) = \begin{pmatrix} 0 & 0 & 0 \\ 0 & 0 & 0 \\ 0 & 0 & 0 \end{pmatrix}$. Verify that $m(P)m(R) = 0 = m(R)m(P)$.

Exercise 9. Show that $\begin{pmatrix} 0 & 0 & 1 \\ 1 & 0 & 0 \\ 0 & 1 & 0 \end{pmatrix}$ has cube = id but a square $\neq$ id. Describe the linear transformation with this matrix.

As in the 2-dimensional case, the distributive law of matrix multiplication over matrix addition is a consequence of this law for linear transformations. The same is true for the associative laws of matrix multiplication and matrix addition. Thus

$$\begin{aligned}(m(T)m(S))m(R) &= m(TS)m(R) = m((TS)R) = m(T(SR)) \\ &= m(T)m(SR) = m(T)(m(S)m(R)), \qquad (5)\end{aligned}$$

$$\begin{aligned}m(T)m(S) + m(T)m(R) &= m(TS) + m(TR) = m(TS + TR) \\ &= m(T(S+R)) = m(T)m(S+R) \\ &= m(T)(m(S) + m(R)). \qquad (6)\end{aligned}$$

Exercise 10. Let $\mathbf{U}$ be a unit vector with coordinates $\begin{pmatrix} u_1 \\ u_2 \\ u_3 \end{pmatrix}$. Show that the projection P to the line along $\mathbf{U}$ has the matrix $\begin{bmatrix} u_1^2 & u_2u_1 & u_3u_1 \\ u_1u_2 & u_2^2 & u_3u_2 \\ u_1u_3 & u_2u_3 & u_3^2 \end{bmatrix}$.

Exercise 11. Show that the matrix for the reflection R through the line along $\mathbf{U}$ is

$$m(R) = \begin{bmatrix} 2u_1^2 - 1 & 2u_2u_1 & 2u_3u_1 \\ 2u_1u_2 & 2u_2^2 - 1 & 2u_3u_2 \\ 2u_1u_3 & 2u_2u_3 & 2u_3^2 - 1 \end{bmatrix}.$$

EXAMPLE 8. If $\mathbf{U} = \begin{bmatrix} 1/\sqrt{3} \\ 1/\sqrt{3} \\ 1/\sqrt{3} \end{bmatrix}$, then, defining P and R as in Exercises 10 and 11,

$$m(P) = \begin{bmatrix} \frac{1}{3} & \frac{1}{3} & \frac{1}{3} \\ \frac{1}{3} & \frac{1}{3} & \frac{1}{3} \\ \frac{1}{3} & \frac{1}{3} & \frac{1}{3} \end{bmatrix} = \frac{1}{3}\begin{bmatrix} 1 & 1 & 1 \\ 1 & 1 & 1 \\ 1 & 1 & 1 \end{bmatrix},$$

while

$$m(R) = 2m(P) - m(I) = \frac{1}{3}\begin{bmatrix} 1 & 2 & 2 \\ 2 & 1 & 2 \\ 2 & 2 & 2 \end{bmatrix}.$$

Exercise 12. Find the matrices of projection to the line along $\mathbf{U} = \frac{1}{3}\begin{pmatrix} 2 \\ 1 \\ 2 \end{pmatrix}$ and of reflection through this line.

§1. Elementary Matrices and Diagonal Matrices

As in the 2×2 case, there are certain simple 3×3 matrices from which we can build up an arbitrary matrix. We consider matrices of three types: shear matrices, permutation matrices, and diagonal matrices.

(1) *Shear Matrices*: For $i \neq j$, let e_{ij}^s = matrix with all 1's on the diagonal and s in the i, j position and 0 otherwise. For example,

$$e_{12}^s = \begin{bmatrix} 1 & s & 0 \\ 0 & 1 & 0 \\ 0 & 0 & 1 \end{bmatrix}, \qquad e_{23}^s = \begin{bmatrix} 1 & 0 & 0 \\ 0 & 1 & s \\ 0 & 0 & 1 \end{bmatrix}.$$

(2) *Permutation Matrices*: p_{ij} = matrix which is obtained from the identity by interchanging the i'th and j'th rows and leaving the remaining row unchanged. For example,

$$p_{13} = \begin{bmatrix} 0 & 0 & 1 \\ 0 & 1 & 0 \\ 1 & 0 & 0 \end{bmatrix}, \qquad p_{23} = \begin{bmatrix} 1 & 0 & 0 \\ 0 & 0 & 1 \\ 0 & 1 & 0 \end{bmatrix}.$$

A matrix of one of these types is called an *elementary matrix*. If e is an elementary matrix and m is any given matrix, it is easy to describe the matrices em and me.

In the following discussion, we set

$$m = \begin{bmatrix} a_1 & a_2 & a_3 \\ b_1 & b_2 & b_3 \\ c_1 & c_2 & c_3 \end{bmatrix}.$$

EXAMPLE 9.

$$e_{21}^{s}m = \begin{pmatrix} 1 & 0 & 0 \\ s & 1 & 0 \\ 0 & 0 & 1 \end{pmatrix} \begin{pmatrix} a_1 & a_2 & a_3 \\ b_1 & b_2 & b_3 \\ c_1 & c_2 & c_3 \end{pmatrix} = \begin{pmatrix} a_1 & a_2 & a_3 \\ sa_1 + b_1 & sa_2 + b_2 & sa_3 + b_3 \\ c_1 & c_2 & c_3 \end{pmatrix},$$

We see that $e_{21}^{s}m$ is the matrix we get, starting with m, by adding s times the first row of m to the second row of m and leaving the other rows alone.

Exercise 13. Show that $e_{12}^{s}m$ is the matrix obtained from m by adding s times the second row to the first row and leaving the other rows alone.

EXAMPLE 10.

$$me_{21}^{s} = \begin{pmatrix} a_1 & a_2 & a_3 \\ b_1 & b_2 & b_3 \\ c_1 & c_2 & c_3 \end{pmatrix} \begin{pmatrix} 1 & 0 & 0 \\ s & 1 & 0 \\ 0 & 0 & 1 \end{pmatrix} = \begin{pmatrix} a_1 + sa_2 & a_2 & a_3 \\ b_1 + sb_2 & b_2 & b_3 \\ c_1 + sc_2 & c_2 & c_3 \end{pmatrix},$$

which is the matrix obtained from m by adding s times the second column to the first column and leaving the other columns alone.

Exercise 14. Describe the matrix me_{13}^{s}.

The preceding calculations work in all cases, and so we have:

Proposition 1. *For each i, j with $i \neq j$, $e_{ij}^{s}m$ is the matrix obtained from m by adding s times the j'th row to the i'th row of m and leaving the other rows alone. Also, me_{ij}^{s} is the matrix obtained from m by adding s times the i'th column to the j'th column.*

Note: Recall that $\text{id} = \begin{pmatrix} 1 & 0 & 0 \\ 0 & 1 & 0 \\ 0 & 0 & 1 \end{pmatrix}$, so $e_{ij}^{s}\text{id} = e_{ij}^{s}$.

Thus we can instantly remember what multiplication by e_{ij}^{s} does to an arbitrary matrix by thinking of e_{ij}^{s} itself as the result of multiplying the identity matrix by e_{ij}^{s}. For instance, $e_{31}^{s} = \begin{pmatrix} 1 & 0 & 0 \\ 0 & 1 & 0 \\ s & 0 & 1 \end{pmatrix}$, so the product $e_{31}^{s}\text{id}$ = the matrix obtained by adding s times the first row of the identity to its last row. Hence, this is what multiplication by e_{31}^{s} does to *any* matrix: it adds s times the first row to the last row.

EXAMPLE 11.

$$p_{13}m = \begin{pmatrix} 0 & 0 & 1 \\ 0 & 1 & 0 \\ 1 & 0 & 0 \end{pmatrix} \begin{pmatrix} a_1 & a_2 & a_3 \\ b_1 & b_2 & b_3 \\ c_1 & c_2 & c_3 \end{pmatrix} = \begin{pmatrix} c_1 & c_2 & c_3 \\ b_1 & b_2 & b_3 \\ a_1 & a_2 & a_3 \end{pmatrix}.$$

Thus $p_{13}m$ is the matrix obtained from m by interchanging the first and last rows and leaving the second row alone.

Exercise 15. Show that mp_{13} is the matrix obtained by interchanging the first and third columns of m and leaving the second column alone.

For every permutation matrix p_{ij}, the situation is similar to that we have just found; so we have:

Proposition 2. *For each i, j with $i \neq j$, $p_{ij}m$ is the matrix obtained from m by interchanging the i'th and j'th rows, and mp_{ij} is the matrix obtained from m by interchanging the i'th and j'th columns.*

Again, to remember how p_{ij} acts on an arbitrary matrix m, we need only look at p_{ij} and remember that $p_{ij} = p_{ij}\text{id}$.

We call a matrix $\begin{bmatrix} t_1 & 0 & 0 \\ 0 & t_2 & 0 \\ 0 & 0 & t_3 \end{bmatrix}$ a *diagonal* matrix.

EXAMPLE 12. Let $d = \begin{bmatrix} t_1 & 0 & 0 \\ 0 & t_2 & 0 \\ 0 & 0 & t_3 \end{bmatrix}$.

$$dm = \begin{bmatrix} t_1 & 0 & 0 \\ 0 & t_2 & 0 \\ 0 & 0 & t_3 \end{bmatrix} \begin{bmatrix} a_1 & a_2 & a_3 \\ b_1 & b_2 & b_3 \\ c_1 & c_2 & c_3 \end{bmatrix} = \begin{bmatrix} t_1a_1 & t_1a_2 & t_1a_3 \\ t_2b_1 & t_2b_2 & t_2b_3 \\ t_3c_1 & t_3c_2 & t_3c_3 \end{bmatrix}.$$

Exercise 16. With d as in the preceding example, calculate md.

By the same calculation as in Example 12 and Exercise 16, we find:

Proposition 3. *If d is any diagonal matrix, dm is the matrix obtained from m by multiplying the i'th row of m by t_i for each i, where t_i is the entry in d in the (i, i)-position. Also, md is obtained in a similar way, the i'th column of m being multiplied by t_i.*

Exercise 17.

(a) Show that $(p_{12})^2 = \text{id}$ and similarly for p_{13} and p_{23}.
(b) Show that $p_{12}p_{23} \neq p_{23}p_{12}$.

Exercise 18. Show that if

$$d_t = \begin{bmatrix} t_1 & 0 & 0 \\ 0 & t_2 & 0 \\ 0 & 0 & t_3 \end{bmatrix} \quad \text{and} \quad d_s = \begin{bmatrix} s_1 & 0 & 0 \\ 0 & s_2 & 0 \\ 0 & 0 & s_3 \end{bmatrix},$$

then $d_sd_t = d_td_s$.

Exercise 19. Calculate the following matrices:

(a) $m = \begin{pmatrix} 1 & 0 & 0 \\ s & 1 & 0 \\ 0 & 0 & 1 \end{pmatrix} \begin{pmatrix} 4 & 0 & 0 \\ 0 & 5 & 0 \\ 0 & 0 & 6 \end{pmatrix} \begin{pmatrix} 1 & 0 & 0 \\ 0 & 1 & 0 \\ 0 & t & 0 \end{pmatrix}$,

(b) $m = \begin{pmatrix} 1 & 0 & 0 \\ 0 & 0 & 1 \\ 0 & 1 & 0 \end{pmatrix} \begin{pmatrix} 2 & 0 & 0 \\ 0 & -2 & 0 \\ 0 & 0 & 0 \end{pmatrix} \begin{pmatrix} 0 & 0 & 1 \\ 0 & 1 & 0 \\ 1 & 0 & 0 \end{pmatrix}$.

By multiplying elementary matrices together, several at a time, we can build up *all* matrices.

Theorem 3.3. *Let m be a* 3×3 *matrix. We can find a diagonal matrix d and elementary matrices* $e_1 e_2, \ldots, e_n$ *and* $f_1, \ldots, f_m$ *so that*

$$m = e_1 e_2 \ldots e_n d f_1 f_2 \ldots f_m .$$

We shall give the proof of this theorem in the next chapter.

CHAPTER 3.4

Inverses and Systems of Equations

Let T be a linear transformation of $\mathbb{R}^3$. As in the case of two dimensions, we say that the linear transformation S is an inverse of T if

$$ST = I \quad \text{and} \quad TS = I. \tag{1}$$

Can a linear transformation have more than one inverse? The answer is no and the proof is the same as in $\mathbb{R}^2$. (See Chapter 2.4, page 52.)

EXAMPLE 1. If $t \neq 0$, then $D_{1/t}$ is the inverse of D_t.

EXAMPLE 2. The inverse of R_θ^1, rotation by θ radians around the x_1-axis, then, is $R_{-\theta}^1 = R_{2\pi-\theta}^1$.

EXAMPLE 3. Let A be the transformation with matrix $\begin{bmatrix} 1 & 0 & 0 \\ 0 & 2 & 0 \\ 0 & 0 & 3 \end{bmatrix}$. The transformation B with matrix $\begin{bmatrix} 1 & 0 & 0 \\ 0 & \frac{1}{2} & 0 \\ 0 & 0 & \frac{1}{3} \end{bmatrix}$ is the inverse of A, because

$$AB\begin{bmatrix} x_1 \\ x_2 \\ x_3 \end{bmatrix} = A\left[\begin{bmatrix} 1 & 0 & 0 \\ 0 & \frac{1}{2} & 0 \\ 0 & 0 & \frac{1}{3} \end{bmatrix}\begin{bmatrix} x_1 \\ x_2 \\ x_3 \end{bmatrix}\right] = A\left[\begin{bmatrix} x_1 \\ \frac{1}{2}x_2 \\ \frac{1}{3}x_3 \end{bmatrix}\right] = \begin{bmatrix} 1 & 0 & 0 \\ 0 & 2 & 0 \\ 0 & 0 & 3 \end{bmatrix}\begin{bmatrix} x_1 \\ \frac{1}{2}x_2 \\ \frac{1}{3}x_3 \end{bmatrix} = \begin{bmatrix} x_1 \\ x_2 \\ x_3 \end{bmatrix},$$

so $AB = I$, and, similarly, $BA = I$.

EXAMPLE 4. Let π be a plane and let P be projection on the plane π. We claim P does not have an inverse, and we shall give two proofs for this statement.

Suppose Q is a transformation satisfying $PQ = QP = I$.

(a) Choose a vector $\mathbf{X}$ which is not in the plane π. Then $\mathbf{X} = I(\mathbf{X}) = PQ(\mathbf{X}) = P(Q(\mathbf{X}))$.

But $\mathbf{X}$ is not in π, while $P(Q(\mathbf{X}))$ is in π, so we have a contradiction. Thus, no such Q exists.

(b) Choose a vector $\mathbf{X} \neq \mathbf{0}$ with $\mathbf{X} \perp \pi$. Then $P(\mathbf{X}) = \mathbf{0}$, and so $Q(P(\mathbf{X})) = \mathbf{0}$. Thus $\mathbf{X} = I(\mathbf{X}) = Q(P(\mathbf{X})) = \mathbf{0}$. This is a contradiction. Therefore, no such Q exists.

EXAMPLE 5. Let T have matrix

$$\begin{bmatrix} 1 & -1 & 0 \\ 0 & 1 & -1 \\ -1 & 0 & 1 \end{bmatrix}.$$

We claim T has no inverse. Suppose S satisfies $ST = TS = I$. Let $\mathbf{X} = \begin{bmatrix} x_1 \\ x_2 \\ x_3 \end{bmatrix}$ be a vector in $\mathbb{R}^3$. Set $S(\mathbf{X}) = \begin{bmatrix} y_1 \\ y_2 \\ y_2 \end{bmatrix}$. Then

$$\mathbf{X} = I(\mathbf{X}) = T(S(\mathbf{X})) = T\begin{bmatrix} y_1 \\ y_2 \\ y_3 \end{bmatrix}$$

$$= \begin{bmatrix} 1 & -1 & 0 \\ 0 & 1 & -1 \\ -1 & 0 & 1 \end{bmatrix}\begin{bmatrix} y_1 \\ y_2 \\ y_3 \end{bmatrix} = \begin{bmatrix} y_1 - y_2 \\ y_2 - y_3 \\ -y_1 + y_3 \end{bmatrix}.$$

Since $(y_1 - y_2) + (y_2 - y_3) + (-y_1 + y_3) = 0$, $\mathbf{X}$ lies on the plane: $x_1 + x_2 + x_3 = 0$. Thus every vector in $\mathbb{R}^3$ lies on this plane. This is false, so S does not exist, and the claim is proved.

Exercise 1. Find an inverse for the transformation T which reflects each vector in the plane $x_3 = 0$.

Exercise 2. Find conditions on the numbers a, b, c so that the transformation T with matrix

$$\begin{pmatrix} a & 0 & 0 \\ 0 & b & 0 \\ 0 & 0 & c \end{pmatrix}$$

has an inverse S. Calculate S when it exists.

Exercise 3. Find an inverse S for the transformation T whose matrix is

$$\begin{pmatrix} 0 & 1 & 0 \\ 1 & 0 & 0 \\ 0 & 0 & 1 \end{pmatrix}.$$

Hint: $T\begin{pmatrix} x_1 \\ x_2 \\ x_3 \end{pmatrix} = \begin{pmatrix} x_2 \\ x_1 \\ x_3 \end{pmatrix}$ for every $\begin{pmatrix} x_1 \\ x_2 \\ x_3 \end{pmatrix}$.

Exercise 4. Find an inverse S for the transformation with matrix $\begin{pmatrix} 0 & 1 & 0 \\ 0 & 0 & 1 \\ 1 & 0 & 0 \end{pmatrix}$.

Exercise 5. Show that the transformation T with matrix $\begin{pmatrix} a & b & c \\ 2a & 2b & 2c \\ 3a & 3b & 3c \end{pmatrix}$ has no inverse for any values of a, b, c.

Exercise 6. Show that the transformation T with matrix $\begin{bmatrix} a & b & c \\ d & e & f \\ a+d & b+e & c+f \end{bmatrix}$ has no inverse for any values of a, b, c, d, e, f.

Exercise 7. Let T be the transformation with matrix $\begin{bmatrix} a_1 & a_2 & a_3 \\ b_1 & b_2 & b_3 \\ c_1 & c_2 & c_3 \end{bmatrix}$. Suppose that there exist scalars t_1, t_2, t_3, not all 0, such that

$$t_1 \begin{bmatrix} a_1 \\ a_2 \\ a_3 \end{bmatrix} + t_2 \begin{bmatrix} b_1 \\ b_2 \\ b_3 \end{bmatrix} + t_3 \begin{bmatrix} c_1 \\ c_2 \\ c_3 \end{bmatrix} = 0.$$

(a) Show that there is a plane π such that for every vector $\mathbf{X}$, $T(\mathbf{X})$ lies in π.
(b) Conclude that T has no inverse.

Let T be a linear transformation of $\mathbb{R}^3$. For a given vector $\mathbf{Y}$, we may try to solve the equation

$$T(\mathbf{X}) = \mathbf{Y} \tag{1}$$

by some vector $\mathbf{X}$. Suppose S is an inverse of T. Then

$$T(S(\mathbf{Y})) = TS(\mathbf{Y}) = I(\mathbf{Y}) = \mathbf{Y},$$

so $\mathbf{X} = S(\mathbf{Y})$ solves (2).

Now let $\mathbf{X}$ be a solution of (2). Then

$$S(\mathbf{Y}) = S(T(\mathbf{X})) = ST(\mathbf{X}) = I(\mathbf{X}) = \mathbf{X}.$$

So $S(\mathbf{Y})$ is the only solution of (2).

We have shown: if T has an inverse, then Eq. (2) has exactly one solution $\mathbf{X}$ for each given $\mathbf{Y}$.

Conversely, let T be a linear transformation which has the property that Eq. (2) possesses exactly one solution for each $\mathbf{Y}$. We shall show that it follows that T has an inverse.

Denote by $S(\mathbf{Y})$ the solution of (2). Then $T(S(\mathbf{Y})) = \mathbf{Y}$, and if $T(\mathbf{X}) = \mathbf{Y}$, then $\mathbf{X} = S(\mathbf{Y})$. We have thus defined a transformation S which sends each vector $\mathbf{Y}$ into $S(\mathbf{Y})$. By the definition of S, $TS(\mathbf{Y}) = T(S(\mathbf{Y})) = \mathbf{Y}$, for each $\mathbf{Y}$ in $\mathbb{R}^3$. So $TS = I$. Also, if $\mathbf{X}$ is any vector, set $\mathbf{Y} = T(\mathbf{X})$. Then $\mathbf{X} = S(\mathbf{Y})$, by the definition of S, and so $\mathbf{X} = S(T(\mathbf{X})) = ST(\mathbf{X})$. Hence $I = ST$. Thus we have shown that S is an inverse of T.

Our conscience should bother us about one point. S is a transformation of $\mathbb{R}^3$; but is it a *linear* transformation? To answer this question, choose two vectors, **X** and **Y**.

$$T(S(\mathbf{X}) + S(\mathbf{Y})) = T(S(\mathbf{X})) + T(S(\mathbf{Y})) \qquad \text{(since } T \text{ is linear)}$$
$$= TS(\mathbf{X}) + TS(\mathbf{Y}) = I(\mathbf{X}) + I(\mathbf{Y}) = \mathbf{X} + \mathbf{Y},$$

so

$$S(\mathbf{X}) + S(\mathbf{Y})$$

solves (2) with $\mathbf{X} + \mathbf{Y}$ as the right-hand side. By the definition of S, it follows that $S(\mathbf{X}) + S(\mathbf{Y}) = S(\mathbf{X} + \mathbf{Y})$. Similarly, $S(t\mathbf{X}) = tS(\mathbf{X})$ whenever t is a scalar.

Exercise 8. Prove the last statement.

We have proved:

Proposition 1. *Let T be a linear transformation of $\mathbb{R}^3$. Then T has an inverse if and only if the equation*

$$T(\mathbf{X}) = \mathbf{Y} \tag{2}$$

has, for each **Y**, *one and only one solution* **X**.

Corollary. *If T has an inverse, then*

$$T(\mathbf{X}) = \mathbf{0} \qquad \textit{implies that} \quad \mathbf{X} = \mathbf{0}. \tag{3}$$

PROOF. Set $\mathbf{Y} = \mathbf{0}$ in Proposition 1.

Suppose, conversely, that T is a linear transformation for which (3) holds, i.e., **0** is the only vector which T sends into **0**. It follows that the equation $T(\mathbf{X}) = \mathbf{Y}$ has *at most* one solution for each **Y**. To see this, suppose $T(\mathbf{U}) = \mathbf{Y}$ and $T(\mathbf{V}) = \mathbf{Y}$. Then

$$T(\mathbf{U} - \mathbf{V}) = T(\mathbf{U}) - T(\mathbf{V}) = \mathbf{Y} - \mathbf{Y} = \mathbf{0},$$

so by (3), $\mathbf{U} - \mathbf{V} = \mathbf{0}$, or $\mathbf{U} = \mathbf{V}$.

We proceed to show that (3) also implies the *existence* of a solution of $T(\mathbf{X}) = \mathbf{Y}$ for each **Y**. We saw, in Chapter 3.2, that the image of $\mathbb{R}^3$ under any linear transformation is either a plane through **0**, a line through **0**, $\mathbb{R}^3$, or the origin. Suppose the image of $\mathbb{R}^3$ under T is a plane π through **0**. The vectors $T(\mathbf{E}_1)$, $T(\mathbf{E}_2)$, $T(\mathbf{E}_3)$ all then lie in π. Three vectors in a plane are linearly dependent. Thus we can find scalars t_1, t_2, t_3, not all 0, such that $t_1T(\mathbf{E}_1) + t_2T(\mathbf{E}_2) + t_3T(\mathbf{E}_3) = \mathbf{0}$. Therefore,

$$T(t_1\mathbf{E}_1 + t_2\mathbf{E}_2 + t_3\mathbf{E}_3) = t_1T(\mathbf{E}_1) + t_2T(\mathbf{E}_2) + t_3T(\mathbf{E}_3) = \mathbf{0},$$

while $t_1\mathbf{E}_1 + t_2\mathbf{E}_2 + t_3\mathbf{E}_3 = \begin{bmatrix} t_1 \\ t_2 \\ t_3 \end{bmatrix} \neq \mathbf{0}$. This contradicts (3). Hence, the image of $\mathbb{R}^3$ under T cannot be a plane. In the same way, we conclude that the image cannot be a line or the origin, and so the image must be all of $\mathbb{R}^3$. This means that for each $\mathbf{Y}$, there is some $\mathbf{X}$ with $T(\mathbf{X}) = \mathbf{Y}$. So the existence of a solution is established. We have just proved: If T is a linear transformation of $\mathbb{R}^3$ such that (3) holds, then Eq. (2) has one and only one solution for each $\mathbf{Y}$.

By Proposition 1, it follows that T has an inverse. Thus we have:

Proposition 2. *Let T be a linear transformation of $\mathbb{R}^3$. If* (3) *holds, i.e., if $T(\mathbf{X}) = \mathbf{0}$ implies that $\mathbf{X} = \mathbf{0}$, then T has an inverse. Conversely, if T has an inverse, then* (3) *holds.*

We now need a practical test to decide whether or not (3) holds, given the matrix of some linear transformation T. In two dimensions, the test was this: If $\begin{pmatrix} a & b \\ c & d \end{pmatrix}$ is the matrix of a transformation A, then A has an inverse if and only if $ad - bc \neq 0$. We seek a similar test in $\mathbb{R}^3$.

Let T be the linear transformation whose matrix is

$$\begin{bmatrix} a_1 & a_2 & a_3 \\ b_1 & b_2 & b_3 \\ c_1 & c_2 & c_3 \end{bmatrix}.$$

We call the vectors

$$\mathbf{A} = \begin{bmatrix} a_1 \\ a_2 \\ a_3 \end{bmatrix}, \qquad \mathbf{B} = \begin{bmatrix} b_1 \\ b_2 \\ b_3 \end{bmatrix}, \qquad \mathbf{C} = \begin{bmatrix} c_1 \\ c_2 \\ c_3 \end{bmatrix}$$

the *row vectors* of this matrix. The vector $\mathbf{X} = \begin{bmatrix} x_1 \\ x_2 \\ x_3 \end{bmatrix}$ satisfies $T(\mathbf{X}) = \mathbf{0}$ if and only if

$$\begin{cases} a_1x_1 + a_2x_2 + a_3x_3 = 0, \\ b_1x_1 + b_2x_2 + b_3x_3 = 0, \\ c_1x_1 + c_2x_2 + c_3x_3 = 0. \end{cases} \tag{4}$$

On p. 124, Chapter 3.0, we showed that (4) has a nonzero solution $\mathbf{X} = \begin{bmatrix} x_1 \\ x_2 \\ x_3 \end{bmatrix}$ if and only if

$$\mathbf{A} \cdot (\mathbf{B} \times \mathbf{C}) = 0. \tag{5}$$

Combining this last result with Proposition 2, we obtain:

Theorem 3.4. *Let T be a linear transformation of $\mathbb{R}^3$ and denote by* **A**, **B**, **C** *the row vectors of the matrix of T. Then T has an inverse if and only if* $\mathbf{A} \cdot (\mathbf{B} \times \mathbf{C}) \neq 0$.

Let T be a linear transformation of $\mathbb{R}^3$ and assume that $\mathbf{A} \cdot (\mathbf{B} \times \mathbf{C}) \neq 0$. We wish to calculate the matrix of T^{-1}. Set

$$m(T^{-1}) = \begin{bmatrix} u_1 & v_1 & w_1 \\ u_2 & v_2 & w_2 \\ u_3 & v_3 & w_3 \end{bmatrix},$$

where the u_i, v_i, w_i are to be found. Then $T^{-1}\begin{bmatrix} 1 \\ 0 \\ 0 \end{bmatrix} = \begin{bmatrix} u_1 \\ u_2 \\ u_3 \end{bmatrix}$, so $\begin{bmatrix} 1 \\ 0 \\ 0 \end{bmatrix} = T\begin{bmatrix} u_1 \\ u_2 \\ u_3 \end{bmatrix}$,

which we can write

$$\begin{aligned} a_1u_1 + a_2u_2 + a_3u_3 &= 1, \\ b_1u_1 + b_2u_2 + b_3u_3 &= 0, \\ c_1u_1 + c_2u_2 + c_3u_3 &= 0. \end{aligned} \tag{6}$$

Set $\Delta = \mathbf{A} \cdot (\mathbf{B} \times \mathbf{C})$. Then we have

$$\begin{aligned} \mathbf{A} \cdot (\mathbf{B} \times \mathbf{C}) &= \Delta, \\ \mathbf{B} \cdot (\mathbf{B} \times \mathbf{C}) &= 0, \\ \mathbf{C} \cdot (\mathbf{B} \times \mathbf{C}) &= 0, \end{aligned}$$

and hence, using the hypothesis $\Delta \neq 0$,

$$\begin{aligned} \mathbf{A} \cdot \frac{\mathbf{B} \times \mathbf{C}}{\Delta} &= 1, \\ \mathbf{B} \cdot \frac{\mathbf{B} \times \mathbf{C}}{\Delta} &= 0, \\ \mathbf{C} \cdot \frac{\mathbf{B} \times \mathbf{C}}{\Delta} &= 0. \end{aligned}$$

Thus $\begin{bmatrix} u_1 \\ u_2 \\ u_3 \end{bmatrix} = \mathbf{B} \times \mathbf{C}/\Delta$ satisfies (6), so

$$T\left(\frac{\mathbf{B} \times \mathbf{C}}{\Delta}\right) = \begin{bmatrix} 1 \\ 0 \\ 0 \end{bmatrix},$$

and so

$$\frac{\mathbf{B} \times \mathbf{C}}{\Delta} = T^{-1}\begin{bmatrix} 1 \\ 0 \\ 0 \end{bmatrix}. \tag{7}$$

Now recall that by Exercise 6, Chapter 3.0,

$$\mathbf{A} \cdot (\mathbf{B} \times \mathbf{C}) = \mathbf{B} \cdot (\mathbf{C} \times \mathbf{A}) = \mathbf{C} \cdot (\mathbf{A} \times \mathbf{B}).$$

By a calculation like the one which led to (7), we get

$$\frac{\mathbf{C}\times\mathbf{A}}{\Delta} = T^{-1}\begin{bmatrix}0\\1\\0\end{bmatrix}$$

and

$$\frac{\mathbf{A}\times\mathbf{B}}{\Delta} = T^{-1}\begin{bmatrix}0\\0\\1\end{bmatrix}.$$

By the definition of the cross product,

$$\mathbf{B}\times\mathbf{C} = \begin{bmatrix}\begin{vmatrix}b_2 & b_3\\ c_2 & c_3\end{vmatrix}\\[2mm] \begin{vmatrix}b_3 & b_1\\ c_3 & c_1\end{vmatrix}\\[2mm] \begin{vmatrix}b_1 & b_2\\ c_1 & c_2\end{vmatrix}\end{bmatrix} \quad \text{and}$$

$$\mathbf{C}\times\mathbf{A} = \begin{bmatrix}\begin{vmatrix}c_2 & c_3\\ a_2 & a_3\end{vmatrix}\\[2mm] \begin{vmatrix}c_3 & c_1\\ a_3 & a_1\end{vmatrix}\\[2mm] \begin{vmatrix}c_1 & c_2\\ a_1 & a_2\end{vmatrix}\end{bmatrix} \quad \text{and}$$

$$\mathbf{A}\times\mathbf{B} = \begin{bmatrix}\begin{vmatrix}a_2 & a_3\\ b_2 & b_3\end{vmatrix}\\[2mm] \begin{vmatrix}a_3 & a_1\\ b_3 & b_1\end{vmatrix}\\[2mm] \begin{vmatrix}a_1 & a_2\\ b_1 & b_2\end{vmatrix}\end{bmatrix}.$$

We conclude that

$$m(T^{-1}) = \frac{1}{\Delta}\begin{bmatrix}\begin{vmatrix}b_2 & b_3\\ c_2 & c_3\end{vmatrix} & \begin{vmatrix}c_2 & c_3\\ a_2 & a_3\end{vmatrix} & \begin{vmatrix}a_2 & a_3\\ b_2 & b_3\end{vmatrix}\\[2mm] \begin{vmatrix}b_3 & b_1\\ c_3 & c_1\end{vmatrix} & \begin{vmatrix}c_3 & c_1\\ a_3 & a_1\end{vmatrix} & \begin{vmatrix}a_3 & a_1\\ b_3 & b_1\end{vmatrix}\\[2mm] \begin{vmatrix}b_1 & b_2\\ c_1 & c_2\end{vmatrix} & \begin{vmatrix}c_1 & c_2\\ a_1 & a_2\end{vmatrix} & \begin{vmatrix}a_1 & a_2\\ b_1 & b_2\end{vmatrix}\end{bmatrix}. \tag{8}$$

Note that

$$\Delta = \mathbf{A}\cdot(\mathbf{B}\times\mathbf{C}) = a_1\begin{vmatrix}b_2 & b_3\\ c_2 & c_3\end{vmatrix} + a_2\begin{vmatrix}b_3 & b_1\\ c_3 & c_1\end{vmatrix} + a_3\begin{vmatrix}b_1 & b_2\\ c_1 & c_2\end{vmatrix}.$$

EXAMPLE 6. If $m(T)=\begin{bmatrix}1 & 2 & 3\\ 1 & 0 & -1\\ 0 & 1 & 1\end{bmatrix}$, find $m(T^{-1})$.

$$\Delta = 1\begin{vmatrix}0 & -1\\ 1 & 1\end{vmatrix} + 2\begin{vmatrix}-1 & 1\\ 1 & 0\end{vmatrix} + 3\begin{vmatrix}1 & 0\\ 0 & 1\end{vmatrix}$$
$$= 1\cdot 1 + 2(-1) + 3\cdot 1 = 2.$$

By (8),

$$m(T^{-1}) = \tfrac{1}{2}\begin{bmatrix}1 & -(-1) & -2\\ -1 & -(-1) & 4\\ 1 & -(1) & -2\end{bmatrix} = \begin{bmatrix}\frac{1}{2} & \frac{1}{2} & -1\\ -\frac{1}{2} & \frac{1}{2} & 2\\ \frac{1}{2} & -\frac{1}{2} & -1\end{bmatrix}.$$

Exercise 9. Verify that the matrix just obtained for $m(T^{-1})$ satisfies $m(T)m(T^{-1}) = m(I)$ and $m(T^{-1})m(T) = m(I)$.

Exercise 10. Use (8) to find inverses for

(a) $\begin{pmatrix}0 & 1 & 0\\ 1 & 0 & 0\\ 0 & 0 & 1\end{pmatrix}$,

(b) $\begin{pmatrix}1 & 0 & 0\\ 0 & 0 & -1\\ 0 & 1 & 0\end{pmatrix}$,

(c) $\begin{pmatrix}5 & 5 & 1\\ 1 & 0 & 0\\ 0 & 0 & 1\end{pmatrix}$.

Let m be a matrix which possesses an inverse matrix m^{-1}, i.e.,

$$mm^{-1} = m^{-1}m = \text{identity matrix} = \text{id}.$$

Let $m = \begin{bmatrix}a_1 & a_2 & a_3\\ b_1 & b_2 & b_3\\ c_1 & c_2 & c_3\end{bmatrix}$. We shall give a new approach to the problem of finding the entries in the matrix m^{-1}.

Given a vector $\begin{bmatrix}y_1\\ y_2\\ y_3\end{bmatrix}$, consider the following system of equations for the unknowns x_1, x_2, x_3:

$$\begin{aligned} a_1x_1 + a_2x_2 + a_3x_3 &= y_1,\\ b_1x_1 + b_2x_2 + b_3x_3 &= y_2,\\ c_1x_1 + c_2x_2 + c_3x_3 &= y_3. \end{aligned} \tag{9}$$

Suppose we can solve the system (9) be setting

$$\begin{aligned} x_1 &= d_1y_1 + d_2y_2 + d_3y_3,\\ x_2 &= e_1y_1 + e_2y_2 + e_3y_3,\\ x_3 &= f_1y_1 + f_2y_2 + f_3y_3, \end{aligned} \tag{10}$$

where d_i, e_i, f_i are certain numbers which do not depend on y_1, y_2, y_3. We define a matrix n by

$$n = \begin{bmatrix} d_1 & d_2 & d_3 \\ e_1 & e_2 & e_3 \\ f_1 & f_2 & f_3 \end{bmatrix}. \tag{11}$$

Then (9) and (10) state that, setting

$$\mathbf{X} = \begin{bmatrix} x_1 \\ x_2 \\ x_3 \end{bmatrix} \quad \text{and} \quad \mathbf{Y} = \begin{bmatrix} y_1 \\ y_2 \\ y_3 \end{bmatrix},$$

$m(\mathbf{X}) = \mathbf{Y}$ and $n(\mathbf{Y}) = \mathbf{X}$. So

$$nm(\mathbf{X}) = n(m(\mathbf{X})) = n(\mathbf{Y}) = \mathbf{X}.$$

This holds for each vector $\mathbf{X}$. Hence

$$nm = \text{id}.$$

It follows that

$$n = (nm)(m^{-1}) = \text{id}(m^{-1}) = m^{-1}.$$

We thus have the following result:

The matrix n defined by (11) *is the inverse of the matrix m.*

Example 7. $m = \begin{bmatrix} 1 & 2 & 3 \\ 1 & 0 & -1 \\ 0 & 1 & 1 \end{bmatrix}$. Find m^{-1} by the preceding method.

The system (9) here is

$$\begin{aligned} x_1 + 2x_2 + 3x_3 &= y_1, \\ x_1 \qquad\quad - x_3 &= y_2, \\ x_2 + x_3 &= y_3. \end{aligned}$$

We can solve this by eliminating x_3 from the last equation by setting

$$x_3 = y_3 - x_2.$$

Inserting this expression in the first two equations yields

$$\begin{aligned} x_1 + 2x_2 + 3(y_3 - x_2) &= y_1, \\ x_1 - (y_3 - x_2) &= y_2, \end{aligned}$$

or

$$\begin{aligned} x_1 - x_2 &= y_1 - 3y_3, \\ x_1 + x_2 &= y_2 + y_3. \end{aligned}$$

Solving this system for x_1 and x_2, we find

$$\begin{aligned} 2x_1 &= y_1 + y_2 - 2y_3, \\ 2x_2 &= (y_2 + y_3) - (y_1 - 3y_3) = -y_1 + y_2 + 4y_3, \end{aligned}$$

so

$$x_1 = \tfrac{1}{2} y_1 + \tfrac{1}{2} y_2 - y_3,$$
$$x_2 = -\tfrac{1}{2} y_1 + \tfrac{1}{2} y_2 + 2y_3,$$

and

$$x_3 = \tfrac{1}{2} y_1 - \tfrac{1}{2} y_2 - y_3,$$

since

$$x_3 = y_3 - x_2 = \tfrac{1}{2} y_1 - \tfrac{1}{2} y_2 - y_3.$$

Thus, here,

$$m^{-1} = \begin{bmatrix} d_1 & d_2 & d_3 \\ e_1 & e_2 & e_3 \\ f_1 & f_2 & f_3 \end{bmatrix} = \begin{bmatrix} \tfrac{1}{2} & \tfrac{1}{2} & -1 \\ -\tfrac{1}{2} & \tfrac{1}{2} & 2 \\ \tfrac{1}{2} & -\tfrac{1}{2} & -1 \end{bmatrix}.$$

To our great relief, the answer we found agrees with our earlier answer to Example 6.

Exercise 11. Using the method just described, find inverses for the following matrices:

(a) $\begin{pmatrix} 0 & 1 & 0 \\ 1 & 0 & 0 \\ 0 & 0 & 1 \end{pmatrix}$,

(b) $\begin{pmatrix} 5 & 5 & 1 \\ 1 & 0 & 0 \\ 0 & 0 & 1 \end{pmatrix}$,

(c) $\begin{pmatrix} 1 & -1 & 2 \\ 6 & 0 & 1 \\ 3 & 2 & 1 \end{pmatrix}$.

Exercise 12. Using the methods just described, find an inverse for

$$\begin{pmatrix} 1 & a & b \\ 0 & 1 & c \\ 0 & 0 & 1 \end{pmatrix}.$$

Exercise 13. By computing $\mathbf{A} \cdot (\mathbf{B} \times \mathbf{C})$, show that $\begin{bmatrix} 1 & s & s^2 \\ 1 & t & t^2 \\ 1 & u & u^2 \end{bmatrix}$ has an inverse provided s, t, u are all distinct. *Hint*: Simplify the expression you get for $\mathbf{A} \cdot (\mathbf{B} \times \mathbf{C})$, writing it as a product.

Exercise 14. S and T are linear transformations of $\mathbb{R}^3$ which have inverses. Show that ST has an inverse and that $(ST)^{-1} = T^{-1}S^{-1}$.

Exercise 15. S is an invertible linear transformation of $\mathbb{R}^3$, D is a linear transformation of $\mathbb{R}^3$, and $T = S^{-1}DS$. Show that

$$T^2 = S^{-1}D^2S, \qquad T^3 = S^{-1}D^3S.$$

Exercise 16. n is a 3×3 matrix such that $n^3 = 0$. Show that

$$(\mathrm{id} + n)^{-1} = \mathrm{id} - n + n^2.$$

Exercise 17. Set $n = \begin{pmatrix} 0 & a & b \\ 0 & 0 & c \\ 0 & 0 & 0 \end{pmatrix}$. Show that $n^3 = 0$. Using Exercise 16, find the inverse of $\mathrm{id} + n = \begin{pmatrix} 1 & a & b \\ 0 & 1 & c \\ 0 & 0 & 1 \end{pmatrix}$. Compare your result with the answer to Exercise 12.

Exercise 18. For what values of a, b, c, d does $\begin{pmatrix} 1 & 0 & 0 \\ 0 & a & b \\ 0 & c & d \end{pmatrix}$ have an inverse? Calculate the inverse when it exists.

We defined the inverse S of a linear transformation T by the two conditions:

(i) $ST = I$

and

(ii) $TS = I$.

Suppose that only one of the two conditions is satisfied, say, (i) holds. Does (ii) follow?

Proposition 3. *Let S, T be linear transformations of $\mathbb{R}^3$ such that $ST = I$. Then $TS = I$.*

PROOF. Choose a vector $\mathbf{X}$ such that $T(\mathbf{X}) = \mathbf{0}$. Then $\mathbf{X} = I(\mathbf{X}) = ST(\mathbf{X}) = S(\mathbf{0}) = \mathbf{0}$. Then, by Proposition 2, T has an inverse, T^{-1}, with $T^{-1}T = TT^{-1} = I$. Since $ST = I$, $S = SI = S(TT^{-1}) = (ST)T^{-1} = IT^{-1} = IT^{-1} = T^{-1}$. Hence $TS = T(T^{-1}) = I$.

Exercise 19. Let S, T be two linear transformations. Show that if the product ST has an inverse, then S has an inverse and T has an inverse.

§1. Inverses of Elementary Matrices and Diagonal Matrices

Recall the elementary matrices e_{ij}^s and p_{ij} that we studied in Chapter 3.3. Let us find the inverses of these matrices.

EXAMPLE 8. $e_{31}^s e_{31}^t = \begin{bmatrix} 1 & 0 & 0 \\ 0 & 1 & 0 \\ s & 0 & 1 \end{bmatrix} \begin{bmatrix} 1 & 0 & 0 \\ 0 & 1 & 0 \\ t & 0 & 1 \end{bmatrix} = \begin{bmatrix} 1 & 0 & 0 \\ 0 & 1 & 0 \\ s+t & 0 & 1 \end{bmatrix}$. Choosing $t = -s$, this gives $e_{31}^s e_{31}^{-s} = \mathrm{id}$, and replacing s by $-s$ and t by s, we get $e_{31}^{-s} e_{31}^s = \mathrm{id}$. So $(e_{31}^s)^{-1} = e_{31}^{-s}$.

Exercise 20. Show that $(e_{32}^s)^{-1} = e_{32}^{-s}$.

Exercise 21. Show that $(e_{12}^s)^{-1} = e_{12}^{-s}$.

In the general case, for all i, j, and s, a similar calculation gives us:

Proposition 4. $(e_{ij}^{s})^{-1} = e_{ij}^{-s}$.

EXAMPLE 9. Recall that $p_{31} = \begin{bmatrix} 0 & 0 & 1 \\ 0 & 1 & 0 \\ 1 & 0 & 0 \end{bmatrix}$. We know that for each matrix m, $p_{31}m$ is obtained from m by interchanging the first and third rows and leaving the second row alone. Hence,

$$p_{31}p_{31} = \text{id},$$

so $(p_{31})^{-1} = p_{31}$.

Exercise 22. Find $(p_{12})^{-1}$.

In the general case, we have:

Proposition 5. *For every i, j, $i \neq j$,*

$$(p_{ij})^{-1} = p_{ij}.$$

Next, consider the diagonal matrix

$$d = \begin{bmatrix} t_1 & 0 & 0 \\ 0 & t_2 & 0 \\ 0 & 0 & t_3 \end{bmatrix}.$$

If any of the numbers t_1, t_2, t_3 is 0, then by Theorem 3.4, d has no inverse. If all $t_i \neq 0$, we have

$$\begin{bmatrix} t_1 & 0 & 0 \\ 0 & t_2 & 0 \\ 0 & 0 & t_3 \end{bmatrix} \begin{bmatrix} 1/t_1 & 0 & 0 \\ 0 & 1/t_2 & 0 \\ 0 & 0 & 1/t_3 \end{bmatrix} = \text{id},$$

so $d^{-1} = \begin{bmatrix} 1/t_1 & 0 & 0 \\ 0 & 1/t_2 & 0 \\ 0 & 0 & 1/t_3 \end{bmatrix}$. Thus we have:

Proposition 6. *A diagonal matrix d has an inverse if and only if its diagonal entries are all $\neq 0$, and in that case d^{-1} is the diagonal matrix whose diagonal entries are the reciprocals of those for d.*

We shall now show that, by multiplying a given matrix m by suitable elementary matrices, we can convert m into a diagonal matrix.

Let $m = \begin{bmatrix} a_1 & a_2 & a_3 \\ b_1 & b_2 & b_3 \\ c_1 & c_2 & c_3 \end{bmatrix}$. If $a_1 \neq 0$, we can choose t so that $b_1 + ta_1 = 0$. Then

$$e_{21}^t m = \begin{bmatrix} a_1 & a_2 & a_3 \\ 0 & b_2 + ta_2 & b_3 + ta_3 \\ c_1 & c_2 & c_3 \end{bmatrix}.$$

Similarly, we can choose s so that

$$e_{31}^s e_{21}^t m = \begin{bmatrix} a_1 & a_2 & a_3 \\ 0 & b_2 + ta_2 & b_3 + ta_3 \\ 0 & c_2 + sa_2 & c_3 + sa_3 \end{bmatrix}.$$

Thus $e_{31}^s e_{21}^t m = \begin{bmatrix} a_1 & a_2 & a_3 \\ 0 & x & y \\ 0 & z & w \end{bmatrix}$. Similarly, we can find r, q so that

$$e_{31}^s e_{21} m e_{12}^r e_{13}^q = \begin{bmatrix} a_1 & 0 & 0 \\ 0 & x & y \\ 0 & z & w \end{bmatrix}. \tag{12}$$

If $a_1 = 0$, either m is the zero matrix or some entry of m is $\neq 0$, say, $b_3 \neq 0$. Then

$$mp_{13} = \begin{bmatrix} a_3 & a_2 & a_1 \\ b_3 & b_2 & b_1 \\ c_3 & c_2 & c_1 \end{bmatrix} \quad \text{and} \quad p_{12} m p_{13} = \begin{bmatrix} b_3 & b_2 & b_1 \\ a_3 & a_2 & a_1 \\ c_3 & c_2 & c_1 \end{bmatrix}.$$

Since $b_3 \neq 0$, we can apply formula (12) to the matrix $p_{12} m p_{13}$ and get

$$e_{31}^s e_{21}^t p_{12} m p_{13} e_{12}^r e_{13}^q = \begin{bmatrix} b_3 & 0 & 0 \\ 0 & x & y \\ 0 & z & w \end{bmatrix}. \tag{13}$$

If $x \neq 0$, we can proceed as earlier and find scalars i, j so that (2) yields

$$e_{32}^i e_{31}^s e_{21}^t p_{12} m p_{13} e_{12}^r e_{13}^q e_{23}^j = \begin{bmatrix} b_3 & 0 & 0 \\ 0 & x & 0 \\ 0 & 0 & u \end{bmatrix}.$$

The right-hand side is a diagonal matrix. If $x = 0$, we consider two possibilities. If y, z, w in (13) are all 0, then the right-hand side of (13) is a diagonal matrix. If at least one of y, z, $w \neq 0$, say $w \neq 0$, we can multiply the left-hand side in (13) on the left by p_{23} and on the right by p_{23} and obtain

$$p_{23} e_{31}^s e_{21}^t p_{12} m p_{13} e_{12}^r e_{13}^q p_{23} = \begin{bmatrix} b_3 & 0 & 0 \\ 0 & w & z \\ 0 & y & x \end{bmatrix},$$

where now $w \neq 0$. Proceeding as we did earlier when we had $x \neq 0$, we find that

$$e_{32}^{i} p_{23} e_{31}^{s} e_{21}^{t} p_{12} m p_{13} e_{12}^{r} e_{13}^{q} p_{23} e_{23}^{j} = \begin{bmatrix} b_3 & 0 & 0 \\ 0 & w & 0 \\ 0 & 0 & u \end{bmatrix},$$

which is again a diagonal matrix. So in every case, we can find elementary matrices $e_1, \ldots, e_k, f_1, \ldots, f_l$ so that

$$e_1 e_2 \ldots e_k m f_1 f_2 \ldots f_l = d,$$

where d is a diagonal matrix. It follows, by multiplying the last equation by e_1^{-1}, that

$$e_2 \ldots e_k m f_1 \ldots f_l = e_1^{-1} d.$$

Continuing, we obtain

$$m f_1 \ldots f_l = e_k^{-1} \ldots e_2^{-1} e_1^{-1} d,$$

and

$$m = e_k^{-1} \ldots e_2^{-1} e_1^{-1} d f_l^{-1} f_{l-1}^{-1} \ldots f_1^{-1}.$$

Note that the inverse of an elementary matrix is again an elementary matrix. We have thus proved:

Theorem 3.5. *Let m be a 3×3 matrix. Then we can find elementary matrices $g_1, \ldots, g_k, h_1, \ldots, h_l$ and a diagonal matrix d so that*

$$m = g_1 g_2 \ldots g_k d h_1 h_2 \ldots h_l. \tag{14}$$

EXAMPLE 10. Let us express the matrix

$$m = \begin{bmatrix} 1 & 1 & 0 \\ 3 & 1 & 2 \\ 5 & 2 & 4 \end{bmatrix}$$

in the form (14),

$$e_{21}^{-3} m = \begin{bmatrix} 1 & 1 & 0 \\ 0 & -2 & 2 \\ 5 & 2 & 4 \end{bmatrix},$$

$$e_{31}^{-5} e_{21}^{-3} m = \begin{bmatrix} 1 & 1 & 0 \\ 0 & -2 & 2 \\ 0 & -3 & 4 \end{bmatrix},$$

$$e_{32}^{-3/2} e_{31}^{-5} e_{21}^{-3} m = \begin{bmatrix} 1 & 1 & 0 \\ 0 & -2 & 2 \\ 0 & 0 & 1 \end{bmatrix}.$$

Also,

$$\begin{bmatrix} 1 & 1 & 0 \\ 0 & -2 & 2 \\ 0 & 0 & 1 \end{bmatrix} e_{12}^{-1} = \begin{bmatrix} 1 & 0 & 0 \\ 0 & -2 & 2 \\ 0 & 0 & 1 \end{bmatrix}$$

and

$$\begin{bmatrix} 1 & 0 & 0 \\ 0 & -2 & 2 \\ 0 & 0 & 1 \end{bmatrix} e_{12}^{-1} e_{23}^{1} = \begin{bmatrix} 1 & 0 & 0 \\ 0 & -2 & 0 \\ 0 & 0 & 1 \end{bmatrix}.$$

Setting $d = \begin{bmatrix} 1 & 0 & 0 \\ 0 & -2 & 0 \\ 0 & 0 & 1 \end{bmatrix}$, we thus have $e_{32}^{-3/2} e_{31}^{-5} e_{21}^{-3} m e_{12}^{-1} e_{23}^{1} = d$, so $m = e_{21}^{3} e_{31}^{5} e_{32}^{3/2} d e_{23}^{-1} e_{12}^{1}$ or

$$\begin{bmatrix} 1 & 1 & 0 \\ 3 & 1 & 2 \\ 5 & 2 & 4 \end{bmatrix} = \begin{bmatrix} 1 & 0 & 0 \\ 3 & 1 & 0 \\ 0 & 0 & 1 \end{bmatrix} \begin{bmatrix} 1 & 0 & 0 \\ 0 & 1 & 0 \\ 5 & 0 & 1 \end{bmatrix} \begin{bmatrix} 1 & 0 & 0 \\ 0 & 1 & 0 \\ 0 & 3/2 & 1 \end{bmatrix}$$

$$\times \begin{bmatrix} 1 & 0 & 0 \\ 0 & -2 & 0 \\ 0 & 0 & 1 \end{bmatrix} \begin{bmatrix} 1 & 0 & 0 \\ 0 & 1 & -1 \\ 0 & 0 & 1 \end{bmatrix} \begin{bmatrix} 1 & 1 & 0 \\ 0 & 1 & 0 \\ 0 & 0 & 1 \end{bmatrix}.$$

Exercise 23. Express the following matrices in the form of (14):

(a) $\begin{pmatrix} 0 & 0 & 1 \\ 1 & 0 & 0 \\ 0 & 1 & 0 \end{pmatrix}$,

(b) $\begin{pmatrix} 1 & 3 & 0 \\ 2 & 0 & 0 \\ 0 & 0 & 0 \end{pmatrix}$,

(c) $\begin{pmatrix} 1 & 1 & 1 \\ 0 & 2 & 2 \\ 0 & 0 & 3 \end{pmatrix}$.

§2. Systems of Three Linear Equations in Three Unknowns

We consider the following system of three equations in three unknowns:

$$\begin{aligned} a_1x_1 + a_2x_2 + a_3x_3 &= u_1, \\ b_1x_1 + b_2x_2 + b_3x_3 &= u_2, \\ c_1x_1 + c_2x_2 + c_3x_3 &= u_3. \end{aligned} \tag{15}$$

For each choice of numbers u_1, u_2, u_3, we may ask: Does the system (15) have a solution x_1, x_2, x_3? And if (15) has a solution, is this solution unique?

We may write the above expression in matrix form by introducing the linear transformation T with matrix

$$m = \begin{bmatrix} a_1 & a_2 & a_3 \\ b_1 & b_2 & b_3 \\ c_1 & c_2 & c_3 \end{bmatrix}.$$

Then the system (15) may be written

$$T(\mathbf{X}) = \mathbf{U}, \tag{16}$$

where $\mathbf{X}$ is the vector $\begin{bmatrix} x_1 \\ x_2 \\ x_3 \end{bmatrix}$ and $\mathbf{U} = \begin{bmatrix} u_1 \\ u_2 \\ u_3 \end{bmatrix}$. We call (16) the *nonhomogeneous system with matrix m.*

If the matrix m has an inverse, then the linear transformation T has an inverse T^{-1}. We then have

$$T(T^{-1}(\mathbf{U})) = (TT^{-1})(\mathbf{U}) = \mathbf{U};$$

so $\mathbf{X} = T^{-1}(\mathbf{U})$ is a solution of (16), and, conversely, if $T(\mathbf{X}) = \mathbf{U}$, then $\mathbf{X} = T^{-1}(T(\mathbf{X})) = T^{-1}(\mathbf{U})$.

So there is a unique solution vector $\mathbf{X}$ for each choice of $\mathbf{U}$.

In particular, if $\mathbf{U} = \mathbf{0}$, we find that $\mathbf{X} = \begin{bmatrix} 0 \\ 0 \\ 0 \end{bmatrix} = T^{-1}\begin{bmatrix} 0 \\ 0 \\ 0 \end{bmatrix}$ is the unique solution of the system

$$T(\mathbf{X}) = \mathbf{0}. \tag{17}$$

This system, with the zero vector on the right-hand side, is called the *homogeneous system* associated with the system (15).

No matter what the matrix m is, the homogeneous system has at least one solution, namely, the solution $\mathbf{X} = \begin{bmatrix} 0 \\ 0 \\ 0 \end{bmatrix}$. This is called the *trivial solution* of the homogeneous system, and we have seen above that if m has an inverse, then the trivial solution is the only solution of the homogeneous system.

But what if m does not possess an inverse? Will the system (17) then have a nontrivial solution? Proposition 2 of this chapter tells us that the answer is yes.

As in the 2-dimensional case, we obtain the following three general results.

Proposition 7. *The system* (16) *has a unique solution* $\mathbf{X}$ *for every* $\mathbf{U}$ *if and only if the transformation T has an inverse.*

Proposition 8. *The homogeneous system* (17) *has a non-trivial solution if and only if T fails to have an inverse.*

In the case that T fails to have an inverse, the general solution of (16) is described as follows.

Proposition 9. *If* $\overline{\mathbf{X}}$ *is a particular solution of* (16), *so that* $T(\overline{\mathbf{X}}) = \mathbf{U}$, *we may express every solution of* (16) *in the form* $\overline{\mathbf{X}} + \mathbf{X}^h$, *where* $\mathbf{X}^h$ *is a solution of the homogeneous system* (17).

EXAMPLE 11. Find all solutions of the nonhomogeneous system

$$\begin{cases} x_1 + x_2 + x_3 = 1, \\ 2x_1 - x_2 = 5, \\ 5x_1 + 2x_2 + 3x_3 = 8. \end{cases}$$

If x_1, x_2, x_3 solves the corresponding homogeneous system

$$\begin{cases} x_1 + x_2 + x_3 = 0 \\ 2x_1 - x_2 = 0, \\ 5x_1 + 2x_2 + 3x_3 = 0, \end{cases}$$

then $2x_1 = x_2$ and $x_1 + 2x_1 + x_3 = 0$, so $x_3 = -3x_1$. Hence, the most general solution $\mathbf{X}^h = \begin{bmatrix} x_1 \\ x_2 \\ x_3 \end{bmatrix}$ of the homogeneous system is

$$\mathbf{X}^h = \begin{bmatrix} x_1 \\ 2x_1 \\ -3x_1 \end{bmatrix}.$$

A particular solution $\overline{\mathbf{X}}$ of the nonhomogeneous system is

$$\overline{\mathbf{X}} = \begin{bmatrix} 2 \\ -1 \\ 0 \end{bmatrix}.$$

By Proposition 9, the general solution of the nonhomogeneous system is then

$$\mathbf{X} = \begin{bmatrix} 2 \\ -1 \\ 0 \end{bmatrix} + \begin{bmatrix} x_1 \\ 2x_1 \\ -3x_1 \end{bmatrix} = \begin{bmatrix} 2 + x_1 \\ -1 + 2x_1 \\ -3x_1 \end{bmatrix},$$

where x_1 is an arbitrary real number.

§3. Two Equations in Three Unknowns

In the system (15) under consideration, the number of unknowns equals the number of equations. Suppose we are given a system of two equations in three unknowns:

$$\begin{aligned} a_1x_1 + a_2x_2 + a_3x_3 &= u_1, \\ b_1x_1 + b_2x_2 + b_3x_3 &= u_2. \end{aligned} \tag{18}$$

The homogeneous system corresponding to (18) is

$$\begin{aligned} a_1x_1 + a_2x_2 + a_3x_3 &= 0, \\ b_1x_1 + b_2x_2 + b_3x_3 &= 0. \end{aligned} \tag{19}$$

Setting $\mathbf{X} = \begin{bmatrix} x_1 \\ x_2 \\ x_3 \end{bmatrix}$, $\mathbf{A} = \begin{bmatrix} a_1 \\ a_2 \\ a_3 \end{bmatrix}$, $\mathbf{B} = \begin{bmatrix} b_1 \\ b_2 \\ b_3 \end{bmatrix}$, we may write the system (18) as

$$\mathbf{A} \cdot \mathbf{X} = u_1, \qquad \mathbf{B} \cdot \mathbf{X} = u_2 \tag{18'}$$

and the system (19) as

$$\mathbf{A} \cdot \mathbf{X} = 0, \qquad \mathbf{B} \cdot \mathbf{X} = 0. \tag{19'}$$

Proposition 10. *If* $\overline{\mathbf{X}}$ *is one solution of* (18), *then any solution of* (18) *can be written as* $\overline{\mathbf{X}} + \mathbf{X}^h$, *where* $\mathbf{X}^h$ *is a solution of the homogeneous system* (19).

PROOF. If $\mathbf{A} \cdot \overline{\mathbf{X}} = \mathbf{A} \cdot \mathbf{X} = u_1$ and $\mathbf{B} \cdot \overline{\mathbf{X}} = \mathbf{B} \cdot \mathbf{X} = u_2$, then $\mathbf{A} \cdot (\mathbf{X} - \overline{\mathbf{X}}) = 0$ and $\mathbf{B} \cdot (\mathbf{X} - \overline{\mathbf{X}}) = 0$, so $\mathbf{X} - \overline{\mathbf{X}}$ is a solution $\mathbf{X}^h$ of (19).

If **A** and **B** are linearly independent, then the solution space of (19) is just the line perpendicular to the plane spanned by **A** and **B**, i.e., the line along the nonzero vector $\mathbf{A} \times \mathbf{B}$.

If **A** and **B** are linearly dependent, but not both **0**, then the solution space of (19) will be the plane through the origin perpendicular to the line containing **A** and **B**.

If $\mathbf{A} = \mathbf{B} = \mathbf{0}$, then the solution of (19) is all of $\mathbb{R}^3$.

CHAPTER 3.5

Determinants

Let m be the matrix

$$\begin{bmatrix} a_1 & a_2 & a_3 \\ b_1 & b_2 & b_3 \\ c_1 & c_2 & c_3 \end{bmatrix}.$$

Let **A**, **B**, **C** denote the row vectors of this matrix. The quantity

$$\mathbf{A} \cdot (\mathbf{B} \times \mathbf{C}) = \mathbf{B} \cdot (\mathbf{C} \times \mathbf{A}) = \mathbf{C} \cdot (\mathbf{A} \times \mathbf{B})$$

is called the *determinant of* m and is denoted $\det(m)$ or

$$\begin{vmatrix} a_1 & a_2 & a_3 \\ b_1 & b_2 & b_3 \\ c_1 & c_2 & c_3 \end{vmatrix}.$$

Expressed in these terms, Theorem 3.4 of Chapter 3.4 says that m has an inverse if and only if $\det(m) \neq 0$. In Exercise 6 in Chapter 3.0, we saw that

$$\mathbf{A} \cdot (\mathbf{B} \times \mathbf{C}) = \mathbf{B} \cdot (\mathbf{C} \times \mathbf{A}) = \mathbf{C} \cdot (\mathbf{A} \times \mathbf{B}).$$

Also $\mathbf{A} \times \mathbf{B} = -\mathbf{B} \times \mathbf{A}$.

In what follows, we shall frequently make use of these relations.

(i) *If two rows are interchanged, the determinant changes sign.*

PROOF.

$$\begin{bmatrix} b_1 & b_2 & b_3 \\ a_1 & a_2 & a_3 \\ c_1 & c_2 & c_3 \end{bmatrix} = \mathbf{B} \cdot (\mathbf{A} \times \mathbf{C}) = \mathbf{B} \cdot (-\mathbf{C} \times \mathbf{A}) = -\mathbf{B} \cdot (\mathbf{C} \times \mathbf{A})$$

$$= -\mathbf{A} \cdot (\mathbf{B} \times \mathbf{C}) = -\begin{vmatrix} a_1 & a_2 & a_3 \\ b_1 & b_2 & b_3 \\ c_1 & c_2 & c_3 \end{vmatrix}.$$

Interchanging the last two rows, we get

$$\begin{vmatrix} a_1 & a_2 & a_3 \\ c_1 & c_2 & c_3 \\ b_1 & b_2 & b_3 \end{vmatrix} = \mathbf{A} \cdot (\mathbf{C} \times \mathbf{B}) = -\mathbf{A} \cdot (\mathbf{B} \times \mathbf{C}) = -\begin{vmatrix} a_1 & a_2 & a_3 \\ b_1 & b_2 & b_3 \\ c_1 & c_2 & c_3 \end{vmatrix}.$$

Finally, interchanging the first and third rows,

$$\begin{vmatrix} c_1 & c_2 & c_3 \\ b_1 & b_2 & b_3 \\ a_1 & a_2 & a_3 \end{vmatrix} = \mathbf{C} \cdot (\mathbf{B} \times \mathbf{A}) = -\mathbf{C} \cdot (\mathbf{A} \times \mathbf{B}) = -\begin{vmatrix} a_1 & a_2 & a_3 \\ b_1 & b_2 & b_3 \\ c_1 & c_2 & c_3 \end{vmatrix}.$$

Thus (i) is proved.

(ii) *If a row is* 0, *then* $\det(m) = 0$.

Proof. If $\mathbf{A} = \mathbf{0}$, $\det(m) = \mathbf{0} \cdot (\mathbf{B} \times \mathbf{C}) = 0$, and if $\mathbf{B}$ or $\mathbf{C} = \mathbf{0}$, $\det(m) = \mathbf{A} \cdot (\mathbf{0} \times \mathbf{C})$ or $\mathbf{A} \cdot (\mathbf{B} \times \mathbf{0})$, and so $\det(m) = 0$.

(iii) *If two rows are equal, then* $\det(m) = 0$.

Proof. If $\mathbf{A} = \mathbf{B}$, $\det(m) = \mathbf{A} \cdot (\mathbf{A} \times \mathbf{C}) = 0$, by Chapter 3.0., p. 121. Similarly, if $\mathbf{A} = \mathbf{C}$, $\det(m) = 0$. If $\mathbf{B} = \mathbf{C}$, $\det(m) = \mathbf{A} \cdot (\mathbf{B} \times \mathbf{B}) = \mathbf{A} \cdot \mathbf{0} = 0$. So (iii) is proved.

(iv) *Suppose the three rows* **A**, **B**, **C** *are linearly dependent. Then* $\det(m) = 0$.

Proof. If **B** and **C** are linearly dependent, then $\mathbf{B} \times \mathbf{C} = \mathbf{0}$, so

$$\det(m) = \mathbf{A} \cdot (\mathbf{B} \times \mathbf{C}) = \mathbf{A} \cdot \mathbf{0} = 0.$$

If **B** and **C** are linearly independent, then $\mathbf{A} = c_1 B + c_2 C$, and so

$$\begin{aligned} \det(m) = \mathbf{A} \cdot (\mathbf{B} \times \mathbf{C}) &= (c_1\mathbf{B} + c_2\mathbf{C}) \cdot \mathbf{B} \times \mathbf{C} \\ &= c_1\mathbf{B} \cdot (\mathbf{B} \times \mathbf{C}) + c_2\mathbf{C} \cdot (\mathbf{B} \times \mathbf{C}) = c_1 0 + c_2 0 = 0, \end{aligned}$$

so (iv) is proved.

(v) *Suppose the three rows* **A**, **B**, **C** *are linearly independent. Then* $\det(m) \neq 0$.

Proof. If $\mathbf{X} = \begin{bmatrix} x_1 \\ x_2 \\ x_3 \end{bmatrix}$ is a vector that m sends into $\mathbf{0}$, then

$$\begin{bmatrix} a_1 & a_2 & a_3 \\ b_1 & b_2 & b_3 \\ c_1 & c_2 & c_3 \end{bmatrix} \begin{bmatrix} x_1 \\ x_2 \\ x_3 \end{bmatrix} = \begin{bmatrix} 0 \\ 0 \\ 0 \end{bmatrix},$$

so $\mathbf{A} \cdot \mathbf{X} = 0$, $\mathbf{B} \cdot \mathbf{X} = 0$, $\mathbf{C} \cdot \mathbf{X} = 0$. By Proposition 1, Chapter 3.0, we conclude that $\mathbf{X} \cdot \mathbf{X} = 0$ and so $\mathbf{X} = \mathbf{0}$.

So m sends only $\mathbf{0}$ into $\mathbf{0}$. By Proposition 2 of Chapter 3.4, it follows that m has an inverse, and so by Theorem 3.4 of Chapter 3.4, $\det(m) \neq 0$.

Putting (iv) and (v) together, we have:

(vi) $\det(m) \neq 0$ *if and only if the rows of m are linearly independent.*

Exercise 1. Show that (iii) and (ii) are consequences of (iv).

(vii) *If a scalar multiple of one row of a matrix is added to another row, the determinant is unchanged.*

PROOF.

$$\begin{vmatrix} a_1 & a_2 & a_3 \\ b_1 + ta_1 & b_2 + ta_2 & b_3 + ta_3 \\ c_1 & c_2 & c_3 \end{vmatrix} = \mathbf{A} \cdot [\ (\mathbf{B} + t\mathbf{A}\ ) \times \mathbf{C}]$$

$$= \mathbf{A} \cdot (\mathbf{B} \times \mathbf{C} + t\mathbf{A} \times \mathbf{C})$$

$$= \mathbf{A} \cdot (\mathbf{B} \times \mathbf{C}) + t\mathbf{A} \cdot (\mathbf{A} \times \mathbf{C})$$

$$= \mathbf{A} \cdot (\mathbf{B} \times \mathbf{C}) = \begin{vmatrix} a_1 & a_2 & a_3 \\ b_1 & b_2 & b_3 \\ c_1 & c_2 & c_3 \end{vmatrix}.$$

Similar reasoning gives the result in the other cases.

Exercise 2. Verify (vii) for the case when t times the last row is added to the first row.

§1. The Transpose of a Matrix

Let $m = \begin{bmatrix} a_1 & a_2 & a_3 \\ b_1 & b_2 & b_3 \\ c_1 & c_2 & c_3 \end{bmatrix}$. We call the line through a_1, b_2, c_3 the *diagonal* of m.

The following pairs of entries lie symmetrically placed with respect to the diagonal:

$$(a_2, b_1), \qquad (a_3, c_1), \qquad (b_3, c_2).$$

Let us interchange the elements in each pair, but leave the elements on the diagonal alone, and write down the matrix this gives:

$$\begin{bmatrix} a_1 & b_1 & c_1 \\ a_2 & b_2 & c_2 \\ a_3 & b_3 & c_3 \end{bmatrix}.$$

We call this new matrix the *transpose of m* and denote it by m^*. Note that the columns of m^* are the rows of m and the rows of m^* are the columns of m.

(viii) $\det(m^*) = \det(m)$.

PROOF.

$$\det(m^*) = a_1\begin{vmatrix} b_2 & c_2 \\ b_3 & c_3 \end{vmatrix} - b_1\begin{vmatrix} a_2 & c_2 \\ a_3 & c_3 \end{vmatrix} + c_1\begin{vmatrix} a_2 & b_2 \\ a_3 & b_3 \end{vmatrix}$$

$$= a_1b_2c_3 - a_1c_2b_3 - b_1a_2c_3 + b_1c_2a_3 + c_1a_2b_3 - c_1a_3b_2$$

$$\det(m) = a_1\begin{vmatrix} b_2 & b_3 \\ c_2 & c_3 \end{vmatrix} - a_2\begin{vmatrix} b_1 & b_3 \\ c_1 & c_3 \end{vmatrix} + a_3\begin{vmatrix} b_1 & b_2 \\ c_1 & c_2 \end{vmatrix}$$

$$= a_1b_2c_3 - a_1b_3c_2 - a_2b_1c_3 + a_2b_2c_1 + a_3b_1c_2 - a_3b_2c_1.$$

Thus $\det(m) = \det(m^*)$, as asserted.

We can use this result to give another characterization of matrices with determinant $\neq 0$.

(ix) $\det(m) \neq 0$ *if and only if the columns of m are linearly independent.*

PROOF. If $\det(m) \neq 0$, then $\det(m^*) \neq 0$ by (viii). Hence, the rows of m^* are linearly independent by (vi). But the rows of m^* are the columns of m, so the columns of m are independent.

Conversely, if the columns of m are independent, then the rows of m^* are independent, so $\det(m^*) \neq 0$ and $\det(m) \neq 0$. The statement is proved.

Is the analogue of (vii) true when columns are used instead of rows?

EXAMPLE 1. Fix a matrix

$$m = \begin{pmatrix} a_1 & a_2 & a_3 \\ b_1 & b_2 & b_3 \\ c_1 & c_2 & c_3 \end{pmatrix};$$

let t be a scalar, and set

$$m_t = \begin{vmatrix} a_1 & a_2 + ta_1 & a_3 \\ b_1 & b_2 + tb_1 & b_3 \\ c_1 & c_2 + tc_1 & c_3 \end{vmatrix}.$$

Then

$$m_t^* = \begin{pmatrix} a_1 & b_1 & c_1 \\ a_2 + ta_1 & b_2 + tb_1 & c_2 + tc_1 \\ a_3 & b_3 & c_3 \end{pmatrix}.$$

By (viii),

$$\det(m_t) = \det(m_t^*).$$

By (vii),

$$\det(m_t^*) = \begin{vmatrix} a_1 & b_1 & c_1 \\ a_2 & b_2 & c_2 \\ a_3 & b_3 & b_3 \end{vmatrix} = \det(m^*).$$

Using (viii), we again get $\det(m^*) = \det(m)$, so $\det(m_t) = \det(m)$. Thus

$$\begin{vmatrix} a_1 & a_2 + ta_1 & a_3 \\ b_1 & b_2 + tb_1 & b_2 \\ c_1 & c_2 + tc_1 & c_3 \end{vmatrix} = \begin{vmatrix} a_1 & a_2 & a_3 \\ b_1 & b_2 & b_3 \\ c_1 & c_2 & c_3 \end{vmatrix},$$

and so the analogue of (vii) holds in this case.

Exercise 3. Show that

$$\begin{vmatrix} a_1 + ta_3 & a_2 & a_3 \\ b_1 + tb_3 & b_2 & b_3 \\ c_1 + tc_3 & c_2 & c_3 \end{vmatrix} = \begin{vmatrix} a_1 & a_2 & a_3 \\ b_1 & b_2 & b_3 \\ c_1 & c_2 & c_3 \end{vmatrix}.$$

Reasoning as in Example 1 and Exercise 2, we find that:

(x) *If a scalar multiple of one column of a matrix is added to another column, the determinant is unchanged.*

§2. Elementary Matrices

Recall the elementary matrices e_{ij}^t and p_{ij} we studied in earlier chapters. Let us find their determinants.

EXAMPLE 2.

$$\det(e_{12}^t) = \begin{vmatrix} 1 & t & 0 \\ 0 & 1 & 0 \\ 0 & 0 & 1 \end{vmatrix} = 1\begin{vmatrix} 1 & 0 \\ 0 & 1 \end{vmatrix} - t\begin{vmatrix} 0 & 0 \\ 0 & 1 \end{vmatrix} + 0 = 1.$$

(xi) *For every* i, j, t, $\det(e_{ij}^t) = 1$, *and for every* $1, j$, $\det(p_{ij}) = -1$.

Exercise 4. Prove (xi).

EXAMPLE 3. Let $m = \begin{bmatrix} a_1 & a_2 & a_3 \\ b_1 & b_2 & b_3 \\ c_1 & c_2 & c_3 \end{bmatrix}$. Then

$$e_{12}^t m = \begin{bmatrix} a_1 + tb_1 & a_2 + tb_2 & a_3 + tb_3 \\ b_1 & b_2 & b_3 \\ c_1 & c_2 & c_3 \end{bmatrix}.$$

Hence, by (vii), $\det(e_{12}^t m) = \det(m)$.

EXAMPLE 4. Let m be as before. Then

$$me_{12}^t = \begin{bmatrix} a_1 & a_2 + ta_1 & a_3 \\ b_1 & b_2 + tb_1 & b_3 \\ c_1 & c_2 + tc_1 & c_3 \end{bmatrix}.$$

Hence, by (x), $\det(me_{12}^t) = \det(m)$.

EXAMPLE 5. Let m be as before.

$$p_{13}m = \begin{bmatrix} c_1 & c_2 & c_3 \\ b_1 & b_2 & b_3 \\ a_1 & a_2 & a_3 \end{bmatrix}.$$

Then, by (i), $\det(p_{13}m) = -\det m$. By (xi), $\det(p_{13}) = -1$. So we have

$$\det(p_{13}m) = (\det(p_{13}))\det(m).$$

Exercise 5. Show that for each matrix m,

$$\det(e_{13}^t m) = \det(e_{13}^t)\det(m)$$

and

$$\det(me_{13}^t) = \det(m)\det(e_{13}^t).$$

Exercise 6. Show that for each matrix m,

$$\det(p_{12}m) = \det(p_{12})\det(m).$$

Reasoning as in the preceding examples and exercises, we find:

(xii) *For every i, j, t and every matrix m,*

$$\det(e_{ij}^t m) = \det(e_{ij}^t)\det(m)$$

and

$$\det(me_{ij}^t) = \det(m)\det(e_{ij}^t).$$

Also:

(xiii) *For every i, j,*

$$\det(p_{ij}m) = \det(mp_{ij}) = (\det m)(\det p_{ij}).$$

Let $d = \begin{bmatrix} t_1 & 0 & 0 \\ 0 & t_2 & 0 \\ 0 & 0 & t_3 \end{bmatrix}$ and $m = \begin{bmatrix} a_1 & a_2 & a_3 \\ b_1 & b_2 & b_3 \\ c_1 & c_2 & c_3 \end{bmatrix}$. Then

$$dm = \begin{bmatrix} t_1a_1 & t_1a_2 & t_1a_3 \\ t_2b_2 & t_2b_2 & t_2b_3 \\ t_3c_1 & t_3c_2 & t_3c_3 \end{bmatrix}.$$

If **A**, **B**, **C** are the rows of m, then $t_1\mathbf{A}$, $t_2\mathbf{B}$, $t_3\mathbf{C}$ are the rows of dm. Hence

$$\begin{aligned} \det(dm) &= t_1\mathbf{A} \cdot (t_2\mathbf{B} \times t_3\mathbf{C}) \\ &= t_1t_2t_3\mathbf{A} \cdot \mathbf{B} \times \mathbf{C} = t_1t_2t_3\det(m). \end{aligned}$$

Also, $\det(d) = t_1t_2t_3$. Thus we have proved:

(xiv) *If d is a diagonal matrix and m is any matrix, then* $\det(dm) = \det(d)\det(m)$. *Similarly, we get* $\det(md) = \det(m)\det(d)$.

In Theorem 2.7 of Chapter 2.5, we showed that if a, b are two 2×2 matrices, then $\det(ab) = (\det a)(\det b)$. We now proceed to prove the corresponding relation for 3×3 matrices.

Theorem 3.6. *Let a, b be two* 3×3 *matrices. Then* $\det(ab) = (\det a)(\det b)$.

(*Note*: Theorems 3.5 and 3.6 appear below.)

In the 2×2 case, we proved the corresponding result by direct computation. Although it would be possible to do the same with 3×3 case, we prefer to give a proof based on the properties of elementary matrices.

Proof. By Theorem 3.3 of Chapter 3.4, there exists a diagonal matrix d and elementary matrices e_i, f_j such that

$$a = e_1 \ldots e_k d f_1 \ldots f_l$$

Using relations (xii), (xiii), and (xiv), we see that

$$\det a = (\det e_1) \ldots (\det e_k)(\det d)(\det f_1) \ldots (\det f_l).$$

Similarly, there exists a diagonal matrix d' and elementary matrices g_i, h_j so that

$$b = g_1 \ldots g_r d' h_1 \ldots h_s$$

and

$$\det b = (\det g_1) \ldots (\det g_r)(\det d')(\det h_1) \ldots (\det h_s).$$

Hence

$$ab = e_1 \ldots e_k d f_1 \ldots f_l g_1 \ldots g_r d' h_1 \ldots h_s$$

and

$$\det(ab) = (\det(e_1)) \ldots (\det e_k)(\det d)(\det f_1) \ldots (\det h_s).$$

So

$$\det(ab) = \det(a)\det(b).$$

Note: Even though, in general, $ab \neq ba$, we now see that $\det(ab) = \det(ba)$, because both are equal to $(\det a)(\det b) = (\det b)(\det a)$.

§3. Geometric Meaning of 3 × 3 Determinants

Next we proceed to extend to determinants of 3×3 matrices the results we found in Chapter 2.5 concerning the relations between determinants and *orientation* and between determinants and *area*.

Consider a triplet of vectors $\mathbf{X}_1$, $\mathbf{X}_2$, $\mathbf{X}_3$ regarded as an ordered triplet with $\mathbf{X}_1$ first, $\mathbf{X}_2$ second, and $\mathbf{X}_3$ third. Suppose $\mathbf{X}_1$ and $\mathbf{X}_2$ are linearly independent. Then $\mathbf{X}_1 \times \mathbf{X}_2$ is perpendicular to the plane of $\mathbf{X}_1$ and $\mathbf{X}_2$, and it is so chosen that the rotation about $\mathbf{X}_1 \times \mathbf{X}_2$ which sends $\mathbf{X}_1$ to $\mathbf{X}_2$ is through a positive angle α. The *upper half-space* determined by the ordered pair $\mathbf{X}_1$, $\mathbf{X}_2$ is the set of all vectors $\mathbf{X}$ such that $(\mathbf{X}_1 \times \mathbf{X}_2) \cdot \mathbf{X} > 0$.

The triplet $\mathbf{X}_1$, $\mathbf{X}_2$, $\mathbf{X}_3$ is said to be *positively oriented* if $\mathbf{X}_3$ lies in the upper half-space determined by the ordered pair $\mathbf{X}_1$, $\mathbf{X}_2$ i.e., if $(\mathbf{X}_1 \times \mathbf{X}_2) \cdot \mathbf{X}_3 > 0$. If $(\mathbf{X}_1 \times \mathbf{X}_2) \cdot \mathbf{X}_3 < 0$, the triplet is said to be *negatively oriented*.

Example 6.

(a) The triplet $\mathbf{E}_1$, $\mathbf{E}_2$, $\mathbf{E}_3$ is positively oriented since $(\mathbf{E}_1 \times \mathbf{E}_2) \cdot \mathbf{E}_3 = \mathbf{E}_3 \cdot \mathbf{E}_3 = 1 > 0$.
(b) The triplet $\mathbf{E}_1$, $\mathbf{E}_2$, $-\mathbf{E}_3$ is negatively oriented, since $(\mathbf{E}_1 \times \mathbf{E}_2) \cdot (-\mathbf{E}_3) = -1$.
(c) The triplet $\mathbf{E}_2$, $\mathbf{E}_1$, $\mathbf{E}_3$ is negatively oriented, since $(\mathbf{E}_2 \times \mathbf{E}_1) \cdot \mathbf{E}_3 = (-\mathbf{E}_3) \cdot \mathbf{E}_3 = -1$. (See Fig. 3.11.)

Let

$$\mathbf{X}_1 = \begin{bmatrix} x_{11} \\ x_{21} \\ x_{31} \end{bmatrix}, \qquad \mathbf{X}_2 = \begin{bmatrix} x_{12} \\ x_{22} \\ x_{32} \end{bmatrix}, \qquad \mathbf{X}_3 = \begin{bmatrix} x_{13} \\ x_{23} \\ x_{33} \end{bmatrix}$$

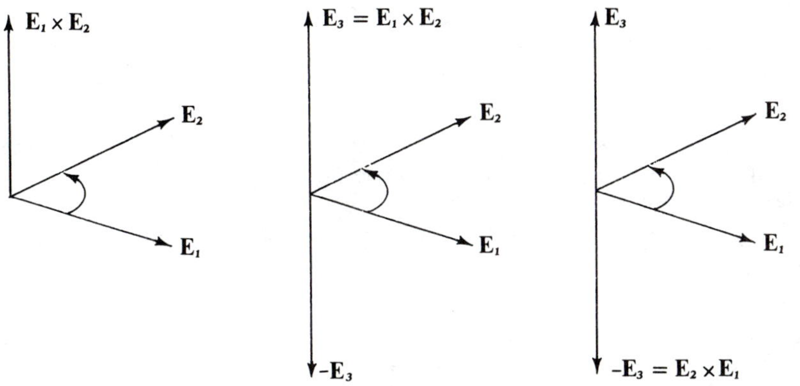

Figure 3.11

be an oriented triplet of vectors. Then

$$(\mathbf{X}_1 \times \mathbf{X}_2) \cdot \mathbf{X}_3 = \begin{vmatrix} x_{11} & x_{21} & x_{31} \\ x_{12} & x_{22} & x_{32} \\ x_{13} & x_{23} & x_{33} \end{vmatrix}.$$

By (viii), the right-hand term equals

$$\begin{vmatrix} x_{11} & x_{12} & x_{13} \\ x_{21} & x_{22} & x_{23} \\ x_{31} & x_{32} & x_{33} \end{vmatrix},$$

which is the determinant of the matrix whose columns are the vectors $\mathbf{X}_1$, $\mathbf{X}_2$, $\mathbf{X}_3$. Let us denote this matrix by $(\mathbf{X}_1 | \mathbf{X}_2 | \mathbf{X}_3)$. Thus

$$(\mathbf{X}_1 \times \mathbf{X}_2) \cdot \mathbf{X}_3 = \det(\mathbf{X}_1 | \mathbf{X}_2 | \mathbf{X}_3),$$

and so we obtain:

Proposition 1. *The triplet* $\mathbf{X}_1$, $\mathbf{X}_2$, $\mathbf{X}_3$ *is positively oriented if and only if* $\det(\mathbf{X}_1 | \mathbf{X}_2 | \mathbf{X}_3) > 0$.

Next let A be a linear transformation which has an inverse. We say that *A preserves orientation* if whenever $\mathbf{X}_1$, $\mathbf{X}_2$, $\mathbf{X}_3$ is a positively oriented triplet, then the triplet $A(\mathbf{X}_1)$, $A(\mathbf{X}_2)$, $A(\mathbf{X}_3)$ of image vectors is also positively oriented.

In Chapter 2.5, Theorem 2.5, we showed that a linear transformation of $\mathbb{R}^2$ preserves orientation if and only if the determinant of its matrix > 0. Is the analogous result true for $\mathbb{R}^3$?

Let A be a linear transformation of $\mathbb{R}^3$ with $m(A) = \begin{pmatrix} a_{11} & a_{12} & a_{13} \\ a_{21} & a_{22} & a_{23} \\ a_{31} & a_{32} & a_{33} \end{pmatrix}$.

Suppose A has an inverse and preserves orientation. Since the triplet $\mathbf{E}_1$, $\mathbf{E}_2$, $\mathbf{E}_3$ is positively oriented, it follows that the triplet $A(\mathbf{E}_1)$, $A(\mathbf{E}_2)$, $A(\mathbf{E}_3)$ is positively oriented. Hence, by Theorem 3.5, the determinant of the matrix $(A(\mathbf{E}_1)|A(\mathbf{E}_2)|A(\mathbf{E}_3))$ is >0. But $A(\mathbf{E}_1) = \begin{pmatrix} a_{11} \\ a_{21} \\ a_{31} \end{pmatrix}$, etc. So $(A(\mathbf{E}_1)|A(\mathbf{E}_2)|A(\mathbf{E}_3)) = m(A)$, and so $\det(m(A)) > 0$. Conversely, let A be a linear transformation and suppose $\det(m(A)) > 0$. Let $\mathbf{X}_1$, $\mathbf{X}_2$, $\mathbf{X}_3$ be a positively oriented triplet of vectors with $\mathbf{X}_1 = \begin{pmatrix} x_{11} \\ x_{21} \\ x_{31} \end{pmatrix}$, with $\mathbf{X}_2$ and $\mathbf{X}_3$ expressed similarly. Then $A(\mathbf{X}_1) = \begin{pmatrix} a_{11}x_{11} + a_{12}x_{21} + a_{13}x_{31} \\ a_{21}x_{11} + a_{22}x_{21} + a_{23}x_{31} \\ a_{31}x_{11} + a_{32}x_{21} + a_{33}x_{31} \end{pmatrix}$, and we have similar expressions for $A(\mathbf{X}_2)$, $A(\mathbf{X}_3)$.

The matrix

$$(A(\mathbf{X}_1)|A(\mathbf{X}_2)|A(\mathbf{X}_3))$$

$$= \begin{pmatrix} a_{11}x_{11} + a_{12}x_{21} + a_{13}x_{31} & a_{11}x_{12} + a_{12}x_{22} + a_{13}x_{32} & a_{11}x_{13} + a_{12}x_{23} + a_{13}x_{33} \\ a_{21}x_{11} + a_{22}x_{21} + a_{23}x_{31} & a_{21}x_{12} + a_{22}x_{22} + a_{23}x_{32} & a_{21}x_{13} + a_{22}x_{23} + a_{23}x_{33} \\ a_{31}x_{11} + a_{32}x_{21} + a_{33}x_{31} & a_{31}x_{12} + a_{32}x_{22} + a_{33}x_{32} & a_{31}x_{13} + a_{32}x_{23} + a_{33}x_{33} \end{pmatrix}$$

$$= \begin{pmatrix} a_{11} & a_{12} & a_{13} \\ a_{21} & a_{22} & a_{23} \\ a_{31} & a_{32} & a_{33} \end{pmatrix} \begin{pmatrix} x_{11} & x_{12} & x_{13} \\ x_{21} & x_{22} & x_{23} \\ x_{31} & x_{32} & x_{33} \end{pmatrix}.$$

Then, by Theorem 3.7,

$$\det(A(\mathbf{X}_1)|A(\mathbf{X}_2)|A(\mathbf{X}_3)) = \begin{vmatrix} a_{11} & a_{12} & a_{13} \\ a_{21} & a_{22} & a_{23} \\ a_{31} & a_{32} & a_{33} \end{vmatrix} \begin{vmatrix} x_{11} & x_{12} & x_{13} \\ x_{21} & x_{22} & x_{23} \\ x_{31} & x_{32} & x_{33} \end{vmatrix}.$$

Since the triplet $\mathbf{X}_1$, $\mathbf{X}_2$, $\mathbf{X}_3$ is positively oriented, the second determinant on the right-hand side is >0, and since by hypothesis $\det(m(A)) > 0$, the first determinant on the right-hand side is also >0. Hence $A(\mathbf{X}_1)$, $A(\mathbf{X}_2)$, $A(\mathbf{X}_3)$ is a positively oriented triplet. Thus A preserves orientation. We have proved:

Theorem 3.7. *A linear transformation A on $\mathbb{R}^3$ preserves orientation if and only if* $\det(m(A)) > 0$.

We now proceed to describe the relation between 3×3 determinants and volume. Let

$$\mathbf{X}_1 = \begin{pmatrix} x_{11} \\ x_{21} \\ x_{31} \end{pmatrix}, \quad \mathbf{X}_2 = \begin{pmatrix} x_{12} \\ x_{22} \\ x_{32} \end{pmatrix}, \quad \mathbf{X}_3 = \begin{pmatrix} x_{13} \\ x_{23} \\ x_{33} \end{pmatrix}$$

be a positively oriented triplet of vectors. Denote by Π the parallelepiped with edges along these vectors, i.e., Π consists of all vectors

$$\mathbf{X} = t_1\mathbf{X}_1 + t_2\mathbf{X}_2 + t_3\mathbf{X}_3,$$

where t_1, t_2, t_3 are scalars between 0 and 1. By (ix) of Chapter 3.0,

$$\text{volume}\,(\Pi) = \mathbf{X}_3 \cdot (\mathbf{X}_1 \times \mathbf{X}_2) = \begin{vmatrix} x_{13} & x_{23} & x_{33} \\ x_{11} & x_{21} & x_{31} \\ x_{12} & x_{22} & x_{32} \end{vmatrix} = \begin{vmatrix} x_{11} & x_{21} & x_{31} \\ x_{12} & x_{22} & x_{32} \\ x_{13} & x_{23} & x_{33} \end{vmatrix}.$$

By (viii), the right-hand side equals

$$\begin{vmatrix} x_{11} & x_{12} & x_{13} \\ x_{21} & x_{22} & x_{23} \\ x_{31} & x_{32} & x_{33} \end{vmatrix} = \det(\mathbf{X}_1 \,|\, \mathbf{X}_2 \,|\, \mathbf{X}_3).$$

So

$$\text{volume}\,(\Pi) = \det(\mathbf{X}_1 \,|\, \mathbf{X}_2 \,|\, \mathbf{X}_3). \tag{1}$$

Now let T be a linear transformation having an inverse. Denote by $T(\Pi)$ the image of Π under T. $T(\Pi)$ is the parallelepiped determined by the vectors $T(\mathbf{X}_1)$, $T(\mathbf{X}_2)$, $T(\mathbf{X}_3)$. Hence, by formula (1) (see Fig. 3.12), we have

$$\text{volume}(T(\Pi)) = \det(T(\mathbf{X}_1), T(\mathbf{X}_2), T(\mathbf{X}_3)). \tag{2}$$

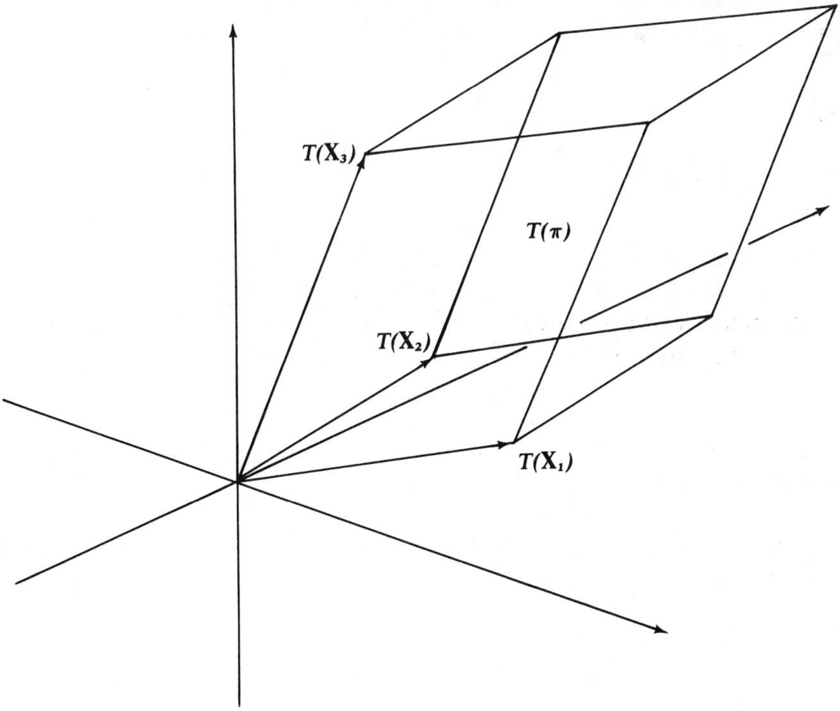

Figure 3.12

On the other hand, the calculation that led us to Theorem 3.5 gives

$$\begin{aligned}\det(T(\mathbf{X}_1)\,|\,T(\mathbf{X}_2)\,|\,T(\mathbf{X}_3)) &= (\det(m(T)))\det(\mathbf{X}_1\,|\,\mathbf{X}_2\,|\,\mathbf{X}_3)\\ &= \det(m(T))\cdot \text{volume}(\Pi).\end{aligned}$$

By (2), this gives

$$\text{volume}(T(\Pi)) = (\det(m(T)))(\text{volume}(\Pi)). \tag{3}$$

We have thus found the following counterpart to Theorem 2.6 of Chapter 2.5.

Theorem 3.8. *Let T be an orientation-preserving linear transformation of $\mathbb{R}^3$. If Π is any parallelepiped, then*

$$\text{volume}(T(\Pi)) = (\det(m(T)))(\text{volume}(\Pi)). \tag{4}$$

Note: If T reverses orientation, the same argument yields formula (4) with a minus sign on the right-hand side, so for every invertible transformation T, we have

$$\text{volume}(T(\Pi)) = |\det(m(T))|\text{volume}(\Pi). \tag{5}$$

If T has *no inverse*, $T(\Pi)$ degenerates into a figure lying in a plane, so the left-hand side is 0 while the right-hand side is 0, since $\det(m(T)) = 0$ in this case. Thus, formula (5) is valid for every linear transformation of $\mathbb{R}^3$.

CHAPTER 3.6

Eigenvalues

EXAMPLE 1. Let π be a plane through the origin and let S be the transformation which reflects each vector through π. If $\mathbf{Y}$ is a vector on π, then $S(\mathbf{Y}) = \mathbf{Y}$, and if $\mathbf{U}$ is a vector perpendicular to π, then $S(\mathbf{U}) = -\mathbf{U}$. Thus for $t = 1$ and $t = -1$, there exist nonzero vectors $\mathbf{X}$ satisfying $S(\mathbf{X}) = t\mathbf{X}$. If $\mathbf{X}$ is any vector which is neither on π nor perpendicular to π, then $S(\mathbf{X})$ is not a multiple of $\mathbf{X}$.

Let T be a linear transformation of $\mathbb{R}^3$ and let t be a real number. We say that t is an *eigenvalue* of T if there is some nonzero vector $\mathbf{X}$ such that

$$T(\mathbf{X}) = t\mathbf{X} \quad \text{and} \quad \mathbf{X} \neq \mathbf{0}.$$

If t is an eigenvalue of T, then we call a vector $\mathbf{Y}$ an *eigenvector of T corresponding to t* if $T(\mathbf{Y}) = t\mathbf{Y}$.

For example, the eigenvalues of S are 1 and -1. The eigenvectors of S corresponding to 1 are all the vectors in π and the eigenvectors of S corresponding to -1 are all the vectors perpendicular to π.

EXAMPLE 2. Fix λ in $\mathbb{R}$. Let D_λ be stretching by λ. Then for every vector $\mathbf{X}$, $D_\lambda(\mathbf{X}) = \lambda\mathbf{X}$. Hence λ is an eigenvalue of D_λ. Every vector $\mathbf{X}$ in $\mathbb{R}^3$ is an eigenvector of D_λ corresponding to the eigenvalue λ.

EXAMPLE 3. Let D be the linear transformation with the diagonal matrix

$$\begin{bmatrix} \lambda_1 & 0 & 0 \\ 0 & \lambda_2 & 0 \\ 0 & 0 & \lambda_3 \end{bmatrix}.$$

If $\mathbf{X} = \begin{pmatrix} x_1 \\ x_2 \\ x_3 \end{pmatrix}$, then $D(\mathbf{X}) = \begin{pmatrix} \lambda_1 & 0 & 0 \\ 0 & \lambda_2 & 0 \\ 0 & 0 & \lambda_3 \end{pmatrix} \begin{pmatrix} x_1 \\ x_2 \\ x_3 \end{pmatrix} = \begin{pmatrix} \lambda_1 x_1 \\ \lambda_2 x_2 \\ \lambda_3 x_3 \end{pmatrix}$. It follows that setting $\mathbf{E}_1 = \begin{pmatrix} 1 \\ 0 \\ 0 \end{pmatrix}$, $\mathbf{E}_2 = \begin{pmatrix} 0 \\ 1 \\ 0 \end{pmatrix}$, $\mathbf{E}_3 = \begin{pmatrix} 0 \\ 0 \\ 1 \end{pmatrix}$, we have

$$D(\mathbf{E}_1) = \lambda_1 \mathbf{E}_1, \qquad D(\mathbf{E}_2) = \lambda_2 \mathbf{E}_2, \qquad D(\mathbf{E}_3) = \lambda_3 \mathbf{E}_3.$$

Thus $\lambda_1, \lambda_2, \lambda_3$ are eigenvalues of D and $\mathbf{E}_1, \mathbf{E}_2, \mathbf{E}_3$ are eigenvectors. Does D have any other eigenvalues? Suppose $D(\mathbf{X}) = t\mathbf{X}$, where $\mathbf{X} = \begin{pmatrix} x_1 \\ x_2 \\ x_3 \end{pmatrix}$, $\mathbf{X} \neq 0$, and t is in $\mathbb{R}$. Then $\begin{pmatrix} tx_1 \\ tx_2 \\ tx_3 \end{pmatrix} = \begin{pmatrix} \lambda_1 x_1 \\ \lambda_2 x_2 \\ \lambda_3 x_3 \end{pmatrix}$, so $tx_i = \lambda_i x_i$ for $i = 1, 2, 3$. Since $x_i \neq 0$ for some i, $t = \lambda_i$. Therefore, $\lambda_1, \lambda_2, \lambda_3$ are all the eigenvalues of D.

Exercise 1. With D as in Example 3, find all the eigenvectors of D.

Exercise 2. Let T be a linear transformation of $\mathbb{R}^3$ satisfying $T^2 = I$. Let t be an eigenvalue of T and $\mathbf{X}$ an eigenvector corresponding to t with $\mathbf{X} \neq \mathbf{0}$.

(a) Show that $T^2(\mathbf{X}) = t^2\mathbf{X}$.
(b) Show that $t = 1$ or $t = -1$.
(c) Apply what you have found to the reflection S in Example 1.

Exercise 3. Let T be a linear transformation of $\mathbb{R}^3$ such that $T^2 = 0$.

(a) Show that 0 is an eigenvalue of T.
(b) Show that 0 is the only eigenvalue of T.

Exercise 4. Let T be the linear transformation with matrix

$$m(T) = \begin{pmatrix} 0 & 0 & 1 \\ 0 & 0 & 0 \\ 0 & 0 & 0 \end{pmatrix}.$$

(a) Show that $T^2 = 0$.
(b) Apply Exercise 3 to determine the eigenvalues of T.
(c) Find all eigenvectors of T.

§1. Characteristic Equation

Given a transformation T with matrix $\begin{pmatrix} a_1 & a_2 & a_3 \\ b_1 & b_2 & b_3 \\ c_1 & c_2 & c_3 \end{pmatrix}$, how can we determine the eigenvalues of T? We proceed as we did for the corresponding problem in two dimensions.

Assume t is an eigenvalue of T and $\begin{bmatrix} x_1 \\ x_2 \\ x_3 \end{bmatrix}$ is a corresponding eigenvector with $\begin{bmatrix} x_1 \\ x_2 \\ x_3 \end{bmatrix} \neq \begin{bmatrix} 0 \\ 0 \\ 0 \end{bmatrix}$. Then $T\begin{bmatrix} x_1 \\ x_2 \\ x_3 \end{bmatrix} = t\begin{bmatrix} x_1 \\ x_2 \\ x_3 \end{bmatrix}$, so

$$\begin{bmatrix} a_1 & a_2 & a_3 \\ b_1 & b_2 & b_3 \\ c_1 & c_2 & c_3 \end{bmatrix}\begin{bmatrix} x_1 \\ x_2 \\ x_3 \end{bmatrix} = t\begin{bmatrix} x_1 \\ x_2 \\ x_3 \end{bmatrix}$$

or

$$\begin{cases} a_1x_1 + a_2x_2 + a_3x_3 = tx_1, \\ b_1x_1 + b_2x_2 + b_3x_3 = tx_2, \\ c_1x_1 + c_2x_2 + c_3x_3 = tx_3. \end{cases} \tag{1}$$

Transposing the right-hand terms, we get

$$\begin{cases} (a_1 - t)x_1 + a_2x_2 + a_3x_3 = 0, \\ b_1x_1 + (b_2 - t)x_2 + b_3x_3 = 0, \\ c_1x_1 + c_2x_2 + (c_3 - t)x_3 = 0. \end{cases} \tag{2}$$

Thus x_1, x_2, x_3 is a nonzero solution of the homogeneous system (2). By Proposition 8 and Theorem 3.4 of Chapter 3.4, it follows that the determinant

$$\begin{vmatrix} a_1 - t & a_2 & a_3 \\ b_1 & b_2 - t & b_3 \\ c_1 & c_2 & c_3 - t \end{vmatrix} = 0. \tag{3}$$

If the left-hand side is expanded, this equation has the form

$$-t^3 + u_1t^2 + u_2t + u_3 = 0, \tag{4}$$

where u_1, u_2, u_3 are certain constants.

Equation (3) is called the *characteristic equation* for the transformation T.

We just saw that if t is an eigenvalue of T, then t is a root of the characteristic equation of T. Conversely, if t is a root of the characteristic equation, then (3) holds. Hence, by Proposition 8 of Chapter 3.4, the system (2) has a nonzero solution x_1, x_2, x_3, and so (1) also has this solution. Therefore,

$$T\begin{bmatrix} x_1 \\ x_2 \\ x_3 \end{bmatrix} = t\begin{bmatrix} x_1 \\ x_2 \\ x_3 \end{bmatrix}.$$

Hence t is an eigenvalue of T. We have proved:

Theorem 3.8. *A real number t is an eigenvalue of the transformation T of $\mathbb{R}^3$ if and only if t is a root of the characteristic equation* (3).

Theorem 3.8 appears later.

EXAMPLE 4. Let T have matrix

$$\begin{bmatrix} 1 & 0 & 0 \\ -5 & 2 & 0 \\ 2 & 3 & 7 \end{bmatrix}.$$

The characteristic equation of T is

$$\begin{vmatrix} 1-t & 0 & 0 \\ -5 & 2-t & 0 \\ 2 & 3 & 7-t \end{vmatrix} = 0$$

or

$$(1-t)\begin{vmatrix} 2-t & 0 \\ 3 & 7-t \end{vmatrix} = (1-t)(2-t)(7-t) = 0.$$

The roots of this equation are 1, 2, and 7, and so these are the eigenvalues of T. Let us calculate the eigenvectors corresponding to the eigenvalue 2. If $\begin{bmatrix} x_1 \\ x_2 \\ x_3 \end{bmatrix}$ is such a vector, then

$$\begin{bmatrix} 1 & 0 & 0 \\ -5 & 2 & 0 \\ 2 & 3 & 7 \end{bmatrix}\begin{bmatrix} x_1 \\ x_2 \\ x_3 \end{bmatrix} = 2\begin{bmatrix} x_1 \\ x_2 \\ x_3 \end{bmatrix} = \begin{bmatrix} 2x_1 \\ 2x_2 \\ 2x_3 \end{bmatrix}.$$

Then

$$x_1 = 2x_1,$$
$$-5x_1 + 2x_2 = 2x_2,$$
$$2x_1 + 3x_2 + 7x_3 = 2x_3.$$

The first equation gives $x_1 = 0$. The second equation puts no restriction on x_2. The third equation yields

$$5x_3 = -3x_2 \quad \text{or} \quad x_3 = -\tfrac{3}{5}x_2.$$

Thus an eigenvector of T corresponding to the eigenvalue 2 must have the form $\begin{bmatrix} 0 \\ x_2 \\ -\frac{3}{5}x_2 \end{bmatrix} = \begin{bmatrix} 0 \\ 5y \\ -3y \end{bmatrix}$, if we set $y = \frac{1}{5}x_2$. Is every vector of this form an eigenvector?

$$\begin{bmatrix} 1 & 0 & 0 \\ -5 & 2 & 0 \\ 2 & 3 & 7 \end{bmatrix}\begin{bmatrix} 0 \\ 5y \\ -3y \end{bmatrix} = \begin{bmatrix} 0 \\ 10y \\ 15y - 21y \end{bmatrix} = \begin{bmatrix} 0 \\ 10y \\ -6y \end{bmatrix} = 2\begin{bmatrix} 0 \\ 5y \\ -3y \end{bmatrix}.$$

Thus the answer is yes and we have: a vector is an eigenvector of T with eigenvalue 2 if and only if it has the form $\begin{bmatrix} 0 \\ 5y \\ -3y \end{bmatrix}$. Note that the vectors of this form fill up the line $\left\{ y\begin{bmatrix} 0 \\ 5 \\ -3 \end{bmatrix} \middle| y \text{ in } \mathbb{R} \right\}$, which passes through the origin.

Exercise 5. Find all eigenvectors corresponding to the eigenvalue 1 for the transformation T of Example 4. Show that these vectors fill up a line through the origin.

EXAMPLE 5. Fix θ with $0 \leqslant \theta < 2\pi$. R_θ^3 is the transformation of $\mathbb{R}^3$ which rotates each vector by θ degrees around the positive x_3-axis. Find all eigenvalues of R_θ^3. We have

$$m(R_\theta^3) = \begin{bmatrix} \cos\theta & -\sin\theta & 0 \\ \sin\theta & \cos\theta & 0 \\ 0 & 0 & 1 \end{bmatrix},$$

so the characteristic equation is

$$\begin{vmatrix} \cos\theta - t & -\sin\theta & 0 \\ \sin\theta & \cos\theta - t & 0 \\ 0 & 0 & 1-t \end{vmatrix} = -\begin{vmatrix} 0 & 0 & 1-t \\ \sin\theta & \cos\theta - t & 0 \\ \cos\theta - t & -\sin\theta & 0 \end{vmatrix}$$
$$= -(1-t)\left[-(\sin\theta)^2 - (\cos\theta - t)^2\right]$$
$$= 0.$$

If t is a real root, then either $1 - t = 0$ or $(\sin\theta)^2 + (\cos\theta - t)^2 = 0$. The second equation implies that $\sin\theta = 0$ and $\cos\theta = t$.

Case 1: $\theta \neq 0, \pi$. In this case, $t = 1$ is the only root of the characteristic equation, and so 1 is the only eigenvalue. If $\mathbf{X}$ lies on the x_3-axis, evidently $T(\mathbf{X}) = \mathbf{X} = 1 \cdot \mathbf{X}$, so the x_3-axis consists of eigenvectors with eigenvalue 1. There are no other eigenvectors.

Case 2: $\theta = 0$. In this case, $R_\theta^3 = I$, so 1 is the only eigenvalue and every vector in $\mathbb{R}^3$ is an eigenvector corresponding to this eigenvalue.

Case 3: $\theta = \pi$. The characteristic equation of R_π^3 is

$$(1-t)(-1-t)^2 = 0$$

with roots $t = 1$ and $t = -1$, so the eigenvalues of R_π^3 are 1 and -1.

Exercise 6. Find the eigenvectors of R_π which correspond to the eigenvalue -1. Describe, in geometrical terms, how $R_\pi^3\mathbf{X}$ is obtained from $\mathbf{X}$ if $\mathbf{X}$ is any vector. Then explain, geometrically, why 1 and -1 occur as eigenvalues of R_π^3.

Now let A be a given linear transformation and let

$$-t^3 + at^2 + bt + c = 0$$

be the characteristic equation of A.

Define $f(t) = -t^3 + at^2 + bt + c$. Then f is a function defined for all real t. The equation $f(t) = 0$ must have at least one real root t_0. To see this, note that $f(t) < 0$ when t is a large positive number, while $f(t) > 0$ when t is a negative number with large absolute value. Therefore, at some point t_0, the graph of f must cross the t-axis. Dividing $f(t)$ by $t - t_0$, we get a quadratic polynomial $-t^2 + dt + e$, where d and e are certain constants. Thus

$$f(t) = (t - t_0)(-t^2 + dt + e).$$

The polynomial $g(t) = -t^2 + dt + e$ may be factored

$$g(t) = -(t - t_1)(t - t_2),$$

where t_1, t_2 are the roots of g, which may be real or conjugate complex numbers. We can distinguish three possibilities.

(i) t_1, t_2 are complex numbers, $t_1 = u + iv$, $t_2 = u - iv$, with $v \neq 0$. Then $f(t) = 0$ has exactly one real root, namely, t_0. In this case, the graph of f appears as in Fig. 3.13.

EXAMPLE 6. For $A = R_\theta^3$, we found in Example 5,

$$f(t) = (t - 1)\left[-(\sin\theta)^2 - (\cos\theta - t)^2\right].$$

Here $g(t) = [-(\sin\theta)^2 - (\cos\theta - t)^2] = -t^2 + 2(\cos\theta)t - 1$. If $\theta \neq 0$ or π, then g has no real roots, so possibility (i) occurs.

(ii) t_1 and t_2 are real and $t_1 = t_2$. If $t_0 = t_1 = t_2$, then $f(t) = 0$ has a triple root at t_0. The graph of f now appears as in Fig. 3.14. If $t_0 \neq t_1$, then f has one simple real root, t_0, and one double real root, $t_1 = t_2$. The graph of f appears as in Fig. 3.15.

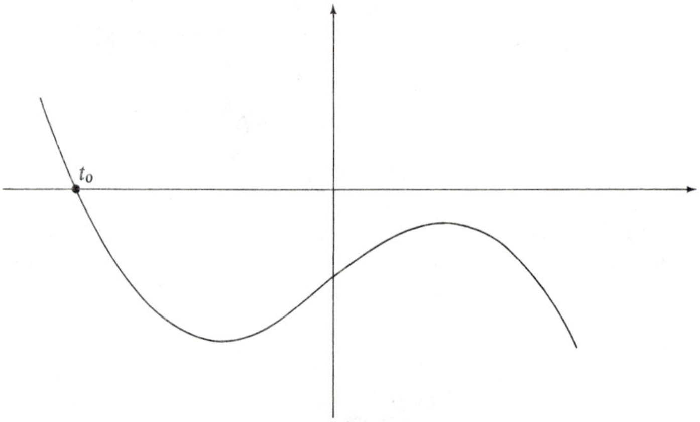

Figure 3.13

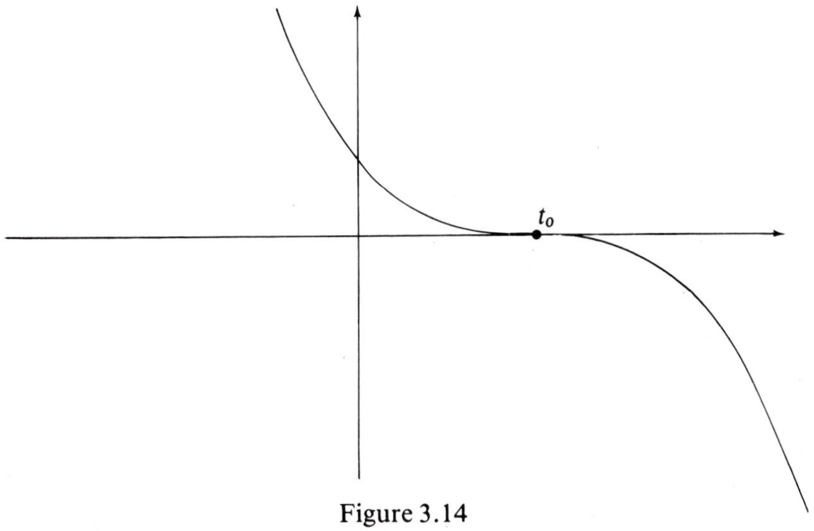

Figure 3.14

EXAMPLE 7. For $A = R_\pi^3$, we found

$$f(t) = (1-t)(-1-t)^2,$$

so $t_0 = 1$, $t_1 = t_2 = -1$, and possibility (ii) occurs.

For $A = D_\lambda$, we have

$$f(t) = \begin{vmatrix} \lambda - t & 0 & 0 \\ 0 & \lambda - t & 0 \\ 0 & 0 & \lambda - t \end{vmatrix} = (\lambda - t)^3,$$

so $t_0 = t_1 = t_2 = \lambda$.

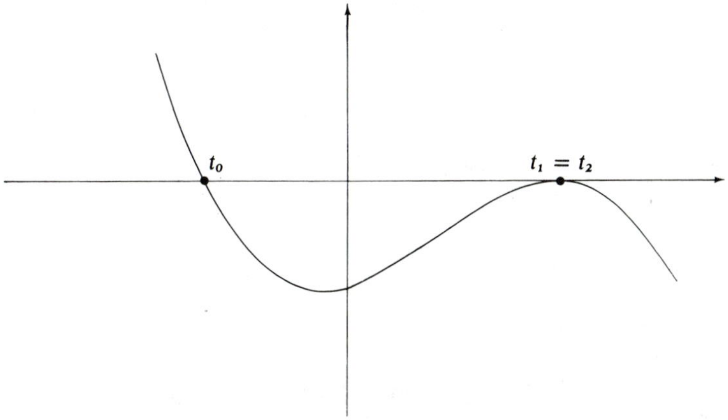

Figure 3.15

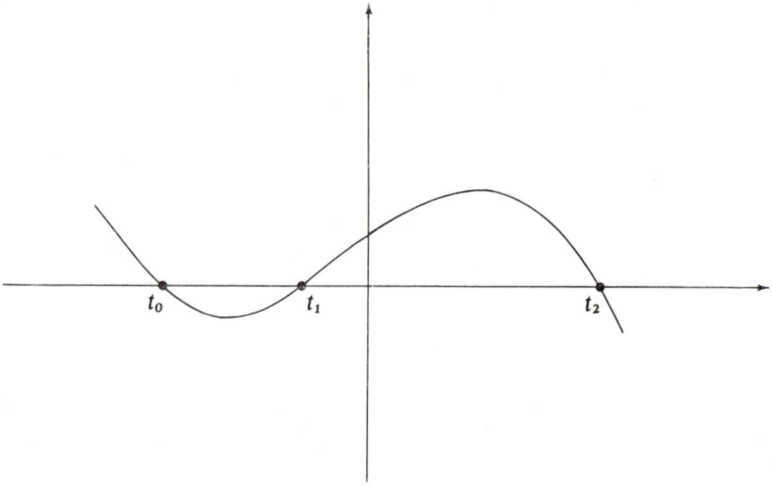

Figure 3.16

(iii) t_1 and t_2 are real and $t_1 \neq t_2$. If $t_1 = t_0$ or $t_2 = t_0$, the situation is as in case (ii). If $t_1 \neq t_0$ and $t_2 \neq t_0$, then the equation $f(t) = 0$ has the three distinct real roots t_0, t_1, and t_2. The graph of f now appears as in Fig. 3.16.

EXAMPLE 8. The transformation T of Example 4 had 1, 2, and 7 as the roots of its characteristic equation and, so, illustrated case (iii).

Let us summarize what we have found. Using Theorem 3.9 we can conclude:

Proposition 1. *If A is a linear transformation of $\mathbb{R}^3$, then A always has at least one eigenvalue and may have one, two, or three distinct eigenvalues.*

Exercise 7. Find all eigenvalues and eigenvectors of the transformation whose matrix is $\begin{pmatrix} 0 & 0 & 2 \\ 0 & 1 & 0 \\ 1 & 0 & 0 \end{pmatrix}$.

Exercise 8. Let T be the transformation with matrix $\begin{pmatrix} c & 0 & b \\ 0 & c & 0 \\ 0 & 0 & c \end{pmatrix}$.

(i) Find the characteristic equation of T.
(ii) Show that c is the only eigenvalue of T.
(iii) Show the eigenvectors of T corresponding to this eigenvalue fill up a plane through 0, and give an equation of this plane.

Exercise 9.

(i) Find the characteristic equation for the transformation with matrix $\begin{pmatrix} 0 & 1 & 0 \\ 0 & 0 & 1 \\ a_1 & a_2 & a_3 \end{pmatrix}$.

(ii) Show that given any three numbers a, b, c, there is some linear transformation of $\mathbb{R}^3$ whose characteristic equation is $-t^3 + at^2 + bt + c = 0$.

Let A be a linear transformation of $\mathbb{R}^3$ and let t be an eigenvalue of A. By the *eigenspace* E_t we mean the collection of all eigenvectors of A which correspond to the eigenvalue t.

EXAMPLE 9. Let P be the linear transformation which projects each vector on the plane π through the origin. Then for each vector $\mathbf{X}$ in π, $P(\mathbf{X}) = \mathbf{X}$, while for each $\mathbf{X}$ perpendicular to π, $P(\mathbf{X}) = \mathbf{0}$. P has no other eigenvectors. Hence, here $E_1 =$ the plane π, $E_0 =$ the line through $\mathbf{0}$ perpendicular to π.

Let A be an arbitrary linear transformation of $\mathbb{R}^3$ and t an eigenvalue of A. If $\mathbf{X}$ is a vector belonging to the eigenspace E_t, then $A(\mathbf{X}) = t\mathbf{X}$. Hence, for each scalar c, $A(c\mathbf{X}) = cA(\mathbf{X}) = ct\mathbf{X} = t(c\mathbf{X})$, and so $c\mathbf{X}$ is also in E_t. Thus E_t contains the line along $\mathbf{X}$. If E_t is not equal to this line, then there is some $\mathbf{Y}$ in E_t such that $\mathbf{X}$ and $\mathbf{Y}$ are linearly independent. For each pair of scalars c_1, c_2, $A(c_1\mathbf{X} + c_2\mathbf{Y}) = c_1A(\mathbf{X}) + c_2A(\mathbf{Y}) = c_1t\mathbf{X} + c_2t\mathbf{Y} = t(c_1\mathbf{X} + c_2\mathbf{Y})$. Thus $c_1\mathbf{X} + c_2\mathbf{Y}$ is in E_t. So the entire plane

$$\{(c_1\mathbf{X} + c_2\mathbf{Y}) \mid c_1, c_2 \text{ in } \mathbb{R}\}$$

is contained in E_t. If E_t does not coincide with this plane, then there is some vector $\mathbf{Z}$ in E_t such that $\mathbf{X}$, $\mathbf{Y}$, $\mathbf{Z}$ are linearly independent. By Proposition 1 of Chapter 3.0, $\{c_1\mathbf{X} + c_2\mathbf{Y} + c_3\mathbf{z} \mid c_1, c_2, c_3 \text{ in } \mathbb{R}\}$ is all of $\mathbb{R}^3$ and this set is contained in E_t. So, in that case, $E_t = \mathbb{R}^3$. We have shown:

Proposition 2. *If A is a linear transformation of $\mathbb{R}^3$, each eigenspace of A is either a line through the origin, or a plane through the origin, or all of $\mathbb{R}^3$.*

§2. Isometries of $\mathbb{R}^3$

In Chapter 2.5 we found all the length-preserving linear transformations of the plane. They turned out to be the rotations and reflections of the plane. Let us try to solve the corresponding problem in 3-space.

A linear transformation T of $\mathbb{R}^3$ which preserves lengths of segments is called an *isometry*. Exactly as in $\mathbb{R}^2$, we find that T is an isometry if and only if

$$|T(\mathbf{X})| = |\mathbf{X}| \qquad \text{for every vector } \mathbf{X}. \tag{5}$$

Proposition 3. *An isometry T preserves the dot product, i.e., for all vectors* $\mathbf{X}$, $\mathbf{Y}$,

$$T(\mathbf{X}) \cdot T(\mathbf{Y}) = \mathbf{X} \cdot \mathbf{Y}.$$

PROOF. Since (5) holds for each vector,

$$|T(\mathbf{X} - \mathbf{Y})|^2 = |\mathbf{X} - \mathbf{Y}|^2$$

or

$$(T(\mathbf{X} - \mathbf{Y})) \cdot (T(\mathbf{X} - \mathbf{Y})) = (\mathbf{X} - \mathbf{Y}) \cdot (\mathbf{X} - \mathbf{Y}).$$

So

$$(T(\mathbf{X}) - T(\mathbf{Y})) \cdot (T(\mathbf{X}) - T(\mathbf{Y})) = (\mathbf{X} - \mathbf{Y}) \cdot (\mathbf{X} - \mathbf{Y})$$

or

$$T(\mathbf{X}) \cdot T(\mathbf{X}) - 2T(\mathbf{X}) \cdot T(\mathbf{Y}) + T(\mathbf{Y}) \cdot T(\mathbf{Y}) = \mathbf{X} \cdot \mathbf{X} - 2\mathbf{X} \cdot \mathbf{Y} + \mathbf{Y} \cdot \mathbf{Y}.$$

Again by (5), $T(\mathbf{X}) \cdot T(\mathbf{X}) = \mathbf{X} \cdot \mathbf{X}$ and similarly for $\mathbf{Y}$, so cancelling we get

$$-2T(\mathbf{X}) \cdot T(\mathbf{Y}) = -2\mathbf{X} \cdot \mathbf{Y},$$

and so

$$T(\mathbf{X}) \cdot T(\mathbf{Y}) = \mathbf{X} \cdot \mathbf{Y}.$$

Proposition 4. *If T is an isometry of $\mathbb{R}^3$, then T has* 1 *or* -1 *as an eigenvalue and has no other eigenvalues.*

PROOF. Every linear transformation T of $\mathbb{R}^3$ has an eigenvalue t, so for some vector $\mathbf{X} \neq \mathbf{0}$, $T(\mathbf{X}) = t\mathbf{X}$. Then

$$|\mathbf{X}| = |T(\mathbf{X})| = |t\mathbf{X}| = |t|\,|\mathbf{X}|, \qquad \text{so} \quad |t| = 1.$$

Hence

$$t = 1 \quad \text{or} \quad t = -1.$$

Proposition 5. *If T is an isometry of $\mathbb{R}^3$, then* $\det(T) = 1$ *or* $\det(T) = -1$.

Note: We write $\det(T)$ for $\det(m(T))$, the determinant of the matrix of T.

PROOF. Consider the cube Q with edges $\mathbf{E}_1$, $\mathbf{E}_2$, $\mathbf{E}_3$. The vectors $T(\mathbf{E}_1)$, $T(\mathbf{E}_2)$, $T(\mathbf{E}_3)$ are edges of the image, $T(Q)$, of Q under T. For each i, $|T(\mathbf{E}_i)| = |\mathbf{E}_i| = 1$, and for each i, j with $i \neq j$, $T(\mathbf{E}_i) \cdot T(\mathbf{E}_j) = \mathbf{E}_i \cdot \mathbf{E}_j = 0$. Thus $T(Q)$ is a cube of side 1. Hence, $\text{vol}(T(Q)) = 1 = \text{vol } Q$. Also, by Theorem 3.6 of Chapter 3.5, $\text{vol } T(Q) = |\det(T)| \cdot \text{vol } Q$. Hence $|\det(T)| = 1$. So $\det T = 1$ or $\det T = -1$.

One example of an isometry is a *rotation about an axis*. Fix a vector $\mathbf{F}$ and denote by π the plane through 0 orthogonal to $\mathbf{F}$. Fix a number θ. We denote by R_θ the transformation of the plane π which rotates each vector in

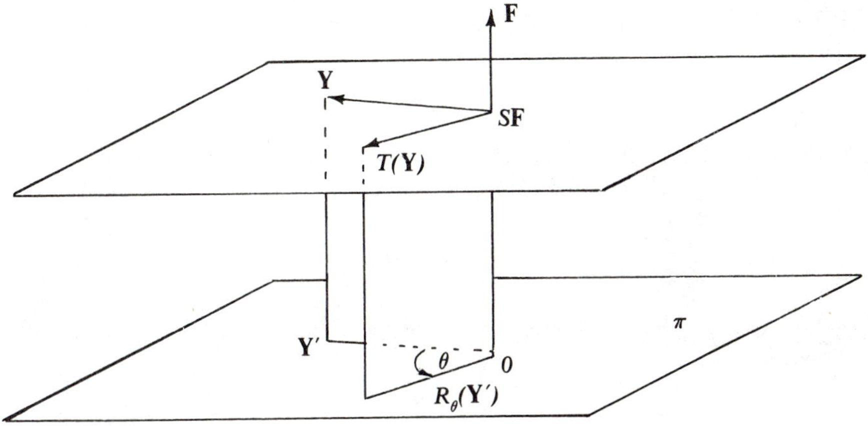

Figure 3.17

π counterclockwise by an angle θ about **F**. Let **Y** be any vector in $\mathbb{R}^3$. We decompose **Y** as

$$\mathbf{Y} = \mathbf{Y}' + s\mathbf{F},$$

where $\mathbf{Y}'$ is the projection of **Y** on π and s is a scalar (see Fig. 3.17). We now define

$$T(\mathbf{Y}) = R_\theta(\mathbf{Y}') + s\mathbf{F}.$$

Note that $T(\mathbf{Y})$ lies in the plane through **Y** perpendicular to **F**. We call the transformation T *a rotation about the axis* **F** *by the angle* θ.

Exercise 10. Prove that T is a linear transformation of $\mathbb{R}^3$ and that $T(\mathbf{F}) = \mathbf{F}$.

For each **Y**, $|T(\mathbf{Y})|^2 = |R_\theta(\mathbf{Y}') + s\mathbf{F}|^2 = (R_\theta(\mathbf{Y}') + s\mathbf{F}) \cdot (R_\theta(\mathbf{Y}') + s\mathbf{F}) = |R_\theta(\mathbf{Y}')|^2 + s^2|\mathbf{F}|^2$, since $R_\theta(\mathbf{Y}')$ is orthogonal to **F**. Also, $|\mathbf{Y}|^2 = |\mathbf{Y}'|^2 + s^2|\mathbf{F}|^2$. Since R_θ is a rotation of π, $|R_\theta(\mathbf{Y}')| = |\mathbf{Y}'|$, and so $|T(\mathbf{Y})|^2 = |\mathbf{Y}|^2$. Thus T is an isometry.

Now choose orthogonal unit vectors $\mathbf{X}_1$, $\mathbf{X}_2$ in π such that $\mathbf{X}_1 \times \mathbf{X}_2 = \mathbf{F}$. Then the triplet $\mathbf{X}_1, \mathbf{X}_2, \mathbf{F}$ is positively oriented. The triplet of vectors $T(\mathbf{X}_1)$, $T(\mathbf{X}_2)$, **F** is also positively oriented, and so T preserves orientation. The proof is contained in Exercise 11.

Exercise 11.

(a) Express $T(\mathbf{X}_1)$ and $T(\mathbf{X}_2)$ as linear combinations of $\mathbf{X}_1$ and $\mathbf{X}_2$ with coefficients depending on θ.
(b) Compute $(T(\mathbf{X}_1) \times T(\mathbf{X}_2)) \cdot \mathbf{F}$ and show it equals $(\mathbf{X}_1 \times \mathbf{X}_2) \cdot \mathbf{F}$ and hence is positive. Thus T preserves orientation.
(c) Using Theorem 3.5 of Chapter 3.5, conclude that $\det T > 0$.

Since T is an isometry, $\det T = \pm 1$ and so, since $\det T > 0$, $\det T = 1$. In sum, we have proved:

Proposition 6. *If T is a rotation about an axis, then T is an isometry and* $\det T = 1$.

What about the converse of this statement? Suppose T is an isometry with $\det T = 1$. Let $\mathbf{F}$ be an eigenvector of T with $|\mathbf{F}| = 1$ and $T(\mathbf{F}) = t\mathbf{F}$. By Proposition 4, we know that $t = \pm 1$. We consider the two cases separately. Let us first suppose $t = 1$. Then

$$T(\mathbf{F}) = \mathbf{F}.$$

Let π be the plane orthogonal to $\mathbf{F}$ and passing through the origin. If $\mathbf{X}$ is a vector in π,

$$T(\mathbf{X}) \cdot \mathbf{F} = T(\mathbf{X}) \cdot T(\mathbf{F}) = \mathbf{X} \cdot \mathbf{F} = 0,$$

so $T(\mathbf{X})$ is orthogonal to $\mathbf{F}$. Hence $T(\mathbf{X})$ lies in π. Thus T transforms π into itself. (See Fig. 3.18) Let us denote by T_π the resulting transformation of the plane π. T_π is evidently a linear transformation of π and an isometry of π since T has these properties on $\mathbb{R}^3$. In Chapter 2.5 we showed that an isometry of the plane is either a rotation or a reflection. Hence either T_π is a rotation of π through some angle θ or T_π is a reflection of π across a line in π through the origin.

Case 1: T_π is a rotation of π through an angle θ. By the discussion following after Proposition 5, we conclude that T is a rotation of $\mathbb{R}^3$ about the axis $\mathbf{F}$.

Case 2: T_π is a reflection across a line in π. In this case there exist nonzero vectors $\mathbf{X}_1$, $\mathbf{X}_2$ in π with $T_\pi(\mathbf{X}_1) = \mathbf{X}_1$, $T_\pi(\mathbf{X}_2) = -\mathbf{X}_2$, and we can choose

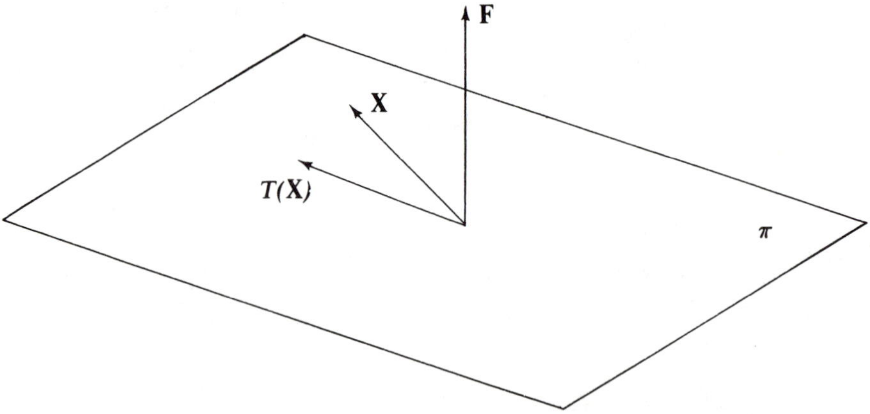

Figure 3.18

these vectors so that the triplet $\mathbf{X}_1$, $\mathbf{X}_2$, $\mathbf{F}$ is positively oriented. Also, $T(\mathbf{X}_1) = T_\pi(\mathbf{X}_1) = \mathbf{X}_1$ and $T(\mathbf{X}_2) = T_\pi(\mathbf{X}_2) = -\mathbf{X}_2$, so the triplet $T(\mathbf{X}_1)$, $T(\mathbf{X}_2)$, $T(\mathbf{F})$ is the triplet $\mathbf{X}_1$, $-\mathbf{X}_2$, $\mathbf{F}$ which is negatively oriented. But $\det T = 1$ by hypothesis, and so we have a contradiction. Thus Case 2 cannot occur. We conclude: If $t = 1$, then T is a rotation about an axis.

Now let us suppose that $t = -1$, so $T(\mathbf{F}) = -\mathbf{F}$. We again form the plane π orthogonal to $\mathbf{F}$ and the transformation T_π of π on itself. As before, T_π is either a rotation of π or reflection in a line of π.

Suppose T_π is a rotation of π by an angle θ. Choose orthogonal unit vectors $\mathbf{X}_1$, $\mathbf{X}_2$ in π such that $\mathbf{X}_1$, $\mathbf{X}_2$, $\mathbf{F}$ is a positively oriented triplet. Then the triplet $T_\pi(\mathbf{X}_1)$, $T_\pi(\mathbf{X}_2)$, $\mathbf{F}$ is again positively oriented.

Exercise 12. Prove this last statement by calculating $(T_\pi(\mathbf{X}_1) \times T_\pi(\mathbf{X}_2)) \cdot \mathbf{F}$ and showing that it is positive.

It follows that the triplet $T_\pi(\mathbf{X}_1)$, $T_\pi(\mathbf{X}_2)$, $-\mathbf{F}$ is negatively oriented. But this is exactly the triplet $T(\mathbf{X}_1)$, $T(\mathbf{X}_2)$, $T(\mathbf{F})$. Since $\mathbf{X}_1$, $\mathbf{X}_2$, $\mathbf{F}$ was a positively oriented triplet, this contradicts the fact that $\det T = 1$. Hence T_π is not a rotation of π so it must be a reflection of π. Therefore we can find orthogonal unit vectors $\mathbf{X}_1$, $\mathbf{X}_2$ in π with $T_\pi(\mathbf{X}_1) = -\mathbf{X}_1$, $T_\pi(\mathbf{X}_2) = \mathbf{X}_2$. Now consider the plane π' determined by the vectors $\mathbf{F}$ and $\mathbf{X}_1$ (see Fig. 3.19). We note that since $T(\mathbf{F}) = -\mathbf{F}$ and $T(\mathbf{X}_1) = T_\pi(\mathbf{X}_1) = -\mathbf{X}_1$, T coincides on the plane π' with minus the identity transformation. Thus T rotates the vectors of π' by 180° about the $\mathbf{X}_2$-axis. Also, $T(\mathbf{X}_2) = T_\pi(\mathbf{X}_2) = \mathbf{X}_2$. Hence T acts on $\mathbb{R}^3$ by rotation by 180° about the $\mathbf{X}_2$-axis.

In summary, we have proved:

Theorem 3.9. *Let T be an isometry of $\mathbb{R}^3$ and let* $\det T = 1$. *Then T is rotation about an axis.*

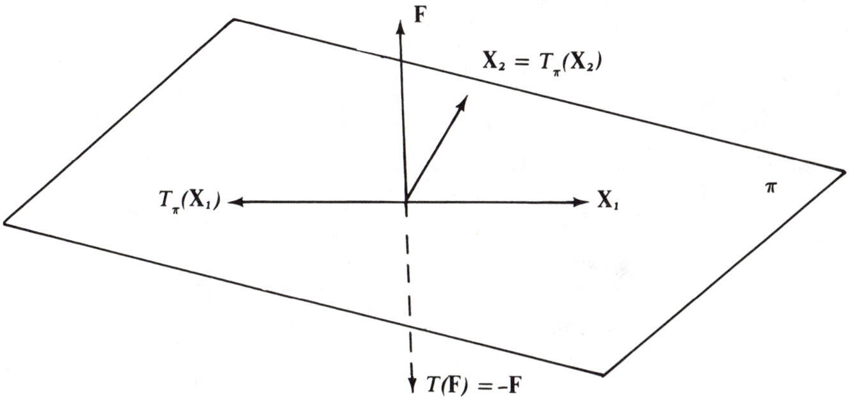

Figure 3.19

Let S and T be two isometries of $\mathbb{R}^3$. What can be said about their product ST?

Exercise 13.

(a) If S and T are isometries, then ST and TS are also isometries.
(b) If S is an isometry, then S^{-1} also is an isometry.

Exercise 14. If S is rotation about an axis and T is rotation about a possibly different axis, then ST is rotation about an axis.

Note: If the axes for S and for T are distinct, our conclusion that ST is again a rotation about some axis, though correct, is by no means evident.

Exercise 15. Let S be rotation by 90° about the x_3-axis and T be rotation by 90° about the x_1-axis. Find the axes for the rotations ST and TS.

Exercise 16. Let T be an isometry with $\det T = -1$.

(a) Show that $-T$ is a rotation.
(b) Conclude that T is the result of first performing a rotation and then reflecting every vector about the origin.

§3. Orthogonal Matrices

In Chapter 2.5, we found that a 2×2 matrix m is the matrix of an isometry if and only if it has one of the following forms:

(i)
$$\begin{pmatrix} \cos\theta & -\sin\theta \\ \sin\theta & \cos\theta \end{pmatrix},$$

(ii)
$$\begin{pmatrix} \cos\theta & \sin\theta \\ \sin\theta & -\cos\theta \end{pmatrix}.$$

Note that in each case the columns are mutually orthogonal unit vectors $\mathbb{R}^2$. It turns out that the analogous statement is true in 3 dimensions. Let

$$m = \begin{bmatrix} a_{11} & a_{12} & a_{13} \\ a_{21} & a_{22} & a_{23} \\ a_{31} & a_{32} & a_{33} \end{bmatrix}.$$

Exercise 17. Assume m is the matrix of an isometry T.

(a) Show $T(\mathbf{E}_i) \cdot T(\mathbf{E}_j) = \begin{cases} 1 & \text{if } i = j, \\ 0 & \text{if } i \neq j, \end{cases} \qquad 1 \leqslant i, j \leqslant 3.$
(b) Show that the columns of m are orthogonal unit vectors in $\mathbb{R}^3$.

Exercise 18. Assume m is the matrix of an isometry. Show that the inverse m^{-1} equals the transpose m^*.

Exercise 19. Show that, conversely, if m is a matrix satisfying $m^{-1} = m^*$, then m is the matrix of an isometry.

A matrix m such that

$$m^{-1} = m^* \tag{6}$$

is called an *orthogonal* matrix. Exercises 18 and 19 together prove the following result:

Proposition 7. *A* 3×3 *matrix* m *is the matrix of an isometry if and only if* m *is an orthogonal matrix.*

Exercise 20. Let R_θ^3 be rotation around the x_3-axis by an angle θ. Let $m = m(R_\theta^3)$. Directly show that m satisfies (6).

Exercise 21. Let T be reflection in the plane $x + y + z = 0$. Let $m = m(T)$. Directly show that m satisfies (6).

Exercise 22. Let S be rotation by 180° around the line: $x = t$, $y = t$, $z = t$. Let $m = m(S)$. Directly show that m satisfies (6).

Exercise 23. Let m be an orthogonal matrix.

(a) Show that m^* is an orthogonal matrix.
(b) Show that the rows of m are orthogonal unit vectors in $\mathbb{R}^3$.

Exercise 24. Show that the product of two orthogonal matrices is an orthogonal matrix.

Exercise 25. Consider the system of equation

$$\begin{cases} a_{11}x_1 + a_{12}x_2 + a_{13}x_3 = y_1, \\ a_{21}x_1 + a_{22}x_2 + a_{23}x_3 = y_2, \\ a_{31}x_1 + a_{32}x_2 + a_{33}x_3 = y_3. \end{cases} \tag{7}$$

Assume the coefficient matrix

$$\begin{pmatrix} a_{11} & a_{12} & a_{13} \\ a_{21} & a_{22} & a_{23} \\ a_{31} & a_{32} & a_{33} \end{pmatrix}$$

is an orthogonal matrix. Show that, given y_1, y_2, y_3, the system (7) is solved by setting

$$\begin{cases} x_1 = a_{11}y_1 + a_{21}y_2 + a_{31}y_3, \\ x_2 = a_{12}y_1 + a_{22}y_2 + a_{32}y_3, \\ x_3 = a_{13}y_1 + a_{23}y_2 + a_{33}y_3. \end{cases} \tag{8}$$

Exercise 26. Let m be a 3×3 matrix. Assume that the column vectors of m are orthogonal unit vectors. Prove that m is an orthogonal matrix.

CHAPTER 3.7

Symmetric Matrices

In the 2-dimensional case, we saw that a special role is played by matrices $\begin{pmatrix} a & b \\ b & d \end{pmatrix}$ which have both off-diagonal elements equal. The corresponding condition in 3 dimensions is symmetry about the diagonal. We say that a matrix is *symmetric* if the entry in the ith position in the jth column is the same as the entry in the jth position in the ith column, i.e., $a_{ij} = a_{ji}$ for all i, j,

$$\begin{bmatrix} a & b & c \\ b & d & e \\ c & e & f \end{bmatrix} = \begin{bmatrix} a_{11} & a_{12} & a_{13} \\ a_{12} & a_{22} & a_{23} \\ a_{13} & a_{23} & a_{33} \end{bmatrix}. \tag{1}$$

Note that this condition does not place any restriction on the diagonal elements themselves, but as soon as we know the elements on the diagonal and above the diagonal, we can fill in the rest of the entries in a symmetric matrix:

If the following matrix is symmetric,

$$\begin{bmatrix} 1 & 2 & -1 \\ x & 7 & 0 \\ y & z & 3 \end{bmatrix}$$

then $x = 2$, $y = -1$, $z = 0$.

We may express the symmetry condition succinctly by using the notion of transpose. Recall that the *transpose* m^* of a matrix m is the matrix whose columns are the rows of m. Thus a matrix m is symmetric if and only if $m^* = m$.

Exercise 1. For which values of the letters x, y, z, will the following matrices be symmetric?

(a) $\begin{pmatrix} 1 & 2 & 1 \\ x & 2 & 1 \\ y & z & 4 \end{pmatrix}$,

(b) $\begin{pmatrix} 2 & 5 & z \\ x & 1 & y \\ 2 & 4 & 0 \end{pmatrix}$,

(c) $\begin{pmatrix} x & 5 & 2 \\ 5 & y & 1 \\ 2 & 1 & z \end{pmatrix}$,

(d) $\begin{pmatrix} 1 & x & 3 \\ y & 2 & 4 \\ 3 & z & 6 \end{pmatrix}$,

(e) $\begin{pmatrix} 1 & y & 4 \\ 3 & x & 5 \\ 4 & 6 & z \end{pmatrix}$.

Exercise 2. Show that every diagonal matrix is symmetric.

Exercise 3. Show that if A and B are symmetric, then $A + B$ is symmetric and cA is symmetric for any c.

Exercise 4. True or false? If a and b are symmetric, then ab is symmetric. (Show that $(ab)^* = b^*a^*$).

Exercise 5. Show that if a is any matrix, then the average $\frac{1}{2}(a + a^*)$ is a symmetric matrix.

Exercise 6. True of false? The square of a symmetric matrix is symmetric.

Exercise 7. Prove that for any 3×3 matrix m, the product $m(m^*)$ is symmetric. (Recall that $(mn)^* = n^*m^*$.)

We shall need a general formula involving the transpose of a matrix. Let a be any matrix.

$$a = \begin{bmatrix} a_{11} & a_{12} & a_{13} \\ a_{21} & a_{22} & a_{23} \\ a_{31} & a_{32} & a_{33} \end{bmatrix}.$$

Then

$$a^* = \begin{bmatrix} a_{11} & a_{21} & a_{31} \\ a_{12} & a_{22} & a_{32} \\ a_{13} & a_{23} & a_{33} \end{bmatrix}.$$

Lemma 1. *Let A be the linear transformation with matrix a and A^* the linear transformation with matrix a^*. Then for every pair of vectors* $\mathbf{X}$, $\mathbf{Y}$,

$$A(\mathbf{X}) \cdot \mathbf{Y} = \mathbf{X} \cdot A^*(\mathbf{Y}). \tag{2}$$

Proof of Lemma 1. Set $\mathbf{X} = \begin{bmatrix} x_1 \\ x_2 \\ x_3 \end{bmatrix}$, $\mathbf{Y} = \begin{bmatrix} y_1 \\ y_2 \\ y_3 \end{bmatrix}$. Then

$$A(\mathbf{X}) \cdot \mathbf{Y} = \begin{bmatrix} a_{11}x_1 + a_{12}x_2 + a_{13}x_3 \\ a_{21}x_1 + a_{22}x_2 + a_{23}x_3 \\ a_{31}x_1 + a_{32}x_2 + a_{33}x_3 \end{bmatrix} \cdot \begin{bmatrix} y_1 \\ y_2 \\ y_3 \end{bmatrix}$$

$$= \begin{matrix} a_{11}x_1y_1 + a_{12}x_2y_1 + a_{13}x_3y_1 \\ + a_{21}x_1y_2 + a_{22}x_2y_2 + a_{23}x_3y_2 \\ + a_{31}x_1y_3 + a_{32}x_2y_3 + a_{33}x_3y_3 \end{matrix}. \tag{3}$$

$$\mathbf{X} \cdot A^*(\mathbf{Y}) = \begin{bmatrix} x_1 \\ x_2 \\ x_3 \end{bmatrix} \cdot \begin{bmatrix} a_{11} & a_{21} & a_{31} \\ a_{12} & a_{22} & a_{32} \\ a_{13} & a_{23} & a_{33} \end{bmatrix} \begin{bmatrix} y_1 \\ y_2 \\ y_3 \end{bmatrix}$$

$$= \begin{bmatrix} x_1 \\ x_2 \\ x_3 \end{bmatrix} \cdot \begin{bmatrix} a_{11}y_1 + a_{21}y_2 + a_{31}y_3 \\ a_{12}y_1 + a_{22}y_2 + a_{32}y_3 \\ a_{13}y_1 + a_{23}y_2 + a_{33}y_3 \end{bmatrix}$$

$$= \begin{matrix} a_{11}x_1y_1 + a_{21}x_1y_2 + a_{31}x_1y_3 \\ + a_{12}x_2y_1 + a_{22}x_2y_2 + a_{32}x_2y_3 \\ + a_{13}x_3y_1 + a_{23}x_3y_2 + a_{33}x_3y_3 \end{matrix}. \tag{4}$$

The first line of the sum (3) is the same as the first column of (4), and similarly for the other two lines. So the sums (3) and (4) consist of the same terms in different arrangements, and thus $A(\mathbf{X}) \cdot \mathbf{Y} = \mathbf{X} \cdot A^*(\mathbf{Y})$.

An immediate consequence of Lemma 1 is:

Lemma 2. *If m is a symmetric matrix, and M is the corresponding linear transformation, then for all vectors* $\mathbf{X}$, $\mathbf{Y}$,

$$M(\mathbf{X}) \cdot \mathbf{Y} = \mathbf{X} \cdot M(\mathbf{Y}). \tag{5}$$

We saw in Theorem 2.10 in Chapter 2.6, that eigenvectors of a symmetric 2×2 matrix corresponding to distinct eigenvalues are orthogonal. We now prove the analogous result in 3 dimensions.

Theorem 3.10. *Let m be a symmetric 3×3 matrix and let M be the corresponding linear transformation. Let t_1, t_2 be distinct eigenvalues of M, and let $\mathbf{X}_1$, $\mathbf{X}_2$ be corresponding eigenvectors. Then $\mathbf{X}_1 \cdot \mathbf{X}_2 = 0$.*

Proof.

$$M(\mathbf{X}_1) \cdot \mathbf{X}_2 = (t_1\mathbf{X}_1) \cdot \mathbf{X}_2 = t_1(\mathbf{X}_1 \cdot \mathbf{X}_2),$$

$$\mathbf{X}_1 \cdot M(\mathbf{X}_2) = X_1 \cdot (t_2\mathbf{X}_2) = t_2(\mathbf{X}_1 \cdot \mathbf{X}_2).$$

Then by (5),

$$t_1(\mathbf{X}_1 \cdot \mathbf{X}_2) = t_2(\mathbf{X}_1 \cdot \mathbf{X}_2).$$

If $\mathbf{X}_1 \cdot \mathbf{X}_2 \neq 0$, we can divide by $\mathbf{X}_1 \cdot \mathbf{X}_2$ and get $t_1 = t_2$, which contradicts our assumption. Therefore $\mathbf{X}_1 \cdot \mathbf{X}_2 = 0$.

Example 1. Let us calculate the eigenvalues and eigenvectors of the linear transformation M with matrix

$$m = \begin{bmatrix} 1 & 1 & -1 \\ 1 & 0 & 0 \\ -1 & 0 & 0 \end{bmatrix}.$$

The characteristic equation of m is

$$\begin{vmatrix} 1-t & 1 & -1 \\ 1 & -t & 0 \\ -1 & 0 & -t \end{vmatrix} = (1-t)t^2 - 1(-t) - 1(-t) = 0$$

or

$$-t^3 + t^2 + 2t = 0,$$

i.e.,

$$-t(t^2 - t - 2) = 0.$$

Since $t^2 - t - 2 = (t-2)(t+1)$, the eigenvalues of m are

$$t_1 = 0, \qquad t_2 = 2, \qquad t_3 = -1.$$

Let $\mathbf{X}_i$ denote an eigenvector corresponding to t_i, for $i = 1, 2, 3$.

$$M(\mathbf{X}_1) = 0\mathbf{X}_1 = \mathbf{0},$$

so setting

$$\mathbf{X}_1 = \begin{bmatrix} x \\ y \\ z \end{bmatrix}, \qquad \begin{bmatrix} 1 & 1 & -1 \\ 1 & 0 & 0 \\ -1 & 0 & 0 \end{bmatrix} \begin{bmatrix} x \\ y \\ z \end{bmatrix} = \begin{bmatrix} 0 \\ 0 \\ 0 \end{bmatrix}$$

or

$$\begin{aligned} x + y - z &= 0, \\ x &= 0, \\ -x &= 0. \end{aligned}$$

Therefore $x = 0$, $y = z$, so $\mathbf{X}_1 = \begin{bmatrix} 0 \\ 1 \\ 1 \end{bmatrix}$ is an eigenvector corresponding to t_1.

Similarly, if

$$\mathbf{X}_2 = \begin{bmatrix} x \\ y \\ z \end{bmatrix}, \qquad \begin{bmatrix} 1 & 1 & -1 \\ 1 & 0 & 0 \\ -1 & 0 & 0 \end{bmatrix} \begin{bmatrix} x \\ y \\ z \end{bmatrix} = 2\begin{bmatrix} x \\ y \\ z \end{bmatrix},$$

so

$$\begin{aligned} x + y - z &= 2x, \\ x &= 2y, \\ -x &= 2z. \end{aligned}$$

Setting $x = 2$, we must take $y = 1$ and $z = -1$. Then $x + y - z = 2 + 1 + 1 = 4 = 2x$, so the first equation is also satisfied. Thus $\mathbf{X}_2 = \begin{bmatrix} 2 \\ 1 \\ -1 \end{bmatrix}$. Similarly we find $\mathbf{X}_3 = \begin{bmatrix} 1 \\ -1 \\ 1 \end{bmatrix}$ is an eigenvector corresponding to $t_3 = -1$. Each pair of two out of our three eigenvectors

$$\begin{bmatrix} 0 \\ 1 \\ 1 \end{bmatrix}, \quad \begin{bmatrix} 2 \\ 1 \\ -1 \end{bmatrix}, \quad \begin{bmatrix} 1 \\ -1 \\ 1 \end{bmatrix}$$

is indeed orthogonal, as stated in Theorem 3.10.

Note: The eigenvectors corresponding to a given eigenvalue fill a line in this case. For instance, the set of eigenvectors of m corresponding to $t_2 = 2$ is the set of all vectors

$$t\mathbf{X}_2 = t\begin{bmatrix} 2 \\ 1 \\ -1 \end{bmatrix} = \begin{bmatrix} 2t \\ t \\ -t \end{bmatrix}, \qquad -\infty < t < \infty.$$

Exercise 8. Find the eigenvalues of the matrix $m = \begin{pmatrix} 3 & 4 & 0 \\ 4 & 3 & 0 \\ 0 & 0 & 1 \end{pmatrix}$ and find all corresponding eigenvectors.

Every 3×3 matrix has at least one eigenvalue, as we showed in Proposition 1 of Chapter 3.6. However, in general, we cannot say more.

Exercise 9. Give an example of a 3×3 matrix having as its only eigenvalue the number 1 and such that the corresponding eigenvectors make up a line.

For symmetric matrices, the situation is much better, as the following fundamental theorem shows:

Spectral Theorem in $\mathbb{R}^3$. *Let m be a symmetric 3×3 matrix and let M be the corresponding linear transformation. Then we can find three orthogonal unit vectors $\mathbf{X}_1$, $\mathbf{X}_2$, and $\mathbf{X}_3$ such that each $\mathbf{X}_i$ is an eigenvector of M.*

Proof. M has at least one eigenvalue t_1, as we know. Let $\mathbf{X}_1$ be a corresponding eigenvector of length 1. Denote by π the plane through the origin which is perpendicular to $\mathbf{X}_1$. We wish to show that we can find two further mutually perpendicular eigenvectors of M lying in π. We claim that π is *invariant* under M, i.e., that if a vector $\mathbf{X}$ is in π, then $M(\mathbf{X})$ also lies in π (See Fig. 3.20). Suppose $\mathbf{X}$ belongs to π. Then by (5),

$$M(\mathbf{X}) \cdot \mathbf{X}_1 = \mathbf{X} \cdot M(\mathbf{X}_1) = \mathbf{X} \cdot t_1\mathbf{X}_1 = t_1(\mathbf{X} \cdot \mathbf{X}_1).$$

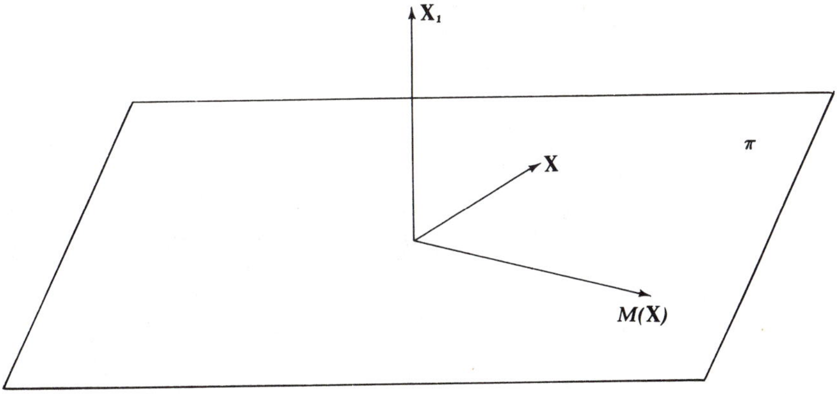

Figure 3.20

But $\mathbf{X} \cdot \mathbf{X}_1 = 0$, since $\mathbf{X}$ lies in π. Hence $M(\mathbf{X}) \cdot \mathbf{X}_1 = 0$, so $M(\mathbf{X})$ is in π, and our claim is proved.

We denote by A the transformation of π defined by

$$A(\mathbf{X}) = M(\mathbf{X}) \qquad \text{for } \mathbf{X} \text{ in } \pi.$$

We shall now go on to show that π may be identified with the plane $\mathbb{R}^2$. Also, when we make this identification, A turns into a linear transformation of $\mathbb{R}^2$ having a symmetric matrix. Using the results we found in Chapter 2.6, we shall then find two eigenvectors for this 2×2 symmetric matrix and these will turn out to give the "missing" eigenvectors in $\mathbb{R}^3$ for our original transformation M.

Let $\mathbf{F}_1$, $\mathbf{F}_2$ be vectors in π which are orthogonal and have length 1. If $\mathbf{X}$ and $\mathbf{Y}$ are vectors in π,

$$A(\mathbf{X}) \cdot \mathbf{Y} = M(\mathbf{X}) \cdot \mathbf{Y} = \mathbf{X} \cdot M(\mathbf{Y}) = \mathbf{X} \cdot A(\mathbf{Y}),$$

so A satisfies

$$A(\mathbf{X}) \cdot \mathbf{Y} = \mathbf{X} \cdot A(\mathbf{Y}) \tag{6}$$

whenever $\mathbf{X}$, $\mathbf{Y}$ lie in π. Each vector $\mathbf{X}$ in π can be expressed as

$$\mathbf{X} = x_1\mathbf{F}_1 + x_2\mathbf{F}_2,$$

where $x_1 = \mathbf{X} \cdot \mathbf{F}_1$, $x_2 = \mathbf{X} \cdot \mathbf{F}_2$. We identify $\mathbf{X}$ with the vector $\begin{pmatrix} x_1 \\ x_2 \end{pmatrix}$ in $\mathbb{R}^2$, and in this way π becomes identified with $\mathbb{R}^2$. Also, since A takes π into itself, A gives rise to a linear transformation A^0 of $\mathbb{R}^2$. For each $\mathbf{X} = x_1\mathbf{F}_1 + x_2\mathbf{F}_2$ in π, $A(\mathbf{X})$ is identified with $A^0\begin{pmatrix} x_1 \\ x_2 \end{pmatrix}$ in $\mathbb{R}^2$ (see Fig. 3.21). What is the matrix of A^0? Since $A(\mathbf{F}_1)$ and $A(\mathbf{F}_2)$ lie in π, we have

$$A(\mathbf{F}_1) = a\mathbf{F}_1 + b\mathbf{F}_2,$$

$$A(\mathbf{F}_2) = c\mathbf{F}_1 + d\mathbf{F}_2,$$

where $a = A(\mathbf{F}_1) \cdot \mathbf{F}_1$, $b = A(\mathbf{F}_1) \cdot \mathbf{F}_2$, $c = A(\mathbf{F}_2) \cdot \mathbf{F}_1$, and $d = A(\mathbf{F}_2) \cdot \mathbf{F}_2$. By

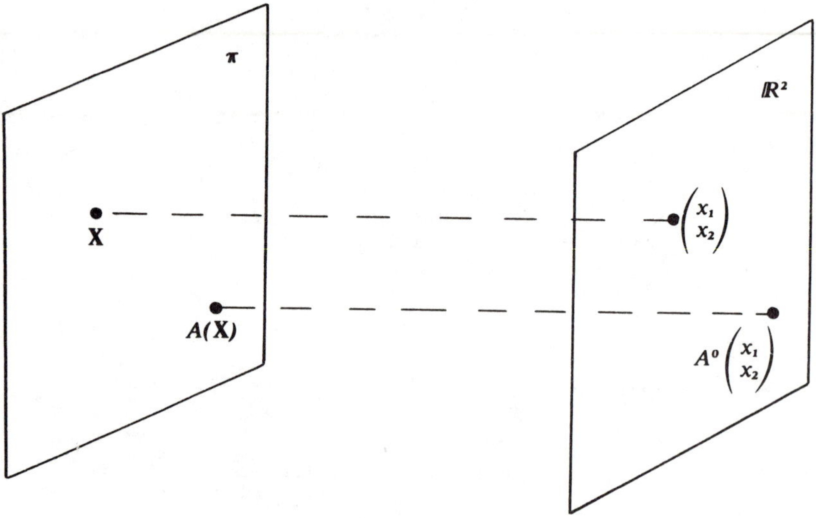

Figure 3.21

(6), we get

$$b = A(\mathbf{F}_1) \cdot \mathbf{F}_2 = \mathbf{F}_1 \cdot A(\mathbf{F}_2) = c. \tag{7}$$

Choose $\mathbf{X} = x_1\mathbf{F}_1 + x_2\mathbf{F}_2$ and set

$$A(\mathbf{X}) = x_1^1\mathbf{F}_1 + x_2^1\mathbf{F}_2 .$$

Then $A^0\begin{pmatrix} x_1 \\ x_2 \end{pmatrix} = \begin{pmatrix} x_1^1 \\ x_2^1 \end{pmatrix}$. Also,

$$\begin{aligned} A(\mathbf{X}) &= x_1A(\mathbf{F}_1) + x_2A(\mathbf{F}_2) = x_1(a\mathbf{F}_1 + b\mathbf{F}_2) + x_2(c\mathbf{F}_1 + d\mathbf{F}_2) \\ &= (ax_1 + cx_2)\mathbf{F}_1 + (bx_1 + dx_2)\mathbf{F}_2 . \end{aligned}$$

Hence,

$$A^0\begin{pmatrix} x_1 \\ x_2 \end{pmatrix} = \begin{pmatrix} x_1^1 \\ x_2^1 \end{pmatrix} = \begin{pmatrix} ax_1 + cx_2 \\ bx_1 + dx_2 \end{pmatrix} = \begin{pmatrix} a & c \\ b & d \end{pmatrix}\begin{pmatrix} x_1 \\ x_2 \end{pmatrix}.$$

Thus the matrix of A^0 is $\begin{pmatrix} a & c \\ b & d \end{pmatrix} = \begin{pmatrix} a & b \\ b & d \end{pmatrix}$, because of (7). We note that A^0 has a symmetric matrix. By Theorem 2.10, Chapter 2.6, there exist two orthogonal nonzero eigenvectors $\begin{pmatrix} u_1 \\ u_2 \end{pmatrix}$ and $\begin{pmatrix} v_1 \\ v_2 \end{pmatrix}$ for A^0. Then, for a certain scalar t, $A^0\begin{pmatrix} u_1 \\ u_2 \end{pmatrix} = t\begin{pmatrix} u_1 \\ u_2 \end{pmatrix}$.

The vector $\mathbf{U} = u_1\mathbf{F}_1 + u_2\mathbf{F}_2$ in π is identified with $\begin{pmatrix} u_1 \\ u_2 \end{pmatrix}$ and $A(\mathbf{U})$ is identified with $A^0\begin{pmatrix} u_1 \\ u_2 \end{pmatrix} = t\begin{pmatrix} u_1 \\ u_2 \end{pmatrix}$ in $\mathbb{R}^2$. Also, $t\mathbf{U}$ is identified with $t\begin{pmatrix} u_1 \\ u_2 \end{pmatrix}$. So $M(\mathbf{U}) = A(\mathbf{U}) = t\mathbf{U}$, and so $\mathbf{U}$ is an eigenvector of M lying in π. Similarly,

$\mathbf{V} = v_1\mathbf{F}_1 + v_2\mathbf{F}_2$ is an eigenvector of M lying in π. Finally,

$$\mathbf{U} \cdot \mathbf{V} = (u_1\mathbf{F}_1 + u_2\mathbf{F}_2) \cdot (v_1\mathbf{F}_1 + v_2\mathbf{F}_2) = u_1v_1 + u_2v_2 = \begin{pmatrix} u_1 \\ u_2 \end{pmatrix} \cdot \begin{pmatrix} v_1 \\ v_2 \end{pmatrix} = 0.$$

It follows that the three vectors

$$\mathbf{X}_1, \quad \frac{\mathbf{U}}{|\mathbf{U}|}, \quad \frac{\mathbf{V}}{|\mathbf{V}|}$$

are an orthonormal set in $\mathbb{R}^3$ consisting of eigenvectors of M.

Note: Although we have just shown that the transformation M has three mutually orthogonal eigenvectors $\mathbf{X}_1$, $\mathbf{X}_2$, $\mathbf{X}_3$, we have not shown that the corresponding eigenvalues are distinct. Indeed, this need not be the case. If M is the linear transformation with matrix $\begin{bmatrix} 1 & 0 & 0 \\ 0 & 1 & 0 \\ 0 & 0 & 2 \end{bmatrix}$, then M has only two distinct eigenvalues, 1 and 2, although it has three orthogonal eigenvectors $\mathbf{E}_1$, $\mathbf{E}_2$, $\mathbf{E}_3$. Here, $\mathbf{E}_1$ and $\mathbf{E}_2$ both correspond to the eigenvalue 1, while $\mathbf{E}_3$ corresponds to the eigenvalue 2.

EXAMPLE 2. Let us find all eigenvalues and eigenvectors for the linear transformation M with matrix

$$m = \begin{bmatrix} 1 & 0 & 5 \\ 0 & 1 & 3 \\ 5 & 3 & 1 \end{bmatrix}.$$

The characteristic equation is

$$\begin{vmatrix} 1-t & 0 & 5 \\ 0 & 1-t & 3 \\ 5 & 3 & 1-t \end{vmatrix} = (1-t)\left[(1-t)^2 - 9\right] + 5(-5(1-t)) = 0$$

or $(1-t)[(1-t)^2 - 9 - 25] = 0$. So the eigenvalues are $t_1 = 1$ and the roots of $(1-t)^2 - 34 = 0$, which are $t_2 = 1 + \sqrt{34}$ and $t_3 = 1 - \sqrt{34}$. We seek an eigenvector $\mathbf{X}_1 = \begin{pmatrix} x \\ y \\ z \end{pmatrix}$ corresponding to $t = 1$. So we must solve

$$M\begin{pmatrix} x \\ y \\ z \end{pmatrix} = \begin{pmatrix} x \\ y \\ z \end{pmatrix}, \text{ i.e., } \begin{bmatrix} 1 & 0 & 5 \\ 0 & 1 & 3 \\ 5 & 3 & 1 \end{bmatrix}\begin{pmatrix} x \\ y \\ z \end{pmatrix} = \begin{pmatrix} x \\ y \\ z \end{pmatrix} \text{ or}$$

$$\begin{aligned} x + 5z &= x, \\ y + 3z &= y, \\ 5x + 3y + z &= z. \end{aligned}$$

The first two equations give $z = 0$. Then the third gives $5x + 3y = 0$. Hence, $x = -3$, $y = 5$, $z = 0$ solves all three equations. So we take

$$\mathbf{X}_1 = \begin{bmatrix} -3 \\ 5 \\ 0 \end{bmatrix}.$$

The plane π through the origin and orthogonal to $\mathbf{X}_1$, which occurred in the proof of the Spectral Theorem, here has the equation

$$\begin{bmatrix} x \\ y \\ z \end{bmatrix} \cdot \begin{bmatrix} -3 \\ 5 \\ 0 \end{bmatrix} = -3x + 5y = 0 \quad \text{or} \quad y = \tfrac{3}{5}x.$$

By the proof of the Spectral Theorem, we can find in π a second eigenvector $\mathbf{X}_2$ of M, corresponding to $t_2 = 1 + \sqrt{34}$. $\mathbf{X}_2 = \begin{pmatrix} x \\ y \\ z \end{pmatrix}$ satisfies

$$y = \tfrac{3}{5}x, \tag{8}$$

since $\mathbf{X}_2$ is on π. Also,

$$M(\mathbf{X}_2) = (1 + \sqrt{34})\mathbf{X}_2 \quad \text{or} \quad \begin{bmatrix} 1 & 0 & 5 \\ 0 & 1 & 3 \\ 5 & 3 & 1 \end{bmatrix} \begin{bmatrix} x \\ y \\ z \end{bmatrix} = (1 + \sqrt{34}) \begin{bmatrix} x \\ y \\ z \end{bmatrix},$$

so

$$x + 5z = (1 + \sqrt{34})x, \tag{9}$$

as well as two further equations. However, Eqs. (8) and (9) suffice to give

$$y = \frac{3}{5}x, \qquad z = \frac{\sqrt{34}}{5}x,$$

so

$$\mathbf{X}_2 = \begin{bmatrix} x \\ y \\ z \end{bmatrix} = \begin{bmatrix} x \\ (3/5)x \\ (\sqrt{34}/5)x \end{bmatrix} = x \begin{bmatrix} 1 \\ 3/5 \\ \sqrt{34}/5 \end{bmatrix}.$$

In particular, taking $x = 5$, we find

$$\mathbf{X}_2 = \begin{bmatrix} 5 \\ 3 \\ \sqrt{34} \end{bmatrix}.$$

Each scalar multiple $t\mathbf{X}_2$ is an eigenvector of M corresponding to $t_2 = 1 + \sqrt{34}$. An eigenvector $\mathbf{X}_3$ corresponding to $t_3 = 1 - \sqrt{34}$ will be orthogonal to both $\mathbf{X}_1$ and $\mathbf{X}_2$, as we know by the Spectral Theorem. Hence, $\mathbf{X}_3$ is a scalar multiple of $\mathbf{X}_1 \times \mathbf{X}_2$.

$$\mathbf{X}_1 \times \mathbf{X}_2 = \begin{bmatrix} -3 \\ 5 \\ 0 \end{bmatrix} \times \begin{bmatrix} 5 \\ 3 \\ \sqrt{34} \end{bmatrix} = \begin{bmatrix} 5\sqrt{34} \\ 3\sqrt{34} \\ -34 \end{bmatrix} = \sqrt{34} \begin{bmatrix} 5 \\ 3 \\ -\sqrt{34} \end{bmatrix}.$$

So we can take $\mathbf{X}_3 = \begin{bmatrix} 5 \\ 3 \\ -\sqrt{34} \end{bmatrix}$ as eigenvector for $t_3 = 1 - \sqrt{34}$.

Note: The eigenvectors $\mathbf{X}_1$, $\mathbf{X}_2$, $\mathbf{X}_3$ do not have unit length. However, the vectors $\mathbf{X}_1/|\mathbf{X}_1|$, etc., are also eigenvectors and do have unit length.

Exercise 10. For each of the following matrices, find an orthonormal set $\mathbf{X}_1$, $\mathbf{X}_2$, $\mathbf{X}_3$ in $\mathbb{R}^3$ consisting of eigenvectors of that linear transformation with matrix m.

(a) $m = \begin{pmatrix} -4 & 0 & 0 \\ 0 & 2 & 3 \\ 0 & 3 & 2 \end{pmatrix}$,

(b) $m = \begin{pmatrix} 1 & 1 & 1 \\ 1 & 1 & 1 \\ 1 & 1 & 1 \end{pmatrix}$,

(c) $m = \begin{bmatrix} \sqrt{2} & 0 & 0 \\ 0 & \sqrt{3} & 0 \\ 0 & 0 & \sqrt{5} \end{bmatrix}$.

In Chapter 2.7, Theorem 2.11, we showed that if a 2×2 matrix m has two linearly independent eigenvectors corresponding to the eigenvalues t_1, t_2, then

$$m = pdp^{-1},$$

where d is the diagonal matrix $\begin{pmatrix} t_1 & 0 \\ 0 & t_2 \end{pmatrix}$ and p is a certain invertible matrix. We shall now prove the corresponding fact in $\mathbb{R}^3$.

Theorem 3.11. *Let M be a linear transformation having linearly independent eigenvectors $\mathbf{X}_1$, $\mathbf{X}_2$, $\mathbf{X}_3$ corresponding to eigenvalues t_1, t_2, t_3. Let m be the matrix of M. Denote by p the matrix $(\mathbf{X}_1 | \mathbf{X}_2 | \mathbf{X}_3)$ whose columns are the vectors $\mathbf{X}_i$. Then p is invertible and, setting* $d = \begin{bmatrix} t_1 & 0 & 0 \\ 0 & t_2 & 0 \\ 0 & 0 & t_3 \end{bmatrix}$,

$$m = pdp^{-1}. \tag{10}$$

Proof. Since $\mathbf{X}_1$, $\mathbf{X}_2$, $\mathbf{X}_3$ are linearly independent by hypothesis, the matrix p has an inverse, by (ix), p. 166. We set $\mathbf{E}_1 = \begin{bmatrix} 1 \\ 0 \\ 0 \end{bmatrix}$, $\mathbf{E}_2 = \begin{bmatrix} 0 \\ 1 \\ 0 \end{bmatrix}$, $\mathbf{E}_3 = \begin{bmatrix} 0 \\ 0 \\ 1 \end{bmatrix}$. Let P be the linear transformation with matrix p. Then

$$P(\mathbf{E}_1) = \mathbf{X}_1, \qquad \text{so} \quad MP(\mathbf{E}_1) = M(\mathbf{X}_1) = t_1\mathbf{X}_1.$$

Similarly, $MP(\mathbf{E}_2) = t_2\mathbf{X}_2$ and $MP(\mathbf{E}_3) = t_3\mathbf{X}_3$. Also, if D is the linear transformation with matrix d,

$$D(\mathbf{E}_1) = t_1\mathbf{E}_1, \qquad \text{so} \quad PD(\mathbf{E}_1) = t_1\mathbf{X}_1.$$

Similarly, $PD(\mathbf{E}_2) = t_2\mathbf{X}_2$ and $PD(\mathbf{E}_3) = t_3\mathbf{X}_3$. Thus the transformations MP and PD give the same results when acting on $\mathbf{E}_1$, $\mathbf{E}_2$, $\mathbf{E}_3$, and so $MP = PD$. Hence, $M = PDP^{-1}$, and $m = pdp^{-1}$.

Corollary 1. *Let m be a symmetric 3×3 matrix and let M be the corresponding linear transformation. Let t_1, t_2, t_3 denote the eigenvalues of M, and set* $d = \begin{bmatrix} t_1 & 0 & 0 \\ 0 & t_2 & 0 \\ 0 & 0 & t_3 \end{bmatrix}$. *Then there exists an orthogonal matrix r of $\mathbb{R}^3$ such that*

$$m = rdr^{-1}. \tag{11}$$

Proof. By the Spectral Theorem, M has eigenvectors $\mathbf{X}_1$, $\mathbf{X}_2$, $\mathbf{X}_3$ with eigenvalues t_1, t_2, t_3 such that $\mathbf{X}_1$, $\mathbf{X}_2$, $\mathbf{X}_3$ form an orthonormal set in $\mathbb{R}^3$. In particular, $\mathbf{X}_1$, $\mathbf{X}_2$, $\mathbf{X}_3$ are linearly independent, and so we can make use of Theorem 3.11. We set $r = (\mathbf{X}_1|\mathbf{X}_2|\mathbf{X}_3)$. By Theorem 3.11, $m = rdr^{-1}$. The only thing left to prove is that r is an orthogonal matrix. But the columns of r are orthonormal vectors, so by Exercise 26 of Chapter 3.6, r is an orthogonal matrix.

Example 3. In Example 2, we studied the matrix $m = \begin{bmatrix} 1 & 0 & 5 \\ 0 & 1 & 3 \\ 5 & 3 & 1 \end{bmatrix}$ and found the eigenvalues $t_1 = 1$, $t_2 = 1 + \sqrt{34}$, $t_3 = 1 - \sqrt{34}$ and corresponding (normalized) eigenvectors

$$\mathbf{X}_1 = \frac{1}{\sqrt{34}}\begin{bmatrix} -3 \\ 5 \\ 0 \end{bmatrix}, \qquad \mathbf{X}_2 = \frac{1}{\sqrt{68}}\begin{bmatrix} 5 \\ 3 \\ \sqrt{34} \end{bmatrix}, \qquad \mathbf{X}_3 = \frac{1}{\sqrt{68}}\begin{bmatrix} 5 \\ 3 \\ -\sqrt{34} \end{bmatrix}.$$

Here $r = (\mathbf{X}_1|\mathbf{X}_2|\mathbf{X}_3) = \begin{bmatrix} -3/\sqrt{34} & 5/\sqrt{68} & 5/\sqrt{68} \\ 5/\sqrt{34} & 3/\sqrt{68} & 3/\sqrt{68} \\ 0 & 1/\sqrt{2} & -1/\sqrt{2} \end{bmatrix}$. Formula (11) gives

$$\begin{bmatrix} 1 & 0 & 5 \\ 0 & 1 & 3 \\ 5 & 3 & 1 \end{bmatrix} = r\begin{bmatrix} 1 & 0 & 0 \\ 0 & 1+\sqrt{34} & 0 \\ 0 & 0 & 1-\sqrt{34} \end{bmatrix}r^{-1}.$$

We note a useful consequence of formula (11).

Corollary 2. *Let m be a symmetric 3×3 matrix with eigenvalues t_1, t_2, t_3. Let r be the matrix occurring in* (11). *Then for each positive integer k,*

$$(m)^k = r \begin{bmatrix} t_1^k & 0 & 0 \\ 0 & t_2^k & 0 \\ 0 & 0 & t_3^k \end{bmatrix} r^{-1}. \tag{12}$$

Exercise 11. Use (12) to find m^5, where $\begin{pmatrix} 1 & 0 & 5 \\ 0 & 1 & 3 \\ 5 & 3 & 1 \end{pmatrix}$ is the matrix of Example 3.

Exercise 12. For each of the matrices m in Exercise 10, obtain the form (11).

CHAPTER 3.8

Classification of Quadric Surfaces

A *quadric surface* is the 3-dimensional generalization of a conic section. Such a surface is determined by an equation in the variables x, y, z so that each term is of second degree; for example,

$$x^2 + 2xy + 3z^2 = 1.$$

The general form of the equation of a quadric surface is

$$ax^2 + 2bxy + 2cxz + dy^2 + 2eyz + fz^2 = 1, \tag{1}$$

where the coefficients a, b, c, d, e, and f are constants. We would like to predict the shape of the quadric surface in terms of the coefficients, much in the same way that we described a conic section in terms of the coefficients of an equation

$$ax^2 + 2bxy + cy^2 = 1$$

in two variables.

As in the 2-dimensional case, we may use the inner product and a symmetric matrix in order to describe the quadric surface. We may then use our analysis of symmetric matrices in order to get a classification of the associated quadric surfaces.

We denote by A the linear transformation with matrix

$$m = \begin{bmatrix} a & b & c \\ b & d & e \\ c & e & f \end{bmatrix} \quad \text{and} \quad \mathbf{X} = \begin{bmatrix} x \\ y \\ z \end{bmatrix}$$

so that

$$A(\mathbf{X}) = \begin{bmatrix} ax + by + cz \\ bx + dy + ez \\ cx + ey + fz \end{bmatrix}$$

and

$$\mathbf{X} \cdot A(\mathbf{X}) = ax^2 + byx + czx + bxy + dy^2 + ezy + cxz + eyz + fz^2$$
$$= ax^2 + 2bxy + 2cxz + dy^2 + 2eyz + fz^2.$$

We can therefore express relation (1) as

$$\mathbf{X} \cdot A(\mathbf{X}) = 1.$$

Observe that m is a symmetric matrix. Consider some examples: If m is a diagonal matrix so that

$$m = \begin{bmatrix} a & 0 & 0 \\ 0 & d & 0 \\ 0 & 0 & f \end{bmatrix},$$

then the equation has the form

$$ax^2 + dy^2 + fz^2 = 1.$$

If $a = d = f = (1/r)^2$ for some $r > 0$, then the equation becomes

$$x^2 + y^2 + z^2 = r^2,$$

and this is a *sphere* of radius r.

If a, d, and f are all positive, then we may write

$$a = \left(\frac{1}{\alpha}\right)^2 = d = \left(\frac{1}{\beta}\right)^2, \quad \text{and} \quad f = \left(\frac{1}{\gamma}\right)^2,$$

and we have the equation

$$\frac{x^2}{\alpha^2} + \frac{y^2}{\beta^2} + \frac{z^2}{\gamma^2} = 1.$$

This gives an *ellipsoid* with axes along the coordinate axes of $\mathbb{R}^3$.

If a and d are positive and f is negative, then

$$a = \left(\frac{1}{\alpha}\right)^2, \qquad d = \left(\frac{1}{\beta}\right)^2,$$

and $f = -1/\gamma^2$ for some a, β, γ; so the equation becomes

$$\frac{x^2}{\alpha^2} + \frac{y^2}{\beta^2} - \frac{z^2}{\gamma^2} = 1.$$

This is a *hyperboloid of one sheet*.

If $a > 0$, but $d < 0$, $f < 0$, then

$$a = \left(\frac{1}{\alpha}\right)^2, \qquad d = -\left(\frac{1}{\beta^2}\right), \qquad f = -\left(\frac{1}{\gamma^2}\right) \qquad \text{for some} \quad a, \beta, \gamma,$$

so the equation becomes

$$\frac{x^2}{\alpha^2} - \frac{y^2}{\beta^2} - \frac{z^2}{\gamma^2} = 1.$$

This is a *hyperboloid of two sheets*.

If $a < 0$, $d < 0$, $f < 0$, there are no solutions, since the sum of three negative numbers can never be 1.

What if one or more of the diagonal entries are zero? If $f = 0$, we have $ax^2 + dy^2 = 1$, and this is either an *elliptical cylinder* (if $a > 0$, $d > 0$), a *hyperbolic cylinder* (if $a > 0$, $d < 0$), or no locus at all if $a < 0$, $d < 0$.

If $f = 0$ and $d = 0$, and $a > 0$, then we have $ax^2 = 1$, and this is a *pair of planes* $x = \pm 1/\sqrt{a}$.

If $a = 0 = d = f$, we have no locus.

This completes the classification of quadric surfaces corresponding to diagonal matrices.

What if the matrix m is not diagonal, or, in other words, if one of the cross-terms in Eq. (1), $2bxy$, $2cxz$, $2eyz$, is nonzero?

In this case, we shall introduce new coordinates $\begin{pmatrix} u \\ v \\ w \end{pmatrix}$ for each vector $\begin{pmatrix} x \\ y \\ z \end{pmatrix}$ in such a way that, expressed in terms of u, v, w, Eq. (1) takes on a simpler form. Recall that by formula (11) of Section 3.7, there exists an orthogonal matrix r such that

$$m = rdr^{-1},$$

where $d = \begin{bmatrix} t_1 & 0 & 0 \\ 0 & t_2 & 0 \\ 0 & 0 & t_3 \end{bmatrix}$ is the diagonal matrix formed with the eigenvalues t_1, t_2, t_3 of A. In other words, we have

$$A = RDR^{-1}, \tag{2}$$

where R and D are the linear transformations whose matrices are r and d. Note that since r is an orthogonal matrix, R is an isometry.

If $\mathbf{X} = \begin{pmatrix} x \\ y \\ z \end{pmatrix}$ is any vector, let $\mathbf{U} = \begin{pmatrix} u \\ v \\ w \end{pmatrix}$ be the vector defined by

$$\mathbf{U} = R^{-1}(\mathbf{X}). \tag{3a}$$

We regard u, v, w as new coordinates of $\mathbf{X}$. Then

$$\mathbf{X} = R(\mathbf{U}). \tag{3b}$$

By (2), $A(\mathbf{X}) = (RDR^{-1})(\mathbf{X}) = RD(\mathbf{U})$, so $\mathbf{X} \cdot A(\mathbf{X}) = R(\mathbf{U}) \cdot R(D(\mathbf{U}))$. Since R is an isometry, the right-hand side equals $\mathbf{U} \cdot D(\mathbf{U}) = \begin{pmatrix} u \\ v \\ w \end{pmatrix} \cdot \begin{bmatrix} t_1 & 0 & 0 \\ 0 & t_2 & 0 \\ 0 & 0 & t_3 \end{bmatrix} \begin{pmatrix} u \\ v \\ w \end{pmatrix} = t_1u^2 + t_2v^2 + t_3w^2$. Expressing $\mathbf{X} \cdot A(\mathbf{X})$ in terms of x, y, z, we get:

Theorem 3.12.

$$ax^2 + 2bxy + 2cxz + dy^2 + 2eyz + fz^2 = t_1u^2 + t_2v^2 + t_3w^2. \tag{4}$$

The quadric surface defined by (1) thus has, as its equation in u, v, w,

$$t_1u^2 + t_2v^2 + t_3w^2 = 1. \tag{5}$$

Note: The new coordinates $\begin{pmatrix} u \\ v \\ w \end{pmatrix}$ of a vector $\mathbf{X} = \begin{pmatrix} x \\ y \\ z \end{pmatrix}$ are actually the coordinates of $\mathbf{X}$ relative to a system of orthogonal coordinate axes. Since R is an isometry, the vectors $R(\mathbf{E}_1)$, $R(\mathbf{E}_2)$, $R(\mathbf{E}_3)$ are three mutually orthogonal unit vectors in $\mathbb{R}^3$.

$$\mathbf{U} = \begin{pmatrix} u \\ v \\ w \end{pmatrix} = u\mathbf{E}_1 + v\mathbf{E}_2 + w\mathbf{E}_3,$$

so

$$\mathbf{X} = R(\mathbf{U}) = uR(\mathbf{E}_1) + vR(\mathbf{E}_2) + wR(\mathbf{E}_3).$$

Thus $\begin{pmatrix} u \\ v \\ w \end{pmatrix}$ are the coordinates of $\mathbf{X}$ in the system whose coordinate axes lie along the vectors $R(\mathbf{E}_1)$, $R(\mathbf{E}_2)$, $R(\mathbf{E}_3)$.

EXAMPLE 1. We wish to classify the quadric surface

$$\Sigma\colon x^2 + 2xy - 2xz = 1.$$

The corresponding symmetric matrix m here is $\begin{bmatrix} 1 & 1 & -1 \\ 1 & 0 & 0 \\ -1 & 0 & 0 \end{bmatrix}$. By Example 1 of Chapter 3.7, the eigenvalues of m are $t_1 = 0$, $t_2 = 2$, $t_3 = -1$. We introduce new coordinates u, v, w as described above. By (5), we find that an equation for Σ in the new coordinates is

$$2v^2 - w^2 = 1. \tag{6}$$

Hence Σ is a hyperbolic cylinder.

Question: How do we express the new coordinates $\begin{pmatrix} u \\ v \\ w \end{pmatrix}$ of a point $\mathbf{X} = \begin{pmatrix} x \\ y \\ z \end{pmatrix}$ in terms of the original coordinates here? We found in Example 1, Chapter 3.7, that the normalized eigenvectors of the matrix $m = \begin{bmatrix} 1 & 1 & -1 \\ 1 & 0 & 0 \\ -1 & 0 & 0 \end{bmatrix}$ are

$$\mathbf{X}_1 = \frac{1}{\sqrt{2}}\begin{bmatrix} 0 \\ 1 \\ 1 \end{bmatrix}, \quad \mathbf{X}_2 = \frac{1}{\sqrt{6}}\begin{bmatrix} 2 \\ 1 \\ -1 \end{bmatrix}, \quad \mathbf{X}_3 = \frac{1}{\sqrt{3}}\begin{bmatrix} 1 \\ -1 \\ 1 \end{bmatrix}.$$

By the way the matrix r occurring in (11) of Chapter 3.7 was obtained, we now have

$$r = \begin{bmatrix} 0 & \frac{2}{\sqrt{6}} & \frac{1}{\sqrt{3}} \\ \frac{1}{\sqrt{2}} & \frac{1}{\sqrt{6}} & -\frac{1}{\sqrt{3}} \\ \frac{1}{\sqrt{2}} & -\frac{1}{\sqrt{6}} & \frac{1}{\sqrt{3}} \end{bmatrix}.$$

Since r is an orthogonal matrix, we have

$$r^{-1} = r^* = \begin{bmatrix} 0 & \frac{1}{\sqrt{2}} & \frac{1}{\sqrt{2}} \\ \frac{2}{\sqrt{6}} & \frac{1}{\sqrt{6}} & -\frac{1}{\sqrt{6}} \\ \frac{1}{\sqrt{3}} & -\frac{1}{\sqrt{3}} & \frac{1}{\sqrt{3}} \end{bmatrix}.$$

If $\mathbf{X}$ is any vector, $\begin{pmatrix} x \\ y \\ z \end{pmatrix}$ are its old coordinates and $\begin{pmatrix} u \\ v \\ w \end{pmatrix}$ its new coordinates, then by (3a),

$$\begin{bmatrix} u \\ v \\ w \end{bmatrix} = R^{-1}\begin{bmatrix} x \\ y \\ z \end{bmatrix} = \begin{bmatrix} 0 & \frac{1}{\sqrt{2}} & \frac{1}{\sqrt{2}} \\ \frac{2}{\sqrt{6}} & \frac{1}{\sqrt{6}} & -\frac{1}{\sqrt{6}} \\ \frac{1}{\sqrt{3}} & -\frac{1}{\sqrt{3}} & \frac{1}{\sqrt{3}} \end{bmatrix}\begin{bmatrix} x \\ y \\ z \end{bmatrix}. \tag{7}$$

Equations (7) allow us to calculate the new coordinates for any given vector $\begin{pmatrix} x \\ y \\ z \end{pmatrix}$ in terms of x, y, and z.

Exercise 1. Classify the quadric surface:

$$x^2 + 10xz + y^2 + 6yz + z^2$$

(see Example 2, p. 197).

Exercise 2. Find an equation in new coordinates of the form

$$\lambda_1 u^2 + \lambda_2 v^2 + \lambda_3 w^2 = 1$$

for the quadric surface $-4x^2 + 2y^2 + 3yz + 2z^2 = 1$.

CHAPTER 4.0

Vector Geometry in 4-Space

§1. Introduction

In the preceding chapters, we have seen how the language and techniques of linear algebra can unify large parts of the geometry of vectors in 2 and 3 dimensions. What begins with an alternative way of treating problems in analytic geometry becomes a powerful tool for investigating increasingly complicated phenomena such as eigenvectors or quadratic forms which would be difficult to approach otherwise.

But now we consider 4-space. In the case of 4 dimensions or higher, linear algebra has to be used almost from the very beginning to define the concepts which correspond to geometric objects in 2 and 3 dimensions.

We cannot visualize these higher-dimensional phenomena directly, but we can use the algebraic intuitions developed in 2 and 3 dimensions to guide us in the study of mathematical ideas that are otherwise almost inaccessible. Many of the algebraic notions which we have used in low dimensions can be transferred almost without change to dimensions 4 and higher and we will therefore continue to use familiar geometric terms such as "vector," "dot product," "linear independence," and "eigenvector" when we study higher-dimensional geometry.

§2. The Algebra of Vectors

A *vector in 4-space* is defined to be a 4-tuple of real numbers $\begin{pmatrix} x_1 \\ x_2 \\ x_3 \\ x_4 \end{pmatrix}$ written in column form with x_i indicating the coordinate in ith place. We denote this vector by a single capital letter, **X**, i.e., we write $\mathbf{X} = \begin{bmatrix} x_1 \\ x_2 \\ x_3 \\ x_4 \end{bmatrix}$. The set of all vectors in 4-space is denoted by $\mathbb{R}^4$.

No longer can we "picture" the vector **X** as an arrow starting at the origin and ending at the point $\begin{bmatrix} x_1 \\ x_2 \\ x_3 \\ x_4 \end{bmatrix}$ in 4-space. The power of linear algebra is that it enables us to manipulate vectors in any dimension by using the same rules for addition and scalar multiplication that we used in dimensions 2 and 3.

We add two vectors by adding their components, so if $\mathbf{X} = \begin{bmatrix} x_1 \\ x_2 \\ x_3 \\ x_4 \end{bmatrix}$ and $\mathbf{U} = \begin{bmatrix} u_1 \\ u_2 \\ u_3 \\ u_4 \end{bmatrix}$, then

$$\mathbf{X} + \mathbf{U} = \begin{bmatrix} x_1 + u_1 \\ x_2 + u_2 \\ x_3 + u_3 \\ x_4 + u_4 \end{bmatrix}.$$

We *multiply* a vector by a scalar r by multiplying each of the coordinates by r, so

$$r\mathbf{X} = r\begin{bmatrix} x_1 \\ x_2 \\ x_3 \\ x_4 \end{bmatrix} = \begin{bmatrix} rx_1 \\ rx_2 \\ rx_3 \\ rx_4 \end{bmatrix}.$$

We set $\mathbf{E}_1 = \begin{bmatrix} 1 \\ 0 \\ 0 \\ 0 \end{bmatrix}$, $\mathbf{E}_2 = \begin{bmatrix} 0 \\ 1 \\ 0 \\ 0 \end{bmatrix}$, $\mathbf{E}_3 = \begin{bmatrix} 0 \\ 0 \\ 1 \\ 0 \end{bmatrix}$, $\mathbf{E}_4 = \begin{bmatrix} 0 \\ 0 \\ 0 \\ 1 \end{bmatrix}$, and we call these four vectors the *basis vectors* of 4-space. The first *coordinate axis* is then obtained by taking all multiples $x_1\mathbf{E}_1 = x_1\begin{bmatrix} 1 \\ 0 \\ 0 \\ 0 \end{bmatrix} = \begin{bmatrix} x_1 \\ 0 \\ 0 \\ 0 \end{bmatrix}$ of $\mathbf{E}_1$, and the ith coordinate axis is defined similarly for each $i = 2, 3, 4$. Any vector **X** may be expressed uniquely as a sum of vectors on the four coordinate axes:

$$\mathbf{X} = \begin{bmatrix} x_1 \\ x_2 \\ x_3 \\ x_4 \end{bmatrix} = \begin{bmatrix} x_1 \\ 0 \\ 0 \\ 0 \end{bmatrix} + \begin{bmatrix} 0 \\ x_2 \\ 0 \\ 0 \end{bmatrix} + \begin{bmatrix} 0 \\ 0 \\ x_3 \\ 0 \end{bmatrix} + \begin{bmatrix} 0 \\ 0 \\ 0 \\ x_4 \end{bmatrix}$$

$$= x_1\mathbf{E}_1 + x_2\mathbf{E}_2 + x_3\mathbf{E}_3 + x_4\mathbf{E}_4$$

In dimension 3, we described $x_1\mathbf{E}_1 + x_2\mathbf{E}_2 + x_3\mathbf{E}_3$ as a diagonal segment in a rectangular prism with edges parallel to the coordinate axes. We drew a picture which was completely determined as soon as we chose a position for each of the coordinate axes. We can do the same thing in the case of a vector in 4 dimensions, although it is not so immediately clear what we mean by the analogue of a 4-dimensional rectangular parallelepiped, and we will have to go further into the algebra of projections before we can interpret the full meaning of the picture (see Fig. 4.1).

The *line through* **X** along the nonzero vector **U** is defined to be the set of all vectors of the form $\mathbf{X} + t\mathbf{U}$ for all real numbers t.

Exercise 1. Let $\mathbf{X} = \begin{bmatrix} 1 \\ 2 \\ 0 \\ -1 \end{bmatrix}$ and $\mathbf{U} = \begin{bmatrix} 1 \\ 1 \\ 1 \\ 2 \end{bmatrix}$. Find the intersection of the line through **X** along **U** and the set of all vectors which have 0 in the fourth coordinate.

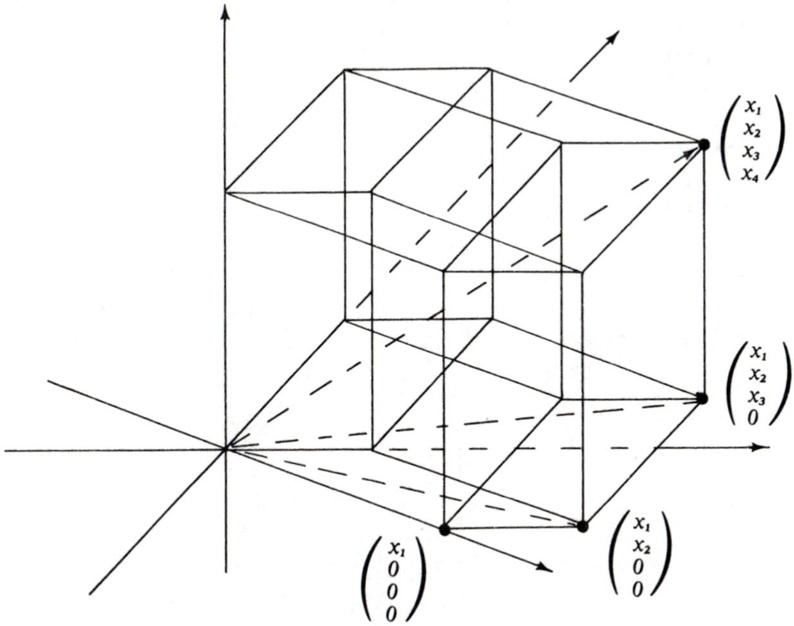

Figure 4.1

Exercise 2. Show that the line of Exercise 1 is the same as the line through $\mathbf{Y} = \begin{bmatrix} 3 \\ 4 \\ 2 \\ 3 \end{bmatrix}$ along $\mathbf{V} = \begin{bmatrix} -2 \\ -2 \\ -2 \\ -4 \end{bmatrix}$. (Show that every vector of the form $\mathbf{X} + t\mathbf{U}$ can be written as $\mathbf{Y} + s\mathbf{V}$ for an appropriate choice of s and, conversely, that every vector $\mathbf{Y} + s\mathbf{V}$ can be written as $\mathbf{X} + t\mathbf{U}$ for an appropriate choice of t.)

Exercise 3. Show that the line of Exercise 1 meets the line through $\mathbf{Z} = \begin{bmatrix} 0 \\ 1 \\ 2 \\ 0 \end{bmatrix}$ along $\mathbf{W} = \begin{bmatrix} -1 \\ -1 \\ -4 \\ -5 \end{bmatrix}$ at exactly one point. (Find t and r such that $\mathbf{X} + t\mathbf{U} = \mathbf{Z} + r\mathbf{W}$, and explain why there is only one such solution.)

By the difference of two vectors $\mathbf{X}$ and $\mathbf{U}$ we mean the vector $\mathbf{X} + (-\mathbf{U})$ which we denote by $\mathbf{X} - \mathbf{U}$.

As in the cases of $\mathbb{R}^2$ and $\mathbb{R}^3$, the vector algebra in $\mathbb{R}^4$ has the following properties: For all vectors $\mathbf{X}, \mathbf{Y}, \mathbf{U}$ and scalars r, s, we have

(i) $(\mathbf{X} + \mathbf{U}) + \mathbf{Y} = \mathbf{X} + (\mathbf{U} + \mathbf{Y})$.
(ii) $\mathbf{X} + \mathbf{U} = \mathbf{U} + \mathbf{X}$.
(iii) There is a vector $\mathbf{0}$ with $\mathbf{X} + \mathbf{0} = \mathbf{X}$ for all $\mathbf{X}$.
(iv) For each $\mathbf{X}$, there is a vector $-\mathbf{X}$ with $\mathbf{X} + (-\mathbf{X}) = \mathbf{0}$.
(v) $(r + s)\mathbf{X} = r\mathbf{X} + s\mathbf{X}$.
(vi) $r(s\mathbf{X}) = (rs)\mathbf{X}$.
(vii) $r(\mathbf{X} + \mathbf{U}) = r\mathbf{X} + r\mathbf{U}$.
(viii) $1\mathbf{X} = \mathbf{X}$ for all $\mathbf{X}$.

As in the previous case, each of these properties can be established componentwise.

In $\mathbb{R}^3$ the set of vectors orthogonal to a fixed nonzero vector forms a plane. In $\mathbb{R}^4$, we call the set of vectors orthogonal to a fixed nonzero vector a *hyperplane*. For example, the vectors orthogonal to $\mathbf{E}_4$ form the hyperplane consisting of all vectors $\begin{bmatrix} x_1 \\ x_2 \\ x_3 \\ x_4 \end{bmatrix}$ with fourth coordinate equal to zero.

We may use the properties of addition and scalar multiplication to define the notion of a *centroid* for sets of vectors in $\mathbb{R}^4$. As before, we define the *midpoint* of a pair of vectors $\mathbf{X}$ and $\mathbf{Y}$ by

$$C(\mathbf{X}, \mathbf{Y}) = \tfrac{1}{2}(\mathbf{X} + \mathbf{Y})$$

and the *centroid* of a triplet of vectors $\mathbf{X}, \mathbf{Y}, \mathbf{U}$ by

$$C(\mathbf{X}, \mathbf{Y}, \mathbf{Z}) = \tfrac{1}{3}(\mathbf{X} + \mathbf{Y} + \mathbf{U}).$$

Similarly, for any 4-tuple or 5-tuple of vectors in $\mathbb{R}^4$, we define the centroid to be the average

$$C(\mathbf{X}, \mathbf{Y}, \mathbf{U}, \mathbf{V}) = \tfrac{1}{4}(\mathbf{X} + \mathbf{Y} + \mathbf{U} + \mathbf{V}),$$

$$C(\mathbf{X}, \mathbf{Y}, \mathbf{Z}, \mathbf{U}, \mathbf{V}) = \tfrac{1}{5}(\mathbf{X} + \mathbf{Y} + \mathbf{Z} + \mathbf{U} + \mathbf{V}),$$

EXAMPLE. $C(\mathbf{E}_1, \mathbf{E}_2, \mathbf{E}_3, \mathbf{E}_4) = \tfrac{1}{4}(\mathbf{E}_1 + \mathbf{E}_2 + \mathbf{E}_3 + \mathbf{E}_4) = \begin{pmatrix} \frac{1}{4} \\ \frac{1}{4} \\ \frac{1}{4} \\ \frac{1}{4} \end{pmatrix}.$

Recall that the centroid of a triangle formed by the endpoints of $\mathbf{X}$, $\mathbf{Y}$, and $\mathbf{Z}$ is two-thirds of the way from $\mathbf{Z}$ to the midpoint of $\mathbf{X}, \mathbf{Y}$. In algebraic language, this means that

$$C(\mathbf{X}, \mathbf{Y}, \mathbf{Z}) = \tfrac{1}{3}\mathbf{Z} + \tfrac{2}{3}C(\mathbf{X}, \mathbf{Y}).$$

A simple substitution for $C(\mathbf{X}, \mathbf{Y})$ shows that this is indeed correct.

Exercise 4. Show that the centroid of the tetrahedron determined by the endpoints of $\mathbf{X}$, $\mathbf{Y}$, $\mathbf{U}$ and $\mathbf{V}$ is three-fourths of the way from $\mathbf{V}$ to the centroid of $\mathbf{X}$, $\mathbf{Y}$, and $\mathbf{U}$.

Exercise 5. Show that the centroid of $\mathbf{X}, \mathbf{Y}, \mathbf{U}, \mathbf{V}$ is the midpoint of the segment joining the midpoint of $\mathbf{X}, \mathbf{Y}$ to the midpoint of $\mathbf{U}, \mathbf{V}$.

Exercise 6. Find a number t such that

$$C(\mathbf{E}_1, \mathbf{E}_2, \mathbf{E}_3, \mathbf{E}_4, t(\mathbf{E}_1 + \mathbf{E}_2 + \mathbf{E}_3 + \mathbf{E}_4)) = \mathbf{0}.$$

Exercise 7. Show that the centroid of $\mathbf{X}, \mathbf{Y}, \mathbf{Z}, \mathbf{U}, \mathbf{V}$ is three-fifths of the way from the midpoint of $\mathbf{U}, \mathbf{V}$ to the centroid of $\mathbf{X}, \mathbf{Y}, \mathbf{Z}$.

§3. Dot Product, Length, and Angle in $\mathbb{R}^4$

In 4-space we may define the length of the vector $\mathbf{X} = \begin{pmatrix} x_1 \\ x_2 \\ x_3 \\ x_4 \end{pmatrix}$ to be $\sqrt{x_1^2 + x_2^2 + x_3^2 + x_4^2}$, denoted by $|\mathbf{X}|$. As before, this number is non-negative and $|\mathbf{X}| = 0$ if and only if $\mathbf{X}$ is the zero vector. Moreover, $|r\mathbf{X}| = |r|\,|\mathbf{X}|$. If $\mathbf{X} \neq \mathbf{0}$, we may write $\mathbf{X} = |\mathbf{X}|\mathbf{U}$, where $\mathbf{U} = (1/|\mathbf{X}|)\mathbf{X}$ is a vector of length 1. The vectors of length 1 in $\mathbb{R}^4$ determine the *unit sphere* in 4-space.

Exercise 8. Show that for any choice of θ and ϕ, the vector $\mathbf{U} = (1/\sqrt{2})\begin{pmatrix} \cos\theta \\ \sin\theta \\ \cos\phi \\ \sin\phi \end{pmatrix}$ is a unit vector in $\mathbb{R}^4$.

Exercise 9. Show that for any θ, ϕ, and α, the vector $\mathbf{V} = \begin{pmatrix} \cos\theta\cos\alpha \\ \sin\theta\cos\alpha \\ \cos\phi\sin\alpha \\ \sin\phi\sin\alpha \end{pmatrix}$ is a unit vector in $\mathbb{R}^4$.

Exercise 10. Show that for any θ, ϕ, and α, the vector $\mathbf{W} = \begin{pmatrix} \cos\theta\cos\phi\cos\alpha \\ \cos\theta\cos\phi\sin\alpha \\ \cos\theta\sin\phi \\ \sin\theta \end{pmatrix}$ is a unit vector in $\mathbb{R}^4$.

Definition. Five points in $\mathbb{R}^4$ are the *vertices* of a *regular* 4-*simplex* if the distance between any two of the points is the same.

Exercise 11. Find a number t so that the endpoints of the vector $\mathbf{E}_1, \mathbf{E}_2, \mathbf{E}_3, \mathbf{E}_4$ and $t(\mathbf{E}_1 + \mathbf{E}_2 + \mathbf{E}_3 + \mathbf{E}_4)$ form the vertices of a regular 4-simplex.

Exercise 12. Show that the vector $s\mathbf{E}_4$ is equidistant from the vectors $\begin{pmatrix} 1 \\ 1 \\ 1 \\ 0 \end{pmatrix}, \begin{pmatrix} 1 \\ -1 \\ -1 \\ 0 \end{pmatrix}, \begin{pmatrix} -1 \\ 1 \\ -1 \\ 0 \end{pmatrix}, \begin{pmatrix} -1 \\ -1 \\ 1 \\ 0 \end{pmatrix}$. For which s do the endpoints of these five vectors form a regular 4-simplex?

Exercise 13. Show that the distance between any two of the six vectors $\mathbf{E}_1 + \mathbf{E}_2$, $\mathbf{E}_1 + \mathbf{E}_3$, $\mathbf{E}_1 + \mathbf{E}_4$, $\mathbf{E}_2 + \mathbf{E}_3$, $\mathbf{E}_2 + \mathbf{E}_4$, $\mathbf{E}_3 + \mathbf{E}_4$ is $\sqrt{2}$ or 2.

As in dimensions 2 and 3, we define a notion of *dot product* in $\mathbb{R}^4$ which enables us to treat many important ideas in linear algebra. We define

$$\mathbf{X} \cdot \mathbf{U} = \begin{pmatrix} x_1 \\ x_2 \\ x_3 \\ x_4 \end{pmatrix} \cdot \begin{pmatrix} u_1 \\ u_2 \\ u_3 \\ u_4 \end{pmatrix} = x_1u_1 + x_2u_2 + x_3u_3 + x_4u_4 .$$

As before, for each vector $\mathbf{X}$, we have $|\mathbf{X}| = \sqrt{\mathbf{X} \cdot \mathbf{X}}$ and $|\mathbf{X}| = 0$ if and only if $\mathbf{X} = \mathbf{0}$. Also, $\mathbf{E}_i \cdot \mathbf{E}_j = 0$ if $i \neq j$.

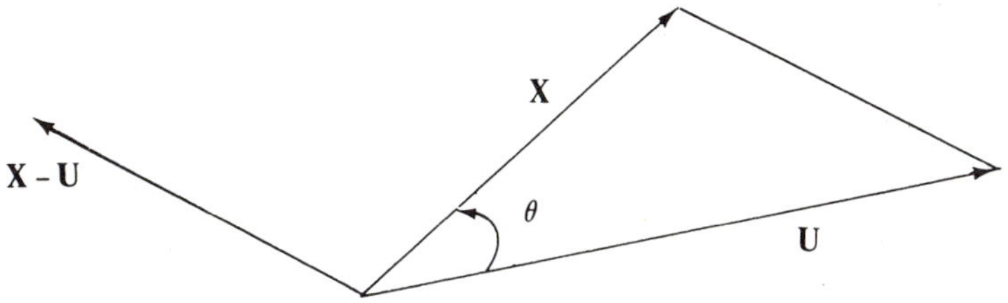

Figure 4.2

Using componentwise arguments, we may establish the following properties of the dot product for any vectors $\mathbf{X}, \mathbf{U}, \mathbf{V}$ and any scalar r:

$$\mathbf{U} \cdot \mathbf{X} = \mathbf{X} \cdot \mathbf{U},$$
$$(r\mathbf{X}) \cdot \mathbf{U} = r(\mathbf{X} \cdot \mathbf{U}),$$
$$\mathbf{X} \cdot (\mathbf{U} + \mathbf{V}) = \mathbf{X} \cdot \mathbf{U} + \mathbf{X} \cdot \mathbf{V}.$$

As in dimensions 2 and 3, we wish to define the angle between two vectors in such a way that the law of cosines will hold, i.e., for any two nonzero vectors $\mathbf{X}$ and $\mathbf{U}$ (see Fig. 4.2), we wish to have

$$|\mathbf{X} - \mathbf{U}|^2 = |\mathbf{X}|^2 + |\mathbf{U}|^2 - 2|\mathbf{X}|\,|\mathbf{U}|\cos\theta.$$

But by the properties of dot product, we have

$$|\mathbf{X} - \mathbf{U}|^2 = (\mathbf{X} - \mathbf{U}) \cdot (\mathbf{X} - \mathbf{U}) = \mathbf{X} \cdot \mathbf{X} - 2\mathbf{X} \cdot \mathbf{U} + \mathbf{U} \cdot \mathbf{U},$$

so

$$|\mathbf{X} - \mathbf{U}|^2 = |\mathbf{X}|^2 + |\mathbf{U}|^2 - 2\mathbf{X} \cdot \mathbf{U}.$$

We would then like to define $\cos\theta$ by the condition

$$|\mathbf{X}|\,|\mathbf{U}|\cos\theta = \mathbf{X} \cdot \mathbf{U} \quad \text{and} \quad 0 \leqslant \theta \leqslant \pi,$$

but to do this, we must have $|\cos\theta| \leqslant 1$, i.e., $\cos^2\theta \leqslant 1$. Thus we must show that for any nonzero $\mathbf{X}$ and $\mathbf{U}$, we have

$$\left(\frac{\mathbf{X} \cdot \mathbf{U}}{|\mathbf{X}| \cdot |\mathbf{U}|}\right)^2 \leqslant 1.$$

(This inequality is known as the *Cauchy–Schwarz Inequality*.)

One case is easy: If $\mathbf{U} = t\mathbf{X}$ for some t, then

$$\frac{\mathbf{X} \cdot \mathbf{U}}{|\mathbf{X}|\,|\mathbf{U}|} = \frac{\mathbf{X} \cdot t\mathbf{X}}{|\mathbf{X}|\,|t\mathbf{X}|} = \frac{t\mathbf{X} \cdot \mathbf{X}}{|t|\,|\mathbf{X}|^2} = \frac{t}{|t|},$$

and $-1 \leqslant t/|t| \leqslant 1$, since $t/|t| = 1$ if $t > 0$ and $t/|t| = -1$ if $t < 0$.

If $\mathbf{U} - t\mathbf{X} \neq \mathbf{0}$ for all t, then we can use the quadratic formula to provide the proof. We have

$$0 < |\mathbf{U} - t\mathbf{X}|^2 = (\mathbf{U} - t\mathbf{X}) \cdot (\mathbf{U} - t\mathbf{X}) = (\mathbf{U} \cdot \mathbf{U}) - 2(\mathbf{U} \cdot \mathbf{X})t + (\mathbf{X} \cdot \mathbf{X})t^2$$

for all t. But if $(\mathbf{U} \cdot \mathbf{X})^2 - (\mathbf{U} \cdot \mathbf{U})(\mathbf{X} \cdot \mathbf{X})$ were positive or zero, we would have solutions t of the equation

$$0 = (\mathbf{U} \cdot \mathbf{U}) - 2(\mathbf{U} \cdot \mathbf{X})t + (\mathbf{X} \cdot \mathbf{X})t^2,$$

given by

$$t = \frac{-(-2\mathbf{U}\cdot\mathbf{X}) \pm \sqrt{4(\mathbf{U}\cdot\mathbf{X})^2 - 4(\mathbf{U}\cdot\mathbf{U})(\mathbf{X}\cdot\mathbf{X})}}{2(\mathbf{X}\cdot\mathbf{X})}.$$

Since we cannot have any such solutions, we must conclude that

$$(\mathbf{U}\cdot\mathbf{X})^2 < (\mathbf{U}\cdot\mathbf{U})(\mathbf{X}\cdot\mathbf{X})$$

i.e.,

$$\frac{(\mathbf{U}\cdot\mathbf{X})^2}{|\mathbf{U}|^2|\mathbf{X}|^2} < 1.$$

We then define θ by the equation

$$\cos\theta = \frac{\mathbf{X}\cdot\mathbf{U}}{|\mathbf{X}|\,|\mathbf{U}|} \qquad \text{for} \quad 0 \leqslant \theta \leqslant \pi.$$

If $\mathbf{U} = t\mathbf{X}$ for $t > 0$, then $\cos\theta = 1$ and $\theta = 0$. If $\mathbf{U} = t\mathbf{X}$ for $t < 0$, then $\cos\theta = -1$ and $\theta = \pi$. If $\mathbf{X}\cdot\mathbf{U} = 0$, then $\theta = \pi/2$ and we say that the vectors $\mathbf{X}$ and $\mathbf{U}$ are *orthogonal*.

Exercise 14. Show that for any θ, the vectors $\cos\theta\mathbf{E}_1 - \sin\theta\mathbf{E}_3$ and $\sin\theta\mathbf{E}_1 + \cos\theta\mathbf{E}_3$ are orthogonal in $\mathbb{R}^4$.

Exercise 15. Find a real number t such that $\begin{pmatrix} 1 \\ 2 \\ 1 \\ -1 \end{pmatrix}$ is orthogonal to $\begin{pmatrix} t \\ -t \\ 1-t \\ 2t-1 \end{pmatrix}$.

We say that a collection of vectors is *orthonormal* if each vector has unit length and if any two distinct vectors in the set are orthogonal. For example, the basis vectors $\{\mathbf{E}_1, \mathbf{E}_2, \mathbf{E}_3, \mathbf{E}_4\}$ form an orthonormal set.

Exercise 16. Show that for any angle θ, the vectors $\{\cos\theta\mathbf{E}_1 + \sin\theta\mathbf{E}_4, \mathbf{E}_2, \mathbf{E}_3, -\sin\theta\mathbf{E}_1 + \cos\theta\mathbf{E}_4,\}$ form an orthonormal set.

Exercise 17. Show that the four vectors $\begin{pmatrix} x \\ y \\ u \\ v \end{pmatrix}, \begin{pmatrix} -y \\ x \\ -v \\ u \end{pmatrix}, \begin{pmatrix} u \\ -v \\ -x \\ y \end{pmatrix}, \begin{pmatrix} v \\ u \\ -y \\ -x \end{pmatrix}$ are mutually orthogonal and all have the same length.

Exercise 18. Show that for all angles θ and ϕ, the vectors

$$\{(\cos\theta\mathbf{E}_1 + \sin\theta\mathbf{E}_2)\cos\phi + \sin\phi\mathbf{E}_3,$$
$$-(\cos\theta\mathbf{E}_1 + \sin\theta\mathbf{E}_2)\sin\phi + \cos\phi\mathbf{E}_3,$$
$$-\sin\theta\mathbf{E}_1 + \cos\theta\mathbf{E}_2, \mathbf{E}_4\}$$

form an orthonormal set.

Exercise 19. Find the angle between the vectors $\mathbf{E}_1 + \mathbf{E}_2$ and $\mathbf{E}_1 + \mathbf{E}_2 + \mathbf{E}_3 + \mathbf{E}_4$.

Exercise 20. Find the angle between the vectors $\mathbf{E}_3 - \frac{1}{2}(\mathbf{E}_1 + \mathbf{E}_2)$ and $\mathbf{E}_4 - \frac{1}{2}(\mathbf{E}_1 + \mathbf{E}_2)$.

Exercise 21. Find the angle between the vectors $\begin{pmatrix} 1 \\ 1 \\ 1 \\ 1 \end{pmatrix}$ and $\begin{pmatrix} 1 \\ 1 \\ 1 \\ -1 \end{pmatrix}$. What are the possible cosines of angles between the vector $\begin{pmatrix} 1 \\ 1 \\ 1 \\ 1 \end{pmatrix}$ and the other vectors which have each coordinate 1 or -1?

Exercise 22. Show that if **U**, **V**, and **W** are distinct vectors with each coordinate 1 or -1 and if **V** and **W** each differ from **U** by exactly one coordinate, then $\mathbf{V} - \mathbf{U}$ and $\mathbf{W} - \mathbf{U}$ are orthogonal and they have the same length.

The collection of vectors $\mathbf{X} = \begin{pmatrix} x_1 \\ x_2 \\ x_3 \\ x_4 \end{pmatrix}$ with $-1 \leqslant x_i \leqslant 1$ for $i = 1, 2, 3, 4$ is called the 4-*cube centered at the origin.*

CHAPTER 4.1

Transformations of 4-Space

By a *transformation of* 4-*space*, we mean a rule T which assigns to each vector X of $\mathbb{R}^4$ some vector $T(X)$ of $\mathbb{R}^4$. The vector $T(X)$ is called the *image* of X under T, and the collection of all images of vectors in $\mathbb{R}^4$ under the transformation T is called the *range* of T. We continue to denote transformations by capital letters such as P, Q, R, S, T.

Examples of transformations are:

(1) Projection to the line along $\mathbf{U} \neq \mathbf{0}$ defined by

$$P(\mathbf{X}) = \left(\frac{\mathbf{X}\cdot\mathbf{U}}{\mathbf{U}\cdot\mathbf{U}}\right)\mathbf{U}.$$

(2) Reflection through the line along $\mathbf{U} \neq \mathbf{0}$ defined by

$$S(\mathbf{X}) = 2P(\mathbf{X}) - \mathbf{X}.$$

(3) Multiplication by a scalar t defined by

$$D_t(\mathbf{X}) = t\mathbf{X}.$$

(4) Projection to the hyperplane perpendicular to $\mathbf{U} \neq \mathbf{0}$ defined by

$$Q(\mathbf{X}) = \mathbf{X} - P(\mathbf{X}).$$

EXAMPLE 1. Let $\mathbf{U} = \begin{bmatrix} 1 \\ 1 \\ 1 \\ 1 \end{bmatrix}$ in $\mathbb{R}^4$; then

$$P(\mathbf{X}) = \left(\frac{x_1 + x_2 + x_3 + x_4}{4}\right)\begin{bmatrix} 1 \\ 1 \\ 1 \\ 1 \end{bmatrix} = \frac{1}{4}\begin{bmatrix} x_1 + x_2 + x_3 + x_4 \\ x_1 + x_2 + x_3 + x_4 \\ x_1 + x_2 + x_3 + x_4 \\ x_1 + x_2 + x_3 + x_4 \end{bmatrix},$$

$$S(\mathbf{X}) = 2P(\mathbf{X}) - \mathbf{X} = \frac{1}{2}\begin{bmatrix} -x_1 + x_2 + x_3 + x_4 \\ x_1 - x_2 + x_3 + x_4 \\ x_1 + x_2 - x_3 + x_4 \\ x_1 + x_2 + x_3 - x_4 \end{bmatrix},$$

$$Q(\mathbf{X}) = \frac{1}{4}\begin{bmatrix} 3x_1 - x_2 - x_3 - x_4 \\ -x_1 + 3x_2 - x_3 - x_4 \\ -x_1 - x_2 + 3x_3 - x_4 \\ -x_1 - x_2 - x_3 + 3x_4 \end{bmatrix}.$$

Exercise 1. In each of the following problems, let P denote projection to the line along $\mathbf{U}$. Find a formula for the coordinates of the image $P(\mathbf{X})$ in terms of the coordinates of $\mathbf{X}$.

(a) $\mathbf{U} = \begin{bmatrix} 1 \\ 0 \\ 0 \\ 0 \end{bmatrix}$,

(b) $\mathbf{U} = \begin{bmatrix} 1 \\ 1 \\ 0 \\ 0 \end{bmatrix}$,

(c) $\mathbf{U} = \begin{bmatrix} 1 \\ -1 \\ 1 \\ -1 \end{bmatrix}$,

(d) $\mathbf{U} = \begin{bmatrix} 1 \\ 0 \\ 2 \\ -1 \end{bmatrix}$.

Exercise 2. For each of the vectors $\mathbf{U}$ in Exercise 1, find a formula for the image of the reflection $S(\mathbf{X})$ through the line along $\mathbf{U}$.

Exercise 3. For each of the vectors $\mathbf{U}$ in Exercise 1, find a formula for $Q(\mathbf{X})$, where Q is the projection to the hyperplane orthogonal to $\mathbf{U}$.

Using the description of a transformation in terms of its coordinates, we can define further transformations, such as:

(5) Rotation in the x_1x_2 plane by angle θ defined by

$$R_\theta^{12}\begin{pmatrix} x_1 \\ x_2 \\ x_3 \\ x_4 \end{pmatrix} = \begin{pmatrix} \cos\theta x_1 - \sin\theta x_2 \\ \sin\theta x_1 + \cos\theta x_2 \\ x_3 \\ x_4 \end{pmatrix}.$$

Exercise 4. In terms of the coordinate of $\mathbf{X} = \begin{pmatrix} x_1 \\ x_2 \\ x_3 \\ x_4 \end{pmatrix}$, calculate the images $R_\pi^{12}(\mathbf{X})$, $R_{\pi/4}^{12}(\mathbf{X})$, $R_{-\pi/4}^{12}(\mathbf{X})$.

Similarly, we have the images of R_θ^{ij} in the x_ix_j plane by setting $x_k' = x_k$ for all $k \neq i, j$ and by defining

$$x_i' = \cos\theta x_i - \sin\theta x_j,$$

$$x_j' = \sin\theta x_i + \cos\theta x_j.$$

Exercise 5. Calculate the images $R_\pi^{34}(\mathbf{X})$, $R_\theta^{34}(R_\phi^{12}(\mathbf{X}))$, $R_\phi^{12}(R_\theta^{34}(\mathbf{X}))$, $R_\theta^{23}(R_\phi^{12}(\mathbf{X}))$, $R_\phi^{12}(R_\theta^{23}(\mathbf{X}))$, where $\mathbf{X} = \begin{pmatrix} x_1 \\ x_2 \\ x_3 \\ x_4 \end{pmatrix}$.

Just as in the case of objects in 3 dimensions, we may picture objects in 4-space by projecting them down to a 2-dimensional plane. The easiest such projection is simply the projection to the first two coordinates in $\mathbb{R}^4$, i.e.,

$$T\begin{pmatrix} x_1 \\ x_2 \\ x_3 \\ x_4 \end{pmatrix} = \begin{pmatrix} x_1 \\ x_2 \end{pmatrix}.$$

We call this the projection to the 1-2-coordinate plane. Even though this transformation takes a vector in $\mathbb{R}^4$ and sends it to a vector in $\mathbb{R}^2$, it possesses the properties of a linear transformation since $T(\mathbf{X} + t\mathbf{U}) =$

$T(\mathbf{X}) + tT(\mathbf{U})$ for any $\mathbf{X}$, $\mathbf{U}$ in $\mathbb{R}^4$ and any real number t. In particular, the images of a line is another line if $T(\mathbf{U}) \neq \mathbf{0}$ and the image is a point if $T(\mathbf{U}) = \mathbf{0}$. This fact makes it easy to draw 2-dimensional pictures of objects in 4-space which are composed of segments—we simply find the images of the vertices of the object and connect the image points by a segment in $\mathbb{R}^2$ if the original vertices are connected by a segment in $\mathbb{R}^4$.

EXAMPLE 2. In $\mathbb{R}^4$, the points $\mathbf{U} = \begin{pmatrix}1\\1\\0\\1\end{pmatrix}$, $\mathbf{V} = \begin{pmatrix}1\\0\\1\\1\end{pmatrix}$, and $\mathbf{W} = \begin{pmatrix}0\\1\\1\\1\end{pmatrix}$ determine an equilateral triangle. The image points are $T(\mathbf{U}) = \begin{pmatrix}1\\1\end{pmatrix}$, $T(\mathbf{V}) = \begin{pmatrix}1\\0\end{pmatrix}$, $T(\mathbf{W}) = \begin{pmatrix}0\\1\end{pmatrix}$ (see Fig. 4.3).

Note that the image itself is not equilateral.

EXAMPLE 3. The tetrahedron in $\mathbb{R}^4$ determined by the vertices $\mathbf{U}_1 = \begin{pmatrix}1\\1\\0\\1\end{pmatrix}$, $\mathbf{U}_2 = \begin{pmatrix}1\\0\\1\\1\end{pmatrix}$, $\mathbf{U}_3 = \begin{pmatrix}0\\1\\1\\1\end{pmatrix}$, $\mathbf{U}_4 = \begin{pmatrix}1\\1\\1\\0\end{pmatrix}$ has the *same image* as in Example 2 since $T(\mathbf{U}_1) = T(\mathbf{U}_4) = \begin{pmatrix}1\\1\end{pmatrix}$.

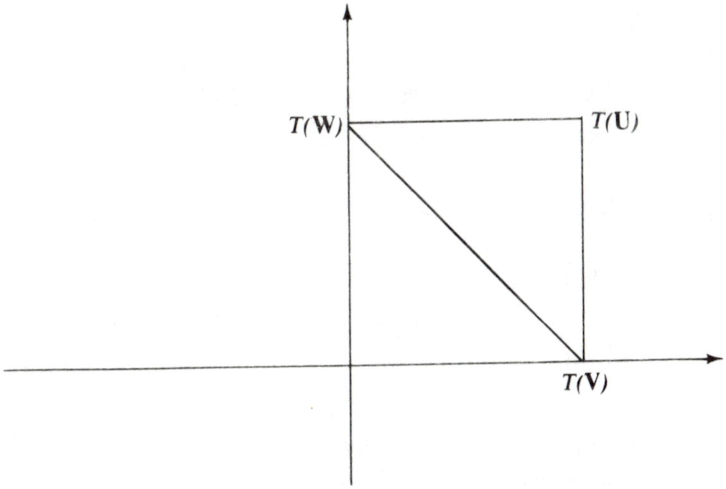

Figure 4.3

EXAMPLE 4. The tetrahedron in $\mathbb{R}^4$ determined by the vertices $\mathbf{V}_1 = \begin{bmatrix} 2 \\ 2 \\ 1 \\ -1 \end{bmatrix}$, $\mathbf{V}_2 = \begin{bmatrix} 1 \\ -1 \\ 0 \\ 3 \end{bmatrix}$, $\mathbf{V}_3 = \begin{bmatrix} -1 \\ 1 \\ 6 \\ 4 \end{bmatrix}$, $\mathbf{V}_4 = \begin{bmatrix} -\frac{1}{2} \\ -\frac{1}{2} \\ 7 \\ 7 \end{bmatrix}$ has the image given by Figure 4.4.

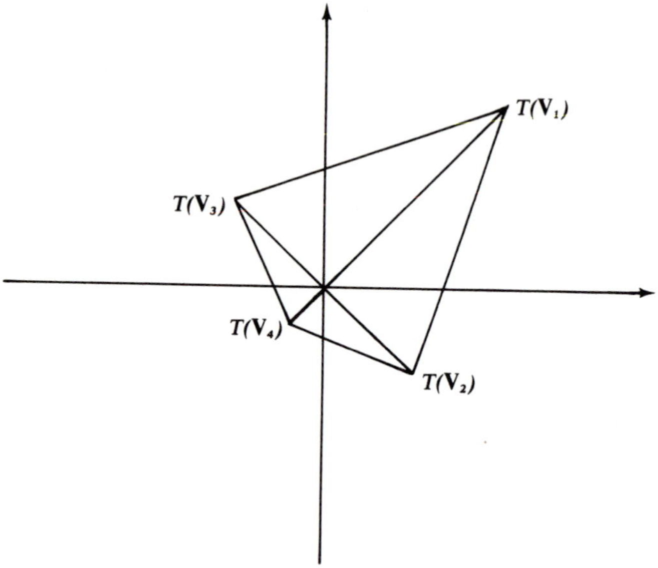

Figure 4.4

EXAMPLE 5. Consider the 4-cube centered at the origin with vertices given by the vectors with all coordinates either 1 or -1. The projection T of this 4-cube to the plane has only four distinct vertices $\begin{pmatrix} 1 \\ 1 \end{pmatrix}$, $\begin{pmatrix} 1 \\ -1 \end{pmatrix}$, $\begin{pmatrix} -1 \\ 1 \end{pmatrix}$, $\begin{pmatrix} -1 \\ -1 \end{pmatrix}$, even though the 4-cube has 16 vertices. For example, the four vertices $\begin{bmatrix} 1 \\ 1 \\ 1 \\ 1 \end{bmatrix}$, $\begin{bmatrix} 1 \\ 1 \\ 1 \\ -1 \end{bmatrix}$, $\begin{bmatrix} 1 \\ 1 \\ -1 \\ -1 \end{bmatrix}$, $\begin{bmatrix} 1 \\ 1 \\ -1 \\ 1 \end{bmatrix}$ are all sent to $\begin{pmatrix} 1 \\ 1 \end{pmatrix}$ under T.

In order to get more useful pictures of an object like the 4-cube, we first rotate the object before projecting to the 1-2-coordinate plane. For example, if we rotate the 4-cube θ degrees in the 1-3-plane, we get

$$R_\theta^{13}\begin{bmatrix}x_1\\x_2\\x_3\\x_4\end{bmatrix}=\begin{bmatrix}\cos\theta x_1-\sin\theta x_3\\x_2\\\sin\theta x_1+\cos\theta x_3\\x_4\end{bmatrix},$$

so

$$TR_\theta^{13}\begin{bmatrix}x_1\\x_2\\x_3\\x_4\end{bmatrix}=\begin{pmatrix}\cos\theta x_1-\sin\theta x_3\\x_2\end{pmatrix}.$$

If $\theta=30°$, we then have $TR_{30}^{13}\begin{bmatrix}1\\1\\1\\1\end{bmatrix}=\begin{pmatrix}\sqrt{3}/2-1/2\\1\end{pmatrix}$.

The picture opens up a certain amount, but we still see only eight distinct vertex images.

If we first rotate in the 1-3-plane by θ degrees and then in the 2-4-plane by ϕ degrees we, get

$$R_\phi^{24}R_\theta^{13}\begin{bmatrix}x_1\\x_2\\x_3\\x_4\end{bmatrix}=\begin{bmatrix}\cos\theta x_1-\sin\theta x_3\\\cos\phi x_2-\sin\phi x_4\\\sin\theta x_1+\cos\theta x_3\\\sin\phi x_2+\cos\phi x_4\end{bmatrix},$$

so

$$TR_\phi^{24}R_\theta^{13}\begin{bmatrix}x_1\\x_2\\x_3\\x_4\end{bmatrix}=\begin{pmatrix}\cos\theta x_1-\sin\theta x_3\\\cos\phi x_2-\sin\phi x_4\end{pmatrix}.$$

Thus

$$TR_{45}^{24}R_{30}^{13}\begin{bmatrix}1\\-1\\-1\\-1\end{bmatrix}=\begin{bmatrix}\sqrt{3}/2+1/2\\-\sqrt{2}/2+\sqrt{2}/2\end{bmatrix},$$

while

$$TR_{30}^{24}R_{30}^{13}\begin{bmatrix}1\\-1\\-1\\-1\end{bmatrix}=\begin{bmatrix}\sqrt{3}/2+1/2\\-\sqrt{3}/2+1/2\end{bmatrix}$$

We get 16 different images for the 16 vertices of the 4-cube, but again it is difficult to interpret the image of the whole 4-cube.

If instead we first rotate in the 1-3-plane, then the 2-4-plane, then the 1-4-plane, we get a general position.

$$TR_{\alpha}^{14}R_{\phi}^{24}R_{\theta}^{13}\begin{pmatrix}x_1\\x_2\\x_3\\x_4\end{pmatrix}=\begin{pmatrix}\cos\alpha(\cos\theta x_1-\sin\theta x_3)-\sin\alpha(\sin\phi x_2+\cos\phi x_4)\\ \cos\phi x_2-\sin\phi x_4\end{pmatrix}.$$

Then

$$TR_{30}^{14}R_{30}^{24}R_{30}^{13}\begin{pmatrix}x_1\\x_2\\x_3\\x_4\end{pmatrix}=\begin{pmatrix}\frac{\sqrt{3}}{2}\left(\frac{\sqrt{3}}{2}x_1-\frac{1}{2}x_3\right)-\frac{1}{2}\left(\frac{1}{2}x_2+\frac{\sqrt{3}}{2}x_4\right)\\ \frac{\sqrt{3}}{2}x_2-\frac{1}{2}x_4\end{pmatrix}$$

$$=\begin{pmatrix}\frac{3}{4}x_1-\frac{1}{4}x_2-\frac{\sqrt{3}}{4}x_3-\frac{\sqrt{3}}{4}x_4\\ \frac{\sqrt{3}}{2}x_2-\frac{1}{2}x_4\end{pmatrix}.$$

Finally, if we rotate by β degrees in the 2-3-plane, we have

$$TR_{\beta}^{23}R_{\alpha}^{14}R_{\phi}^{24}R_{\theta}^{13}\begin{pmatrix}x_1\\x_2\\x_3\\x_4\end{pmatrix}$$

$$=\begin{pmatrix}\cos\alpha\cos\theta x_1-\sin\alpha\sin\phi x_2-\cos\alpha\sin\theta x_3-\sin\alpha\cos\phi x_4\\ -\sin\beta\sin\phi x_1+\cos\beta\cos\phi x_2-\sin\beta\cos\theta x_3-\cos\beta\sin\phi x_4\end{pmatrix}.$$

In particular,

$$TR_{30}^{23}R_{30}^{14}R_{30}^{24}R_{30}^{13}\begin{pmatrix}x_1\\x_2\\x_3\\x_4\end{pmatrix}=\begin{pmatrix}\frac{3}{4}x_1-\frac{1}{4}x_2-\frac{\sqrt{3}}{4}x_3-\frac{\sqrt{3}}{4}x_4\\ -\frac{3}{4}x_1+\frac{1}{4}x_2-\frac{\sqrt{3}}{4}x_3-\frac{\sqrt{3}}{4}x_4\end{pmatrix}.$$

Now we have a picture in "general position" where no two images of coordinate axes are linearly dependent.

These are precisely the sorts of instructions which are used in producing computer graphics images (see Fig. 4.5), for example, in the film *The*

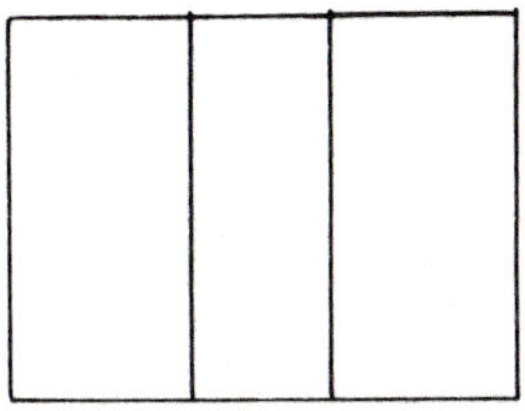

$\theta = 30, \phi = 0, \alpha = 0, \beta = 0$

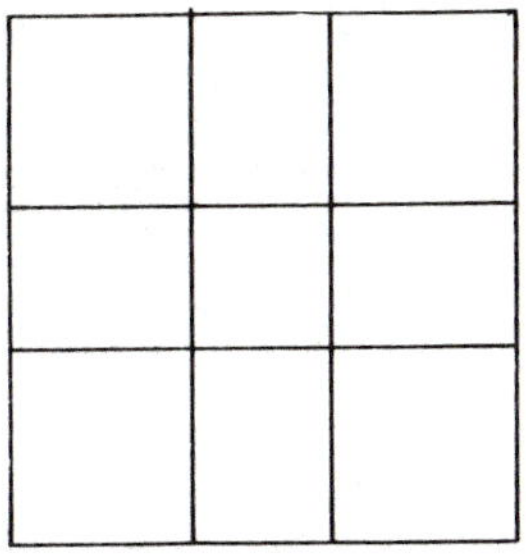

$\theta = 30, \phi = 30, \alpha = 0, \beta = 0$

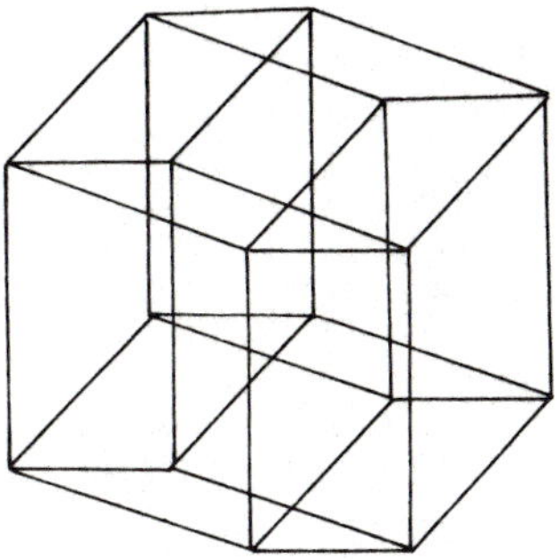

$\theta = 30, \phi = 30, \alpha = 30, \beta = 0$

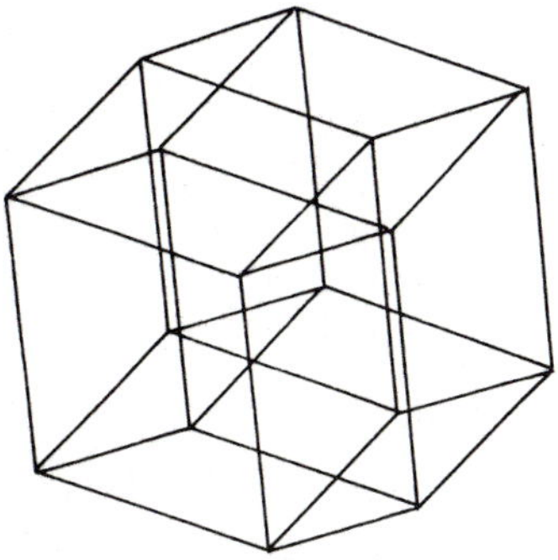

$\theta = 30, \phi = 30, \alpha = 30, \beta = 10$

Figure 4.5

Hypercube: *Projections and Slicing* by Thomas Banchoff and Charles Strauss. We include several different pictures of that object corresponding to other values of θ, ϕ, α, and β.

CHAPTER 4.2

Linear Transformations and Matrices

In Chapter 4.1 we examined a number of transformations T of 4-space all of which have the property that the coordinates of $T(\mathbf{X})$ are given as *linear* functions of the coordinates of $\mathbf{X}$. In each case we have formulas of the sort

$$T\begin{pmatrix} x_1 \\ x_2 \\ x_3 \\ x_4 \end{pmatrix} = \begin{pmatrix} a_{11}x_1 + a_{12}x_2 + a_{13}x_3 + a_{14}x_4 \\ a_{21}x_1 + a_{22}x_2 + a_{23}x_3 + a_{24}x_4 \\ a_{31}x_1 + a_{32}x_2 + a_{33}x_3 + a_{34}x_4 \\ a_{41}x_1 + a_{42}x_2 + a_{43}x_3 + a_{44}x_4 \end{pmatrix}.$$

Any transformation which can be written in this form is called a *linear transformation of* 4-*space*.

The symbol $\begin{pmatrix} a_{11} & a_{12} & a_{13} & a_{14} \\ a_{21} & a_{22} & a_{23} & a_{24} \\ a_{31} & a_{32} & a_{33} & a_{34} \\ a_{41} & a_{42} & a_{43} & a_{44} \end{pmatrix}$ is called the *matrix* of the transformation T and is denoted $m(T)$. We abbreviate $m(T)$ by $((a_{ij}))$, where a_{ij} stands for the entry in the ith *row* and the jth *column*.

We can now list the matrices of the linear transformations in the examples of Chapter 4.1 (with $\mathbf{U} = \begin{pmatrix} 1 \\ 1 \\ 1 \\ 1 \end{pmatrix}$):

$$m(P) = \begin{pmatrix} \frac{1}{4} & \frac{1}{4} & \frac{1}{4} & \frac{1}{4} \\ \frac{1}{4} & \frac{1}{4} & \frac{1}{4} & \frac{1}{4} \\ \frac{1}{4} & \frac{1}{4} & \frac{1}{4} & \frac{1}{4} \\ \frac{1}{4} & \frac{1}{4} & \frac{1}{4} & \frac{1}{4} \end{pmatrix}, \tag{1}$$

$$m(S) = \begin{bmatrix} -\frac{1}{2} & \frac{1}{2} & \frac{1}{2} & \frac{1}{2} \\ \frac{1}{2} & -\frac{1}{2} & \frac{1}{2} & \frac{1}{2} \\ \frac{1}{2} & \frac{1}{2} & -\frac{1}{2} & \frac{1}{2} \\ \frac{1}{2} & \frac{1}{2} & \frac{1}{2} & \frac{1}{2} \end{bmatrix}, \tag{2}$$

$$m(D_t) = \begin{bmatrix} t & 0 & 0 & 0 \\ 0 & t & 0 & 0 \\ 0 & 0 & t & 0 \\ 0 & 0 & 0 & t \end{bmatrix}, \tag{3}$$

$$m(Q) = \begin{bmatrix} \frac{3}{4} & -\frac{1}{4} & -\frac{1}{4} & -\frac{1}{4} \\ -\frac{1}{4} & \frac{3}{4} & -\frac{1}{4} & -\frac{1}{4} \\ -\frac{1}{4} & -\frac{1}{4} & \frac{3}{4} & -\frac{1}{4} \\ -\frac{1}{4} & -\frac{1}{4} & -\frac{1}{4} & \frac{3}{4} \end{bmatrix}, \tag{4}$$

$$m(R_\theta^{12}) = \begin{bmatrix} \cos\theta & -\sin\theta & 0 & 0 \\ \sin\theta & \cos\theta & 0 & 0 \\ 0 & 0 & 1 & 0 \\ 0 & 0 & 0 & 1 \end{bmatrix}. \tag{5}$$

As in dimensions 2 and 3, if T is the linear transformation with matrix $m(T) = ((a_{ij}))$, we then write

$$((a_{ij}))(X) = ((a_{ij}))\begin{bmatrix} x_1 \\ x_2 \\ x_3 \\ x_4 \end{bmatrix} = \begin{bmatrix} a_{11}x_1 + a_{12}x_2 + a_{13}x_3 + a_{14}x_4 \\ a_{21}x_1 + a_{22}x_2 + a_{23}x_3 + a_{24}x_4 \\ a_{31}x_1 + a_{32}x_2 + a_{33}x_3 + a_{34}x_4 \\ a_{41}x_1 + a_{42}x_2 + a_{43}x_3 + a_{44}x_4 \end{bmatrix} = \begin{bmatrix} y_1 \\ y_2 \\ y_3 \\ y_4 \end{bmatrix}$$

and we say that *the matrix* $((a_{ij}))$ *acts on the vector* $\mathbf{X}$ to yield $\mathbf{Y} = \begin{bmatrix} y_1 \\ y_2 \\ y_3 \\ y_4 \end{bmatrix}$. We may then write the equations for the coordinates of $\mathbf{Y}$ as

$$y_i = a_{i1}x_1 + a_{i2}x_2 + a_{i3}x_3 + a_{i4}x_4, \qquad \text{for} \quad i = 1, 2, 3, 4.$$

As in dimensions 2 and 3, we now prove two crucial properties of linear transformations, which show how a matrix acts on sums and scalar products. If $\mathbf{X} = \begin{bmatrix} x_1 \\ x_2 \\ x_3 \\ x_4 \end{bmatrix}$ and $\mathbf{U} = \begin{bmatrix} u_1 \\ u_2 \\ u_3 \\ u_4 \end{bmatrix}$, and if $\mathbf{Y} = ((a_{ij}))(\mathbf{X} + \mathbf{U})$, then

$$\begin{aligned} y_i &= a_{i1}(x_1 + u_1) + a_{i2}(x_2 + u_2) + a_{i3}(x_3 + u_3) + a_{i4}(x_4 + u_4) \\ &= (a_{i1}x_1 + a_{i2}x_2 + a_{i3}x_3 + a_{i4}x_4) + (a_{i1}u_1 + a_{i2}u_2 + a_{i3}u_3 + a_{i4}u_4). \end{aligned}$$

Therefore, $((a_{ij}))(\mathbf{X} + \mathbf{U}) = ((a_{ij}))\mathbf{X} + ((a_{ij}))\mathbf{U}$. It follows that $T(\mathbf{X} + \mathbf{U}) = T(\mathbf{X}) + T(\mathbf{U})$.

Similarly, we may show that $T(r\mathbf{X}) = rT(\mathbf{X})$ for any scalar r.

Conversely, if T is a transformation such that $T(\mathbf{X}+\mathbf{U}) = T(\mathbf{X}) + T(\mathbf{U})$ and $T(r\mathbf{X}) = rT(\mathbf{X})$ for all vectors $\mathbf{X}$, $\mathbf{U}$ and scalars r, then

$$T\begin{bmatrix} x_1 \\ x_2 \\ x_3 \\ x_4 \end{bmatrix} = T\left[\begin{bmatrix} x_1 \\ 0 \\ 0 \\ 0 \end{bmatrix} + \begin{bmatrix} 0 \\ x_2 \\ 0 \\ 0 \end{bmatrix} + \begin{bmatrix} 0 \\ 0 \\ x_3 \\ 0 \end{bmatrix} + \begin{bmatrix} 0 \\ 0 \\ 0 \\ x_4 \end{bmatrix}\right]$$
$$= T(x_1\mathbf{E}_1 + x_2\mathbf{E}_2 + x_3\mathbf{E}_3 + x_4\mathbf{E}_4)$$
$$= x_1T(\mathbf{E}_1) + x_2T(\mathbf{E}_2) + x_3T(\mathbf{E}_3) + x_4T(\mathbf{E}_4).$$

We define a_{ij} by setting $T(\mathbf{E}_j) = \begin{bmatrix} a_{1j} \\ a_{2j} \\ a_{3j} \\ a_{4j} \end{bmatrix}$. Then $T\begin{bmatrix} x_1 \\ x_2 \\ x_3 \\ x_4 \end{bmatrix} = ((a_{ij}))\begin{bmatrix} x_1 \\ x_2 \\ x_3 \\ x_4 \end{bmatrix}$. Hence, T is the linear transformation with matrix $((a_{ij}))$.

In summary, we have:

Theorem 4.1. *A transformation T of 4-space is a linear transformation if and only if $T(\mathbf{X}+\mathbf{U}) = T(\mathbf{X}) + T(\mathbf{U})$ and $T(r\mathbf{X}) = rT(\mathbf{X})$ for any vectors $\mathbf{X}$ and $\mathbf{U}$ and scalars r.*

At this point we conclude our treatment of the linear algebra of 4 dimensions. In much the same way as in dimensions 2 and 3, we may define the notions of products of transformations and of their corresponding matrices, of inverses, determinants, and eigenvalues. These procedures lead to systems of equations in four and more variables which we will take up in the next chapter. We do mention two facts which are important differences between dimension 4 and dimension 3 to help the student in pursuing the subject of linear algebra beyond the material in this book.

First of all, although every linear transformation in $\mathbb{R}^3$ had at least one eigenvalue, this property no longer holds in $\mathbb{R}^4$ (as indeed it did not in $\mathbb{R}^2$). For example, if we consider the double rotation $R_\theta^{34}R_\phi^{12}$, we have

$$R_\theta^{34}R_\phi^{12}\begin{bmatrix} x_1 \\ x_2 \\ x_3 \\ x_4 \end{bmatrix} = \begin{bmatrix} \cos\theta x_1 - \sin\theta x_2 \\ \sin\theta x_1 + \cos\theta x_2 \\ \cos\phi x_3 - \sin\phi x_4 \\ \sin\phi x_3 + \cos\phi x_4 \end{bmatrix},$$

and if $R_\theta^{34}R_\phi^{12}(\mathbf{X}) = \lambda\mathbf{X}$ for some $\lambda \neq 0$, we have, first of all,

$$\cos\theta x_1 - \sin\theta x_2 = \lambda x_1,$$
$$\sin\theta x_1 + \cos\theta x_2 = \lambda x_2;$$

so unless x_1 and $x_2 = 0$,

$$(\cos\theta - \lambda)^2 + \sin^2\theta = 0,$$

and, therefore,

$$1 + \lambda^2 - 2\lambda\cos\theta = 0.$$

The only solutions then are $\lambda = (2\cos\theta \pm \sqrt{4\cos^2\theta - 4})/2$. But this has solutions only if $\cos^2\theta = 1$, $\theta = 0, \pi$. Similarly, the last two equations express the condition that

$$\cos\phi x_3 - \sin\phi x_4 = \lambda x_3,$$
$$\sin\phi x_3 + \cos\phi x_4 = \lambda x_4,$$

which can only occur if $\phi = 0$ or π or if both x_3 and $x_4 = 0$. Therefore, in the case where neither θ nor ϕ is 0 or π, the transformation $R_\theta^{34}R_\phi^{12}$ will have no (real) eigenvalues or eigenvectors.

Also, the definition of the determinant of a matrix in $\mathbb{R}^4$ requires more care, although it is analogous to the definition in $\mathbb{R}^2$ or $\mathbb{R}^3$. We recall that we can express the 3×3 determinant in terms of 2×2 determinants as follows.

$$\begin{vmatrix} a_1 & b_1 & c_1 \\ a_2 & b_2 & c_2 \\ a_3 & b_3 & c_3 \end{vmatrix} = a_1\begin{vmatrix} b_2 & c_2 \\ b_3 & c_3 \end{vmatrix} - a_2\begin{vmatrix} b_1 & c_1 \\ b_3 & c_3 \end{vmatrix} + a_3\begin{vmatrix} b_1 & c_1 \\ b_2 & c_2 \end{vmatrix}.$$

In $\mathbb{R}^4$, we define a 4×4 determinant in terms of 3×3 determinants:

$$\begin{vmatrix} a_1 & b_1 & c_1 & d_1 \\ a_2 & b_2 & c_2 & d_2 \\ a_3 & b_3 & c_3 & d_3 \\ a_4 & b_4 & c_4 & d_4 \end{vmatrix} = a_1\begin{vmatrix} b_2c_2d_2 \\ b_3c_3d_3 \\ b_4c_4d_4 \end{vmatrix} - a_2\begin{vmatrix} b_1c_1d_1 \\ b_3c_3d_3 \\ b_4c_4d_4 \end{vmatrix} + a_3\begin{vmatrix} b_1c_1d_1 \\ b_2c_2d_2 \\ b_4c_4d_4 \end{vmatrix} - a_4\begin{vmatrix} b_1c_1d_1 \\ b_2c_2d_2 \\ b_3c_3d_3 \end{vmatrix}.$$

For example,

$$\begin{vmatrix} \cos\theta - \lambda & -\sin\theta & 0 & 0 \\ \sin\theta & \cos\theta - \lambda & 0 & 0 \\ 0 & 0 & \cos\phi - \lambda & -\sin\phi \\ 0 & 0 & \sin\phi & \cos\phi - \lambda \end{vmatrix}$$

$$= (\cos\theta - \lambda)\begin{vmatrix} \cos\theta - \lambda & 0 & 0 \\ 0 & \cos\phi - \lambda & -\sin\phi \\ 0 & \sin\phi & \cos\phi - \lambda \end{vmatrix}$$

$$- \sin\theta\begin{vmatrix} -\sin\theta & 0 & 0 \\ 0 & \cos\phi - \lambda & -\sin\phi \\ 0 & \sin\phi & \cos\phi - \lambda \end{vmatrix}$$

$$= \left((\cos\theta - \lambda)^2 + \sin^2\theta\right)\left((\cos\phi - \lambda)^2 + \sin^2\phi\right)$$

$$= (\lambda^2 - 2\lambda\cos\theta + 1)(\lambda^2 - 2\lambda\cos\phi + 1).$$

The only cases in which this determinant is zero occur when either θ or ϕ is 0 or π.

CHAPTER 5.1

Homogeneous Systems of Equations

In Chapters 2.4 and 3.4, we studied systems of equations in n unknowns for $n = 2$ and 3. What if $n > 3$?

EXAMPLE 1. Let us look at the system

$$\begin{cases} x_1 + 2x_2 + 3x_3 - x_4 = 0, \\ \qquad\quad x_2 + x_3 + x_4 = 0, \end{cases} \tag{1}$$

in four unknowns. By a solution of (1) we mean a vector $\mathbf{X}$ in $\mathbb{R}^4$,

$$\mathbf{X} = \begin{bmatrix} x_1 \\ x_2 \\ x_3 \\ x_4 \end{bmatrix},$$

which satisfies the two equations in system (1). Thus

$$\begin{bmatrix} -1 \\ -1 \\ 1 \\ 0 \end{bmatrix} \quad \text{and} \quad \begin{bmatrix} 0 \\ 0 \\ 0 \\ 0 \end{bmatrix}$$

are two solutions of (1).

Let us find all solutions of (1). Assume

$$\mathbf{X} = \begin{bmatrix} x_1 \\ x_2 \\ x_3 \\ x_4 \end{bmatrix}$$

is a solution of (1). Subtracting twice the bottom equation from the top, we get

$$(x_1 + 2x_2 + 3x_3 - x_4) - 2(x_2 + x_3 + x_4) = 0 - 2(0)$$

or

$$x_1 + x_3 - 3x_4 = 0.$$

So we have

$$\begin{cases} x_1 + x_3 - 3x_4 = 0, \\ x_2 + x_3 + x_4 = 0, \end{cases} \tag{2}$$

which we rewrite in the form:

$$\begin{cases} x_1 = -x_3 + 3x_4, \\ x_2 = -x_3 - x_4. \end{cases} \tag{3}$$

We just saw that every solution of the system (1) satisfies the system (3). Conversely, retracing our steps, we see that if **X** is a solution of (3), then **X** is also a solution of (1). But now (3) can be solved directly. We give arbitrary values to x_3 and x_4 and then use (3) to calculate x_1 and x_2. For instance, set $x_3 = -10$, $x_4 = 3$. Then, by (3),

$$x_1 = 10 + 3 \cdot 3 = 19,$$

and

$$x_2 = 10 - 3 = 7.$$

Using these values for x_i, $i = 1, 2, 3, 4$, we get

$$\mathbf{X} = \begin{pmatrix} 19 \\ 7 \\ -10 \\ 3 \end{pmatrix}.$$

One can directly check that **X** is a solution of (1).

More generally, fix two numbers u, v. Set $x_3 = u$, $x_4 = v$ and define x_1 and x_2 by (3). Then

$$x_1 = -u + 3v,$$

$$x_2 = -u - v.$$

Set

$$\mathbf{X} = \begin{pmatrix} x_1 \\ x_2 \\ x_3 \\ x_4 \end{pmatrix} = \begin{pmatrix} -u + 3v \\ -u - v \\ u \\ v \end{pmatrix} = u \begin{pmatrix} -1 \\ -1 \\ 1 \\ 0 \end{pmatrix} + v \begin{pmatrix} 3 \\ -1 \\ 0 \\ 1 \end{pmatrix}. \tag{4}$$

Letting u, v take on all possible scalar values, formula (4) then delivers all solutions of (1).

Next, we shall study an arbitrary system of k equations in n unknowns having the following form:

$$\begin{cases} a_{11}x_1 + a_{12}x_2 + \cdots + a_{1n}x_n = 0, \\ a_{21}x_1 + a_{22}x_2 + \cdots + a_{2n}x_n = 0, \\ \qquad\vdots \\ a_{k1}x_1 + a_{k2}x_2 + \cdots + a_{kn}x_n = 0. \end{cases} \tag{H}$$

(H) is called a *homogeneous system of linear equations*.

Here a_{ij}, $1 \leqslant i \leqslant k$, $1 \leqslant j \leqslant n$, are certain given scalars called the *coefficients* of the system (H) and $x_1, \ldots, x_n$ are the unknowns. A *solution* of (H) is an n-tuple of numbers

$$\mathbf{X} = \begin{pmatrix} x_1 \\ x_2 \\ \vdots \\ x_n \end{pmatrix}$$

such that each of the k equations in (H) is satisfied by these n numbers.

By analogy with the discussion in Chapter 4.0, §2, we define an n-tuple $\mathbf{X}$ as a *vector in n-space*, and we denote the totality of all such vectors by $\mathbb{R}^n$. *Addition* of vectors in $\mathbb{R}^n$ and *multiplication of a vector by a scalar* is defined by analogy with the definitions given for the case $n = 4$, and the same basic algebraic properties hold which we noted in that case. Furthermore, the *dot product* of two vectors in $\mathbb{R}^n$ is defined as in Chapter 4.0, §3, by

$$\mathbf{X} \cdot \mathbf{U} = x_1u_1 + x_2u_2 + \cdots + x_nu_n,$$

where

$$\mathbf{X} = \begin{pmatrix} x_1 \\ x_2 \\ \vdots \\ x_n \end{pmatrix}, \qquad \mathbf{U} = \begin{pmatrix} u_1 \\ u_2 \\ \vdots \\ u_n \end{pmatrix}.$$

Using dot-product notation, the system (H) can be written concisely as

$$\begin{cases} \mathbf{A}_1 \cdot \mathbf{X} = 0, \\ \mathbf{A}_2 \cdot \mathbf{X} = 0, \\ \qquad\vdots \\ \mathbf{A}_k \cdot \mathbf{X} = 0, \end{cases} \tag{H}$$

where $\mathbf{X}$ is the unknown vector $\begin{pmatrix} x_1 \\ \vdots \\ x_n \end{pmatrix}$ and

$$\mathbf{A}_1 = \begin{pmatrix} a_{11} \\ a_{12} \\ \vdots \\ a_{1n} \end{pmatrix}, \quad \mathbf{A}_2 = \begin{pmatrix} a_{21} \\ a_{22} \\ \vdots \\ a_{2n} \end{pmatrix}, \quad \text{etc.}$$

Let us now give a procedure for finding all the solutions $\mathbf{X}$ of a given system (H). Consider a second system

$$\begin{cases} a'_{11}x_1 + a'_{12}x_2 + \cdots + a'_{1n}x_n = 0, \\ \qquad\vdots \\ a'_{k1}x_1 + a'_{k2}x_2 + \cdots + a'_{kn}x_n = 0. \end{cases} \tag{H$'$}$$

We say that the systems (H) and (H′) are *equivalent* if every solution of (H) is a solution of (H′), and conversely.

To solve (H) it will be enough if we can find a system (H′) which is equivalent to (H) and which is easy to solve. Note that we did just that in Example 1 when we found the system (2) which was equivalent to (1).

In the next example, $k = 3$ and $n = 4$.

EXAMPLE 2. To solve

$$\begin{cases} x_1 - 2x_2 + 3x_3 \qquad = 0, \\ x_1 + \ x_2 \qquad + x_4 = 0, \\ \qquad\qquad 4x_3 - x_4 = 0, \end{cases} \tag{5}$$

subtract the top line from the middle line and leave the other lines alone. We get the new system

$$\begin{cases} x_1 - 2x_2 + 3x_3 \qquad = 0, \\ \quad 3x_2 - 3x_3 + x_4 = 0, \\ \qquad\qquad 4x_3 - x_4 = 0. \end{cases} \tag{5a}$$

(5a) is equivalent to (5).

Next, add $\frac{2}{3}$ of the middle line in (5a) to the top line. We get

$$\begin{cases} x_1 \qquad + \ x_3 + \tfrac{2}{3}x_4 = 0, \\ \quad 3x_2 - 3x_3 + \ x_4 = 0, \\ \qquad\qquad 4x_3 - \ x_4 = 0. \end{cases} \tag{5b}$$

(5b) is equivalent to (5a), and so it follows that (5b) is equivalent to (5).

Next we add $-\frac{1}{4}$ times the bottom line of (5b) to the top line, getting

$$\begin{cases} x_1 \qquad\qquad\qquad + (\frac{2}{3} + \frac{1}{4})x_4 = 0, \\ \qquad 3x_2 - 3x_3 + \qquad x_4 = 0, \\ \qquad\qquad 4x_3 - \qquad x_4 = 0. \end{cases} \tag{5c}$$

Finally, add $\frac{3}{4}$ of the bottom line of (5c) to the middle line, getting

$$\begin{cases} x_1 \qquad\qquad + \frac{11}{12}x_4 = 0, \\ \qquad 3x_2 \qquad + \frac{1}{4}x_4 = 0, \\ \qquad\qquad 4x_3 - \quad x_4 = 0. \end{cases} \tag{5d}$$

As before, (5c) and (5d) are equivalent to (5). But (5d) can be solved at once. Give an arbitrary value to x_4 and then use (5d) to calculate x_1, x_2, x_3. We find

$$x_1 = -\tfrac{11}{12}x_4,$$
$$x_2 = (-\tfrac{1}{12})x_4,$$
$$x_3 = \tfrac{1}{4}x_4.$$

Hence, we get, as a solution of (5d):

$$\mathbf{X} = \begin{pmatrix} x_1 \\ x_2 \\ x_3 \\ x_4 \end{pmatrix} = \begin{pmatrix} -\frac{11}{12}x_4 \\ -\frac{1}{12}x_4 \\ \frac{1}{4}x_4 \\ x_4 \end{pmatrix} = x_4 \begin{pmatrix} -\frac{11}{12} \\ -\frac{1}{12} \\ \frac{1}{4} \\ 1 \end{pmatrix}. \tag{6}$$

For different choices of x_4, (6) gives all solutions of (5d) and, therefore, all solutions of (5).

The method just used in Example 2 can be applied to any system of the form (H). By a succession of steps in which a scalar times one line of the system is added to some other line, while the remaining lines are left unchanged, and, possibly, by relabeling the unknowns x_i, we finally obtain a system (H′) of the following form:

$$\begin{cases} x_1 \qquad\qquad + b_{11}x_{l+1} + b_{12}x_{l+2} + \cdots + b_{1,n-l}x_n = 0, \\ \quad x_2 \qquad + b_{21}x_{l+1} + b_{22}x_{l+2} + \cdots + b_{2,n-l}x_n = 0, \\ \qquad \ddots \\ \qquad\qquad x_l + b_{l1}x_{l+1} + b_{l2}x_{l+2} + \cdots + b_{l,n-l}x_n = 0, \end{cases} \tag{H′}$$

where l is some integer, depending on the system (H), with $1 \leqslant l \leqslant n$, and b_{ij} are certain constants, such that the original system (H) and this system (H′) are equivalent. To find all solutions of (H′), and, therefore, of (H), we need only fix numbers $u_1, u_2, \ldots, u_{n-l}$, set $x_{l+1} = u_1, \ldots, x_n = u_{n-l}$, and then find $x_1, x_2, \ldots, x_l$ from (H′). We get

$$\begin{aligned} x_1 &= -b_{11}u_1 - b_{12}u_2 - \cdots - b_{1,n-l}u_{n-l}, \\ x_2 &= -b_{21}u_1 - b_{22}u_2 - \cdots - b_{2,n-l}u_{n-l}, \\ &\vdots \\ x_l &= -b_{l1}u_1 - b_{l2}u_2 - \cdots - b_{l,n-l}u_{n-l}. \end{aligned}$$

The solution $\mathbf{X}$ of (H) is then given by

$$\mathbf{X} = \begin{bmatrix} x_1 \\ x_2 \\ \vdots \\ x_l \\ x_{l+1} \\ x_{l+2} \\ \vdots \\ x_n \end{bmatrix} = \begin{bmatrix} -b_{11}u_1 - \cdots - b_{1,n-l}u_{n-l} \\ -b_{21}u_1 - \cdots - b_{2,n-l}u_{n-l} \\ \vdots \\ -b_{l1}u_1 - \cdots - b_{l,n-l}u_{n-l} \\ u_1 \\ u_2 \\ \vdots \\ u_{n-l} \end{bmatrix}.$$

In other words,

$$\mathbf{X} = u_1 \begin{bmatrix} -b_{11} \\ \vdots \\ -b_{l1} \\ 1 \\ 0 \\ \vdots \\ 0 \end{bmatrix} + u_2 \begin{bmatrix} -b_{12} \\ \vdots \\ -b_{l2} \\ 0 \\ 1 \\ \vdots \\ 0 \end{bmatrix} + \cdots + u_{n-l} \begin{bmatrix} -b_{1,n-1} \\ \vdots \\ -b_{l,n-l} \\ 0 \\ 0 \\ \vdots \\ 1 \end{bmatrix}. \tag{7}$$

Letting $u_1, \ldots, u_l$ take on all possible scalar values, (7) *gives us all solutions of* (H), *and each choice of $u_1, \ldots, u_l$ provides a solution of* (H).

EXAMPLE 3. Find a nonzero vector $\begin{bmatrix} x_1 \\ x_2 \\ x_3 \end{bmatrix}$ in $\mathbb{R}^3$ which is orthogonal to each of the vectors $\begin{bmatrix} 2 \\ 3 \\ 0 \end{bmatrix}$ and $\begin{bmatrix} 1 \\ 1 \\ 1 \end{bmatrix}$.

The condition on x_1, x_2, x_3 is

$$\begin{cases} 2x_1 + 3x_2 \phantom{{}+x_3} = 0, \\ x_1 + x_2 + x_3 = 0. \end{cases} \tag{8}$$

So we must solve the system (8), of two equations in three unknowns. Subtracting $\frac{1}{2}$ the top line from the bottom line, we get the equivalent system

$$\begin{cases} 2x_1 + 3x_2 = 0, \\ \quad -\tfrac{1}{2}x_2 + x_3 = 0. \end{cases} \tag{8a}$$

Adding 6 times the bottom line to the top one, we get the equivalent system

$$\begin{cases} 2x_1 + \phantom{-\tfrac{1}{2}x_2} + 6x_3 = 0, \\ \quad -\tfrac{1}{2}x_2 + x_3 = 0. \end{cases} \tag{8b}$$

(8b) can now be solved to give

$$x_1 = -3x_3,$$
$$x_2 = 2x_3.$$

So the solution $\mathbf{X} = \begin{pmatrix} x_1 \\ x_2 \\ x_3 \end{pmatrix}$ of (8) is given by

$$\mathbf{X} = \begin{pmatrix} -3x_3 \\ 2x_3 \\ x_3 \end{pmatrix} = x_3 \begin{pmatrix} -3 \\ 2 \\ 1 \end{pmatrix}.$$

Here x_3 is an arbitrary scalar. In particular, taking $x_3 = -1$, we get: *The vector* $\begin{pmatrix} 3 \\ -2 \\ -1 \end{pmatrix}$ *is orthogonal to* $\begin{pmatrix} 2 \\ 3 \\ 0 \end{pmatrix}$ *and to* $\begin{pmatrix} 1 \\ 1 \\ 1 \end{pmatrix}$.

Note: We could have solved this problem by using the cross product.

Exercise 1. Find all solutions of the system in four unknowns:

$$\begin{cases} x_1 + x_4 = 0, \\ x_1 - x_4 = 0, \\ x_1 + x_2 + x_3 + x_4 = 0. \end{cases} \tag{9}$$

Exercise 2. Find all solutions of the system:

$$\begin{cases} x_1 + 2x_4 = 0, \\ x_1 + x_2 + x_3 + x_4 = 0. \end{cases} \tag{10}$$

Exercise 3. Find all solutions of the system in x_1, x_2, x_3, x_4:

$$\begin{cases} x_1 + 2x_2 = 0, \\ \phantom{x_1 + {}} x_2 + x_3 + x_4 = 0, \\ x_1 + x_2 - x_3 = 0. \end{cases} \tag{11}$$

Exercise 4. Find all solutions of the system consisting of one equation in five variables:

$$2x_1 - x_2 + x_3 - 4x_4 + x_5 = 0.$$

Exercise 5. Find all vectors in $\mathbb{R}^4$ which are orthogonal in $\mathbb{R}^4$:

(a) to the vector $\begin{bmatrix} 1 \\ -1 \\ 1 \\ -1 \end{bmatrix}$;

(b) to the vectors $\begin{bmatrix} 1 \\ -1 \\ 1 \\ -1 \end{bmatrix}, \begin{bmatrix} 0 \\ 1 \\ -1 \\ 0 \end{bmatrix}$;

(c) to the vectors $\begin{bmatrix} 1 \\ -1 \\ 1 \\ -1 \end{bmatrix}, \begin{bmatrix} 0 \\ 1 \\ -1 \\ 0 \end{bmatrix}, \begin{bmatrix} 1 \\ 0 \\ 0 \\ 1 \end{bmatrix}$;

(d) to the vectors $\begin{bmatrix} 1 \\ -1 \\ 1 \\ 1 \end{bmatrix}, \begin{bmatrix} 0 \\ 1 \\ -1 \\ 0 \end{bmatrix}, \begin{bmatrix} 1 \\ 0 \\ 0 \\ 1 \end{bmatrix}$;

(e) to the vectors $\begin{bmatrix} 1 \\ -1 \\ 1 \\ 1 \end{bmatrix}, \begin{bmatrix} 0 \\ 1 \\ 1 \\ 0 \end{bmatrix}, \begin{bmatrix} 1 \\ 0 \\ 0 \\ 1 \end{bmatrix}, \begin{bmatrix} 1 \\ 0 \\ 0 \\ -1 \end{bmatrix}$.

Exercise 6. Find all vectors in $\mathbb{R}^4$ which are orthogonal to the vectors: $\begin{bmatrix} 1 \\ 2 \\ 3 \\ 4 \end{bmatrix}, \begin{bmatrix} 5 \\ 6 \\ 7 \\ 8 \end{bmatrix}$.

CHAPTER 5.2

Subspaces, Linear Dependence, Dimension

§1. Subspaces

Consider a homogeneous system of linear equations in $x_1, \ldots, x_n$ such as we have studied in Chapter 5.1. What "geometric object" in $\mathbb{R}^n$ is described by this system?

EXAMPLE 1. Take $n = 3$. The equation

$$2x_1 - 3x_2 + x_3 = 0$$

describes a plane in $\mathbb{R}^3$, passing through the origin.

Similarly, the equation

$$2x_1 + 5x_2 = 0 \qquad (n = 2)$$

describes a line through the origin in $\mathbb{R}^2$; and the pair of equations

$$\begin{cases} x_1 + x_2 - x_3 = 0, \\ x_1 + 2x_2 + 3x_3 = 0, \end{cases}$$

in x_1, x_2, x_3, describes a line through the origin in $\mathbb{R}^3$.

As analogue of a line through the origin or a plane through the origin, we have the notion of *subspace* of $\mathbb{R}^n$. A subset of S of $\mathbb{R}^n$ is called a *subspace* if for every pair of vectors $\mathbf{X}$ and $\mathbf{Y}$ in S and any scalar t, the sum $\mathbf{X} + \mathbf{Y}$ and the scalar product $t\mathbf{X}$ are in S. Equivalently, S is a subspace if for all $\mathbf{X}$, $\mathbf{Y}$ in S and for all scalars s, t, the vector $s\mathbf{X} + t\mathbf{Y}$ is in S.

EXAMPLE 2. The set consisting of the $\mathbf{0}$ vector alone is a subspace of $\mathbb{R}^n$.

Theorem 5.1. *Let* (H) *be a homogeneous system of linear equations in* $x_1, \ldots, x_n$. *Then the subset* S *of* $\mathbb{R}^n$ *which consists of all solutions of the system* (H) *is a subspace of* $\mathbb{R}^n$.

To prove this, we write (H) in the form:

$$\mathbf{A}_1 \cdot \mathbf{X} = 0, \qquad \mathbf{A}_2 \cdot \mathbf{X} = 0, \ldots, \mathbf{A}_k \cdot \mathbf{X} = 0, \tag{1}$$

where $\mathbf{A}_1, \ldots, \mathbf{A}_k$ are given vectors in $\mathbb{R}^n$ and $\mathbf{X}$ is the unknown vector in $\mathbb{R}^n$. Then the set S of solutions of (H) consists of all $\mathbf{X}$ satisfying (1). Let $\mathbf{X}$ and $\mathbf{Y}$ belong to S and let s, t be scalars. Then

$$\begin{aligned} \mathbf{A}_1 \cdot (t\mathbf{X} + s\mathbf{Y}) &= t(\mathbf{A}_1 \cdot \mathbf{X}) + s(\mathbf{A}_1 \cdot \mathbf{Y}) \\ &= t \cdot 0 + s \cdot 0 = 0. \end{aligned}$$

Similarly, $\mathbf{A}_j \cdot (t\mathbf{X} + s\mathbf{Y}) = 0$ for $j = 2, \ldots, k$. Hence, $t\mathbf{X} + s\mathbf{Y}$ satisfies (1) and therefore belongs to S. Hence, S is a subspace of $\mathbb{R}^n$.

EXAMPLE 3. Let $a_1, a_2, \ldots, a_n$ be n scalars, not all 0. The set S consisting of all $\mathbf{X} = \begin{pmatrix} x_1 \\ x_2 \\ \vdots \\ x_n \end{pmatrix}$ in $\mathbb{R}^n$ such that

$$a_1x_1 + a_2x_2 + \cdots + a_nx_n = 0$$

is a subspace of $\mathbb{R}^n$. Such a subspace, defined by a single equation, is called a *hyperplane* in $\mathbb{R}^n$. A hyperplane in $\mathbb{R}^3$ is a plane through the origin, and a hyperplane in $\mathbb{R}^2$ is a line through the origin. The set of all vectors in $\mathbb{R}^4$ of the form $\begin{pmatrix} u \\ v \\ 0 \\ w \end{pmatrix}$, where u, v, w are scalars, is a hyperplane in $\mathbb{R}^4$.

Exercise 1. Give an equation in x_1, x_2, x_3, x_4 defining this hyperplane.

Exercise 2. Let S be the set of all vectors in $\mathbb{R}^4$ of the form $\begin{pmatrix} t \\ s \\ t \\ s \end{pmatrix}$, where t, s are scalars.

(i) Show that S is a subspace of $\mathbb{R}^4$.
(ii) Give a system of equations in x_1, x_2, x_3, x_4 defining this subspace.

Exercise 3. Let $\mathbf{U}_1$, $\mathbf{U}_2$, $\mathbf{U}_3$ be three vectors in $\mathbb{R}^n$. Denote by S the collection of all vectors of the form: $s_1\mathbf{U}_1 + s_2\mathbf{U}_2 + s_3\mathbf{U}_3$, where s_1, s_2, s_3 are scalars. Show that S is a subspace of $\mathbb{R}^n$.

§2. Linear Dependence

In $\mathbb{R}^2$ and $\mathbb{R}^3$, we had the notion of linear dependence of a set of vectors. For n arbitrary and $\mathbf{Y}_1, \mathbf{Y}_2, \ldots, \mathbf{Y}_l$ a set of l vectors in $\mathbb{R}^n$, we say that the set $\mathbf{Y}_1, \ldots, \mathbf{Y}_l$ is *linearly dependent* if one of the $\mathbf{Y}_i$ is a linear combination of the rest. Equivalently, we can say that the set $\mathbf{Y}_1, \ldots, \mathbf{Y}_l$ is linearly dependent if there exist scalars $s_1, \ldots, s_l$, not all 0, such that

$$s_1\mathbf{Y}_1 + s_2\mathbf{Y}_2 + \cdots + s_l\mathbf{Y}_l = \mathbf{0}. \tag{2}$$

If (2) holds and one of the s_i, say s_2, is not 0, we can solve for $\mathbf{Y}_2$, getting

$$\mathbf{Y}_2 = -\frac{s_1}{s_2}\mathbf{Y}_1 - \frac{s_3}{s_2}\mathbf{Y}_3 - \cdots - \frac{s_l}{s_2}\mathbf{Y}_l,$$

and so $\mathbf{Y}_2$ is a linear combination of the remaining $\mathbf{Y}_i$, and thus the set $\mathbf{Y}_1, \mathbf{Y}_2, \ldots, \mathbf{Y}_l$ is linearly dependent.

Conversely, suppose the set $\mathbf{Y}_1, \ldots, \mathbf{Y}_l$ is linearly dependent. Then for some i,

$$\mathbf{Y}_i = c_1\mathbf{Y}_1 + \cdots + c_{i-1}\mathbf{Y}_{i-1} + c_{i+1}\mathbf{Y}_{i+1} + \cdots + c_l\mathbf{Y}_l,$$

and so we have

$$c_1\mathbf{Y}_1 + \cdots + c_{i-1}\mathbf{Y}_{i-1} + (-1)\mathbf{Y}_i + c_{i+1}\mathbf{Y}_{i+1} + \cdots + c_l\mathbf{Y}_l = \mathbf{0}.$$

Thus (2) can be solved by the set of scalars $c_1, \ldots, c_{i-1}, -1, c_{i+1}, \ldots, c_l$, which are not all 0. Thus, deciding about linear dependence amounts to seeing whether (2) can be satisfied by scalars $s_1, \ldots, s_l$ which are not all 0.

The problem of deciding whether a given set of vectors is linearly dependent or not leads to a homogeneous system of equations of the type we studied in Chapter 5.1.

Example 4. Is the set of vectors in $\mathbb{R}^4$,

$$\mathbf{A}_1 = \begin{pmatrix} 2 \\ 1 \\ 3 \\ 1 \end{pmatrix}, \quad \mathbf{A}_2 = \begin{pmatrix} 1 \\ 0 \\ 2 \\ 1 \end{pmatrix}, \quad \mathbf{A}_3 = \begin{pmatrix} 3 \\ 1 \\ 5 \\ 2 \end{pmatrix},$$

linearly dependent?

We look for scalars s, t, u such that $s\mathbf{A}_1 + t\mathbf{A}_2 + u\mathbf{A}_3 = \mathbf{0}$. This holds if

$$\begin{cases} 2s + t + 3u = 0, \\ s + u = 0, \\ 3s + 2t + 5u = 0, \\ s + t + 2u = 0. \end{cases} \tag{3}$$

Let us solve the system (3). Subtracting multiples of the second line from the other lines, we find that (3) is equivalent to the system

$$\begin{cases} \quad t + u = 0, \\ s \quad\; + u = 0, \\ \quad 2t + 2u = 0, \\ \quad t + u = 0, \end{cases} \tag{3a}$$

which in turn is equivalent to

$$\begin{cases} t + u = 0, \\ s + u = 0, \end{cases} \tag{3b}$$

and therefore has solutions, for arbitrary u:

$$t = -u, \qquad s = -u.$$

Taking $u = 1$, we get: $t = -1$, $s = -1$, $u = 1$ as a solution of (3). Hence,

$$(-1)\mathbf{A}_1 + (-1)\mathbf{A}_2 + (+1)\mathbf{A}_3 = -\mathbf{A}_1 - \mathbf{A}_2 + \mathbf{A}_3 = \mathbf{0}$$

which we can check by direct computation. So the set $\mathbf{A}_1$, $\mathbf{A}_2$, $\mathbf{A}_3$ is linearly dependent, and, in fact, $\mathbf{A}_3 = \mathbf{A}_1 + \mathbf{A}_2$.

Exercise 4. Is the set $\mathbf{B}$, $\mathbf{A}_2$, $\mathbf{A}_3$ linearly dependent, where

$$\mathbf{B} = \begin{bmatrix} 2 \\ 1 \\ 3 \\ 2 \end{bmatrix}, \qquad \mathbf{A}_2 = \begin{bmatrix} 1 \\ 0 \\ 2 \\ 1 \end{bmatrix}, \qquad \mathbf{A}_3 = \begin{bmatrix} 3 \\ 1 \\ 5 \\ 2 \end{bmatrix}?$$

A set of vectors $\mathbf{A}_1, \ldots, \mathbf{A}_l$ in $\mathbb{R}^n$ which is *not* linearly dependent is called *linearly independent*.

Example 5. The set

$$\begin{bmatrix} 1 \\ 0 \\ 0 \\ 0 \end{bmatrix}, \quad \begin{bmatrix} 0 \\ 2 \\ 0 \\ 0 \end{bmatrix}, \quad \begin{bmatrix} 0 \\ 0 \\ 0 \\ 3 \end{bmatrix}$$

is linearly independent.

Exercise 5. For which value of t is the set

$$\begin{bmatrix} 1 \\ 1 \\ -2 \\ 0 \end{bmatrix}, \quad \begin{bmatrix} 0 \\ 3 \\ -3 \\ 0 \end{bmatrix}, \quad \begin{bmatrix} 1 \\ t \\ 4 \\ 0 \end{bmatrix}$$

linearly dependent?

Exercise 6. Is the set

$$\begin{bmatrix} 1 \\ 0 \\ 2 \\ 3 \end{bmatrix}, \quad \begin{bmatrix} 3 \\ 2 \\ 10 \\ 7 \end{bmatrix}, \quad \begin{bmatrix} 1 \\ -1 \\ 0 \\ 4 \end{bmatrix}$$

linearly independent?

Exercise 7. Let $\mathbf{U}_1, \mathbf{U}_2, \ldots, \mathbf{U}_k$ be nonzero vectors in $\mathbb{R}^n$ which are mutually orthogonal, i.e., $\mathbf{U}_i \cdot \mathbf{U}_j = 0$ if $i \neq j$. Show that the set $\mathbf{U}_1, \mathbf{U}_2, \ldots, \mathbf{U}_k$ is linearly independent.

§3. Dimension

We say that a straight line is one-dimensional (has dimension 1), while a plane has dimension 2 and $\mathbb{R}^3$ has dimension 3. Exactly what does this mean? Furthermore, can we define a *dimension* for subspaces of $\mathbb{R}^n$, $n > 3$? What, for instance, shall we call the dimension of the hyperplane in $\mathbb{R}^5$ defined by the equation: $x_1 + x_2 + x_3 + x_4 + x_5 = 0$?

A straight line, through the origin, in $\mathbb{R}^2$ or $\mathbb{R}^3$, consists of all the scalar multiples of *one* nonzero vector. A plane π, through the origin, in $\mathbb{R}^3$ consists of the set of all linear combinations of *two* vectors **A**, **B** in π. In other words, π consists of the set of all vectors

$$\mathbf{X} = s\mathbf{A} + t\mathbf{B}, \qquad \text{where } s, t \text{ are scalars.}$$

The only restriction on the pair **A**, **B** of vectors in π is that it is a linearly independent pair.

Finally, $\mathbb{R}^3$ can be expressed as the set of all linear combinations

$$x\mathbf{E}_1 + y\mathbf{E}_2 + z\mathbf{E}_3$$

of the *three* vectors $\mathbf{E}_1$, $\mathbf{E}_2$, $\mathbf{E}_3$, and this triplet of vectors is linearly independent. These examples suggest a definition: let S be a subspace of $\mathbb{R}^n$. Suppose there exist d vectors $\mathbf{A}_1, \mathbf{A}_2, \ldots, \mathbf{A}_d$ in S such that

$$\begin{cases} \text{every vector } \mathbf{X} \text{ in } S \text{ is a linear combination} \\ \qquad \mathbf{X} = t_1\mathbf{A}_1 + t_2\mathbf{A}_2 + \cdots + t_d\mathbf{A}_d \\ \text{of this set of vectors,} \end{cases} \tag{4}$$

and

$$\text{the set } \mathbf{A}_1, \ldots, \mathbf{A}_d \text{ is linearly independent.} \tag{5}$$

Then we say that the *dimension of S* equals d.

We also say that a d-tuple of vectors $\mathbf{A}_1, \mathbf{A}_2, \ldots, \mathbf{A}_d$ in S satisfying (4) and (5) is a *basis* of S.

Note: To justify the definition of dimension just given, we need to know that the number d we obtain does not depend upon *how we choose* the vectors $\mathbf{A}_1, \ldots, \mathbf{A}_d$ in S. Fortunately we have:

Theorem 5.2 *Suppose* $\mathbf{A}_1, \mathbf{A}_2, \ldots, \mathbf{A}_d$ *and* $\mathbf{B}_1, \mathbf{B}_2, \ldots, \mathbf{B}_e$ *are two sets of vectors in S which each satisfy* (4) *and* (5). *Then* $d = e$.

Note: In other language, the theorem says that every two bases of a given subspace S consist of the same number of vectors. This number is what we call the dimension of S. We shall not give a proof of Theorem 5.2.

Each subspace S has many different bases. For instance, let π be a plane through the origin in $\mathbb{R}^3$. Every pair of vectors $\mathbf{A}$, $\mathbf{B}$ in π which is linearly independent satisfies (4) and (5) with $S = \pi$ and is therefore a basis for π.

EXAMPLE 6. Let S be the hyperplane in $\mathbb{R}^5$ defined by $x_1 + x_2 + x_3 + x_4 + x_5 = 0$.

(a) Find a basis for S.
(b) Find the dimension of S.

If $\mathbf{X} = \begin{pmatrix} x_1 \\ \vdots \\ x_5 \end{pmatrix}$ is in S, then $x_1 = -x_2 - x_3 - x_4 - x_5$, and so

$$\mathbf{X} = \begin{pmatrix} -x_2 - x_3 - x_4 - x_5 \\ x_2 \\ x_3 \\ x_4 \\ x_5 \end{pmatrix}$$

$$= x_2 \begin{pmatrix} -1 \\ 1 \\ 0 \\ 0 \\ 0 \end{pmatrix} + x_3 \begin{pmatrix} -1 \\ 0 \\ 1 \\ 0 \\ 0 \end{pmatrix} + x_4 \begin{pmatrix} -1 \\ 0 \\ 0 \\ 1 \\ 0 \end{pmatrix} + x_5 \begin{pmatrix} -1 \\ 0 \\ 0 \\ 0 \\ 1 \end{pmatrix}$$

$$= x_2\mathbf{A}_2 + x_3\mathbf{A}_3 + x_4\mathbf{A}_4 + x_5\mathbf{A}_5,$$

where we set $\mathbf{A}_2 = \begin{pmatrix} -1 \\ 1 \\ 0 \\ 0 \\ 0 \end{pmatrix}$ and so forth. Each of the vectors $\mathbf{A}_2$, $\mathbf{A}_3$, $\mathbf{A}_4$, $\mathbf{A}_5$ belongs to S, as we verify by checking the equation which defines S. We just saw that every vector in S is a linear combination of the 4-tuple $\mathbf{A}_2$, $\mathbf{A}_3$, $\mathbf{A}_4$, $\mathbf{A}_5$.

Also, we claim that this 4-tuple of vectors is linearly independent. Suppose, on the contrary, that the 4-tuple is linearly dependent. Then there exist scalars s, t, u, v *not all* 0, so that

$$s\mathbf{A}_2 + t\mathbf{A}_3 + u\mathbf{A}_4 + v\mathbf{A}_5 = 0$$

or

$$s \begin{pmatrix} -1 \\ 1 \\ 0 \\ 0 \\ 0 \end{pmatrix} + t \begin{pmatrix} -1 \\ 0 \\ 1 \\ 0 \\ 0 \end{pmatrix} + u \begin{pmatrix} -1 \\ 0 \\ 0 \\ 1 \\ 0 \end{pmatrix} + v \begin{pmatrix} -1 \\ 0 \\ 0 \\ 0 \\ 1 \end{pmatrix} = \begin{pmatrix} 0 \\ 0 \\ 0 \\ 0 \\ 0 \end{pmatrix}.$$

It follows that

$$-s-t-u-v=0, \qquad s=0, \qquad t=0, \qquad u=0, \qquad v=0.$$

This contradicts the fact that s, t, u, v are not all zero. So the 4-tuple $\mathbf{A}_2$, $\mathbf{A}_3$, $\mathbf{A}_4$, $\mathbf{A}_5$ is linearly dependent. Hence, this 4-tuple satisfies conditions (4) and (5). Thus, the set $\mathbf{A}_2$, $\mathbf{A}_3$, $\mathbf{A}_4$, $\mathbf{A}_5$ is a basis of the hyperplane S, and so the dimension of this hyperplane is 4.

§4. Exercises

Exercise 8. Find a basis for the plane in $\mathbb{R}^3$:

$$x_3 - 2x_1 = 0.$$

Exercise 9. Find a basis for the hyperplane in $\mathbb{R}^4$:

$$4x_1 + 3x_2 - x_4 = 0.$$

Exercise 10. The subspace S of $\mathbb{R}^4$ is defined by the system of equations:

$$\begin{cases} x_1 - 2x_3 = 0, \\ x_2 + x_4 = 0. \end{cases}$$

(a) Find a basis for S.
(b) What is the dimension of S?

Exercise 11. Let S be the subspace of $\mathbb{R}^5$ defined by the equations

$$\begin{cases} 2x_1 - x_2 = 0, \\ 2x_2 - x_3 = 0, \\ 2x_3 - x_4 = 0, \\ 2x_4 - x_5 = 0. \end{cases}$$

Find a basis for S, and give the dimension of S.

Exercise 12. Let $\mathbf{Z}$ be a nonzero vector in $\mathbb{R}^n$. Let S be the subset of $\mathbb{R}^n$ consisting of all vectors $t\mathbf{Z}$ with t a scalar. Show that S is a one-dimensional subspace of $\mathbb{R}^n$.

Exercise 13. Fix n. Show that

(a) $\mathbb{R}^n$ is a subspace of $\mathbb{R}^n$.
(b) Find a basis of $\mathbb{R}^n$.

Exercise 14. Find a basis for the subspace of $\mathbb{R}^4$ defined by the system

$$\begin{cases} 2x_1 + x_2 - 3x_3 + x_4 = 0, \\ 2x_1 - x_2 - x_3 - x_4 = 0. \end{cases}$$

Exercise 15. Find a basis for the subspace S of $\mathbb{R}^4$ defined by the system

$$\begin{cases} 2x_1 + x_2 - 3x_3 + x_4 = 0, \\ 2x_1 - x_2 - x_3 - x_4 = 0, \\ \qquad x_2 + x_3 + 3x_4 = 0. \end{cases} \tag{6}$$

Show that S is one-dimensional.

Exercise 16. Let S be a subspace of $\mathbb{R}^3$. Prove the following statements:

(a) If S has dimension 1, then S is a line.
(b) If S has dimension 2, then S is a plane.
(c) If S has dimension 3, then S is all of $\mathbb{R}^3$.
(d) S cannot have dimension greater than 3.

Exercise 17. Let S be a subspace of $\mathbb{R}^n$. Show that the vector $\mathbf{0}$ belongs to S.

By analogy with the cases $n = 2, 3, 4$, for any n, we call a transformation T of $\mathbb{R}^n$ linear if $T(\mathbf{X} + \mathbf{U}) = T(\mathbf{X}) + T(\mathbf{U})$ and $T(r\mathbf{X}) = rT(\mathbf{X})$ for any vectors $\mathbf{X}$, $\mathbf{U}$ in $\mathbb{R}^n$ and any scalar r. If T is a linear transformation of $\mathbb{R}^n$, then by the *range* of T we mean the collection of all vectors $T(\mathbf{X})$, where $\mathbf{X}$ is in $\mathbb{R}^n$. By the *null space* of T we mean the collection of all vectors $\mathbf{X}$ in $\mathbb{R}^n$ such that $T(\mathbf{X}) = \mathbf{0}$.

Exercise 18. Show that the range of T is a subspace of $\mathbb{R}^n$.

Exercise 19. Show that the null space of T is a subspace of $\mathbb{R}^n$.

Exercise 20. Let T be the transformation of $\mathbb{R}^4$ which assigns to each vector $\mathbf{X} = \begin{pmatrix} x_1 \\ x_2 \\ x_3 \\ x_4 \end{pmatrix}$ the vector $\begin{pmatrix} x_1 \\ x_2 \\ x_3 \\ 0 \end{pmatrix}$.

(a) Show that T is a linear transformation of $\mathbb{R}^4$.
(b) Describe the range of T.
(c) Describe the null space of T.

CHAPTER 5.3

Inhomogeneous Systems of Equations

§1. Solutions

Let a_{ij}, $1 \leqslant i \leqslant k$, $1 \leqslant j \leqslant n$ be a set of constants, and fix k constants $u_1, u_2, \ldots, u_k$. The system

$$\begin{cases} a_{11}x_1 + a_{12}x_2 + \cdots + a_{1n}x_n = u_1, \\ a_{21}x_1 + a_{22}x_2 + \cdots + a_{2n}x_n = u_2, \\ \qquad\qquad\vdots \\ a_{k1}x_1 + a_{k2}x_2 + \cdots + a_{kn}x_n = u_k \end{cases} \tag{I}$$

is called an *inhomogeneous system of linear equations*. If all the $u_i = 0$, (I) turns into the homogeneous system (H) which we studied in Chapter 5.1. We set

$$\mathbf{U} = \begin{bmatrix} u_1 \\ u_2 \\ \vdots \\ u_k \end{bmatrix}, \qquad \mathbf{X} = \begin{bmatrix} x_1 \\ x_2 \\ \vdots \\ x_n \end{bmatrix}.$$

X is a *solution* of (I) if the equations in (I) are satisfied. How can we solve such an inhomogeneous system? For $n = k = 2$ and for $n = k = 3$, we have studied such systems in earlier chapters. In the general case of arbitrary k and n, we can proceed as in our solution of homogeneous systems in Chapter 5.1 to find a succession of systems that are *equivalent* to (I), until we reach a system (I′) of the following form, where we may have relabeled the x_i:

$$\begin{cases} x_1 \qquad\qquad + b_{11}x_{l+1} + b_{12}x_{l+2} + \cdots + b_{1,n-l}x_n = v_1, \\ \quad x_2 \qquad + b_{21}x_{l+1} + b_{22}x_{l+2} + \cdots + b_{2,n-l}x_n = v_2, \\ \qquad \ddots \qquad\qquad\qquad \vdots \\ \qquad\qquad x_l + b_{l1}x_{l+1} + b_{l2}x_{l+2} + \cdots + b_{l,n-l}x_n = v_l, \end{cases} \tag{I'}$$

where $v_1, \ldots, v_l$ is a new sequence of constants, constructed out of the u_i. We then solve the system (I′) directly by choosing $x_{l+1}, \ldots, x_n$ *arbitrarily* and solving for $x_1, x_2, \ldots, x_l$, using (I′). In this way, we find all solutions of (I′), and hence all solutions of (I).

Example 1. Solve the system

$$\begin{cases} 2x + 3y + z = u, \\ x - y - z = v, \\ 3x + 2y \qquad = w, \end{cases} \tag{1}$$

where u, v, w are given numbers. Subtracting twice the middle line from the top line, and then three times the middle line from the bottom line, we get the equivalent system:

$$\begin{cases} x - y - z = v, \\ \quad 5y + 3z = u - 2v, \\ \quad 5y + 3z = w - 3v. \end{cases} \tag{1'}$$

By a similar procedure, we get the following system (1″), equivalent to (1′), and, hence, also equivalent to (1):

$$\begin{cases} x - y - z = v, \\ \quad 5y + 3z = u - 2v, \\ \qquad\quad 0 = (w - 3v) - (u - 2v) = w - u - v. \end{cases} \tag{1''}$$

Observe that (1″) does not have solutions for every choice of u, v, w. The bottom line in (1″) implies that if (1″) has a solution, then $w = v + u$. One more step, adding $\frac{1}{5}$ times the middle of line (1″) to the top line, gives

$$\begin{cases} x \quad - \frac{2}{5}z = v + \frac{1}{5}(u - 2v) = \frac{1}{5}u + \frac{3}{5}v, \\ \quad y + \frac{3}{5}z = \frac{1}{5}u - \frac{2}{5}v, \\ \qquad\quad 0 = w - u - v. \end{cases} \tag{1'''}$$

(1‴) has the form (I′) discussed above. To solve (1‴), we must have $w = u + v$. Under this assumption, we give z an arbitrary value t and find

$$\begin{cases} x = \frac{1}{5}u + \frac{3}{5}v + \frac{2}{5}t, \\ y = \frac{1}{5}u - \frac{2}{5}v - \frac{3}{5}t, \\ z = \qquad\qquad\quad t. \end{cases} \tag{2}$$

For different choices of t, (2) provides us with all solutions of (1‴) and, hence, all solutions of our original system (1). In particular, take $u = 5$, $v = 10$, $w = 15$. Then the system

$$\begin{cases} 2x + 3y + z = 5, \\ x - y - z = 10, \\ 3x + 2y \quad = 15 \end{cases} \tag{3}$$

is solved by fixing a value t and setting

$$\begin{aligned} x &= 1 + 6 + \tfrac{2}{5}t = 7 + \tfrac{2}{5}t, \\ y &= 1 - 4 - \tfrac{3}{5}t = -3 - \tfrac{3}{5}t, \\ z &= \qquad\quad = t. \end{aligned}$$

One can check this by inserting these values in the system (3).

Exercise 1. Find all solutions of the system

$$\begin{aligned} 2x + 3y + z &= 5, \\ x - y - z &= 10. \end{aligned}$$

Exercise 2. Find conditions on u_1, u_2, u_3, u_4 under which there exists a solution x_1, x_2, x_3, x_4 of the system:

$$\begin{aligned} x_1 - x_2 &= u_1, \\ 2x_1 + x_3 &= u_2, \\ x_1 - x_4 &= u_3, \\ x_2 - x_4 &= u_4. \end{aligned}$$

Assuming these conditions are satisfied, find all solutions of the system.

Example 2. Solve the system in four unknowns:

$$\begin{cases} x_1 + \ x_2 = u_1, \\ x_2 + \ x_3 = u_2, \\ x_3 + \ x_4 = u_3, \\ 2x_1 - 3x_2 = u_4. \end{cases} \tag{4}$$

We can easily see that this system is equivalent to

$$\begin{cases} \quad - 5x_2 = u_4 - 2u_1, \\ x_2 + \ x_3 = u_2, \\ x_3 + \ x_4 = u_3, \\ 2x_1 - 3x_2 = u_4, \end{cases} \tag{4'}$$

and (4′), in turn, is equivalent to

$$\begin{cases} -5x_2 = u_4 - 2u_1, \\ \quad x_3 = \tfrac{1}{5}(u_4 - 2u_1) + u_2 = -\tfrac{2}{5}u_1 + u_2 + \tfrac{1}{5}u_4, \\ \quad 2x_1 = u_4 - \tfrac{3}{5}(u_4 - 2u_1) = \tfrac{6}{5}u_1 + \tfrac{2}{5}u_4, \\ x_3 + x_4 = u_3, \end{cases} \tag{4''}$$

and, at last, (4″) is equivalent to the system obtained from (4″) by keeping the three top lines and replacing the bottom line by

$$\begin{aligned} x_4 &= u_3 - \left(-\tfrac{2}{5}u_1 + u_2 + \tfrac{1}{5}u_4\right) \\ &= \tfrac{2}{5}u_1 - u_2 + u_3 - \tfrac{1}{5}u_4 . \end{aligned}$$

Thus the solution of (4) is *unique*, with given u_i, and is as follows:

$$\begin{cases} x_1 = \tfrac{6}{10}u_1 + \tfrac{1}{5}u_4 , \\ x_2 = \tfrac{2}{5}u_1 - \tfrac{1}{5}u_4 , \\ x_3 = -\tfrac{2}{5}u_1 + u_2 + \tfrac{1}{5}u_4 , \\ x_4 = \tfrac{2}{5}u_1 - u_2 + u_3 - \tfrac{1}{5}u_4 . \end{cases} \tag{5}$$

inserting these values for the x_i in (4), we can check that our solution is correct.

Note: In Example 1, the solution of the system (1) was not unique. Also, in Example 1, the solution exists only for certain choices of u, v, w. In Example 2, the solution was unique and exists for every choice of u_1, u_2, u_3, u_4. What can be said about the *existence* and *uniqueness* of the solutions for the system (I)? We state, without proof, the following basic result for the case $k = n$. Let us denote by (H) the homogeneous system corresponding to (I), obtained by setting $u_i = 0$, $i = 1, \ldots, k$, in (I).

Theorem 5.3. *Let $k = n$. We distinguish two cases*:

(i) (H) *has only the trivial solution* **0**. *Then the inhomogeneous system* (I) *has a unique solution for every choice of the* u_i.

(ii) (H) *has a non-trivial solution. Then for certain* u_i, (I) *has no solution. Also, the solution of* (I) *is never unique.*

The following example will illustrate how Theorem 5.3 can be used in proofs.

EXAMPLE 3. Given three points in the plane: (x_1, y_1), (x_2, y_2), (x_3, y_3), which are not collinear, show that there exists a circle which passes through the three points.

The circle C with center (x_0, y_0) and radius R has the equation

$$(x - x_0)^2 + (y - y_0)^2 = R^2$$

or

$$x^2 - 2xx_0 + x_0^2 + y^2 - 2yy_0 + y_0^2 = R^2 .$$

We can rewrite this in the form

$$x^2 + y^2 + ax + by + c = 0,$$

where a, b, c are certain constants. This circle passes through our three given points if and only if

$$x_i^2 + y_i^2 + ax_i + by_i + c = 0, \qquad i = 1, 2, 3.$$

Note that x_i and y_i are given numbers and a, b, and c are numbers to be found. We can rewrite this system in the form:

$$\begin{cases} x_1 a + y_1 b + c = u_1, \\ x_2 a + y_2 b + c = u_2, \\ x_3 a + y_3 b + c = u_3, \end{cases} \tag{6}$$

where $u_i = -x_i^2 - y_i^2$. We regard (6) as an inhomogeneous system of equations in the unknowns a, b, c. The corresponding homogeneous system is the following:

$$\begin{cases} x_1 a + y_1 b + c = 0, \\ x_2 a + y_2 b + c = 0, \\ x_3 a + y_3 b + c = 0, \end{cases} \tag{7}$$

Suppose that this system has a nonzero solution a, b, c. Then the line defined by

$$ax + by + c = 0$$

passes through each of our three points. This contradicts the assumption that the points are not collinear. Hence, (7) has only the trivial solution. Thus we have case (i) in Theorem 5.3 for the system (6), and so (6) has a solution a, b, c. It follows that

$$x_i^2 + y_i^2 + x_i a + y_i b + c = 0, \qquad i = 1, 2, 3.$$

The equation

$$x^2 + y^2 + xa + yb + c = 0$$

is thus satisfied by (x_i, y_i), $i = 1, 2, 3$. This equation can be written

$$\left(x + \frac{a}{2}\right)^2 + \left(y + \frac{b}{2}\right)^2 = -c + \frac{a^2}{4} + \frac{b^2}{4},$$

which represents a circle which passes through each of the three points.

Note: The existence of this circle could also be shown by elementary geometry. However, the method we used, based on Theorem 5.3, is applicable to a wide variety of situations, some of which are given as exercises at the end of this chapter.

§2. Geometric Interpretation

What is the geometric meaning of the collection of solutions of an inhomogeneous system (I)? In the case of a homogeneous system, the solutions form a subspace.

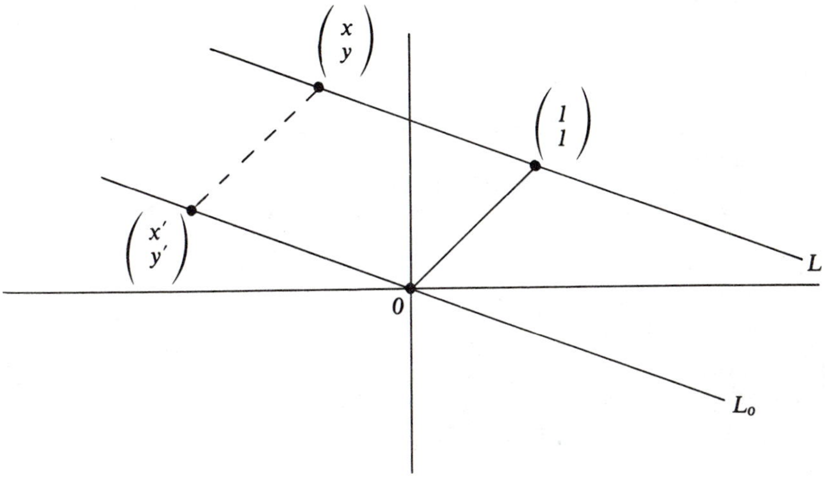

Figure 5.1

EXAMPLE 4. Let L consist of all vectors $\mathbf{X} = \begin{pmatrix} x \\ y \end{pmatrix}$ in $\mathbb{R}^2$ such that

$$2x + 3y = 5. \tag{8}$$

L is clearly a straight line, but also L is not a subspace of $\mathbb{R}^2$. (How do we know that?) Let L_0 consist of all solutions of the homogeneous equation:

$$2x + 3y = 0. \tag{9}$$

L_0 is a subspace of $\mathbb{R}^2$. How are L_0 and L related (see Fig. 5.1)? We can construct L by moving L_0 parallel to itself, *translating* L_0 by a fixed vector. In fact, $\begin{pmatrix} 1 \\ 1 \end{pmatrix}$ satisfies (8). If $\begin{pmatrix} x \\ y \end{pmatrix}$ belongs to L, then $2(x-1) + 3(y-1) = 2x + 3y - 5 = 0$, and so $\begin{pmatrix} x-1 \\ y-1 \end{pmatrix} = \begin{pmatrix} x \\ y \end{pmatrix} - \begin{pmatrix} 1 \\ 1 \end{pmatrix}$ lies on L_0. Thus $\begin{pmatrix} x \\ y \end{pmatrix} = \begin{pmatrix} x' \\ y' \end{pmatrix} + \begin{pmatrix} 1 \\ 1 \end{pmatrix}$, where $\begin{pmatrix} x' \\ y' \end{pmatrix}$ lies on L_0. So L is obtained by translating L_0 by the vector $\begin{pmatrix} 1 \\ 1 \end{pmatrix}$.

Proposition 1. *Let T be the set of solutions in $\mathbb{R}^n$ of the inhomogeneous system (I). Then there is a subspace S of $\mathbb{R}^n$ and a vector $\mathbf{X}^0$ in $\mathbb{R}^n$, such that T is the translate of S by $\mathbf{X}^0$. In other words, T consists of all vectors $\mathbf{X} = \mathbf{Y} + \mathbf{X}^0$, where $\mathbf{Y}$ is in S.*

To show this, we write the system (I) in the form:

$$\mathbf{A}_1 \cdot \mathbf{X} = u_1, \ldots, \mathbf{A}_k \cdot \mathbf{X} = u_k, \tag{I}$$

where $\mathbf{A}_1 = \begin{pmatrix} a_{11} \\ a_{12} \\ \vdots \\ a_{1n} \end{pmatrix}, \ldots, \mathbf{A}_k = \begin{pmatrix} a_{k1} \\ a_{k2} \\ \vdots \\ a_{kn} \end{pmatrix}$. Choose $\mathbf{X}^0 = \begin{pmatrix} x_1^0 \\ \vdots \\ x_n^0 \end{pmatrix}$ satisfying (I). Then if $\mathbf{X}$ satisfies (I),

$$\mathbf{A}_1 \cdot (\mathbf{X} - \mathbf{X}^0) = \mathbf{A}_1 \cdot \mathbf{X} - \mathbf{A}_1 \cdot \mathbf{X}^0 = \mathbf{A}_1 \cdot \mathbf{X} - u_1 = u_1 - u_1 = 0.$$

Hence,

$$\mathbf{A}_1 \cdot (\mathbf{X} - \mathbf{X}^0) = 0.$$

Similarly, $\mathbf{A}_j \cdot (\mathbf{X} - \mathbf{X}^0) = 0$, $j = 2, \ldots, k$. Thus $\mathbf{X} - \mathbf{X}^0$ satisfies the homogeneous system (H) which corresponds to (I).

We denote by S the subspace of $\mathbb{R}^n$ defined by (H). If $\mathbf{X}$ is a solution of (I), then $\mathbf{Y} = \mathbf{X} - \mathbf{X}^0$ is in S and $\mathbf{X} = \mathbf{Y} + \mathbf{X}^0$. Conversely, if $\mathbf{X}$ has this form, then

$$\mathbf{A}_1 \cdot \mathbf{X} = \mathbf{A}_1 \cdot (\mathbf{Y} + \mathbf{X}^0) = \mathbf{A}_1 \cdot \mathbf{Y} + \mathbf{A}_1 \cdot \mathbf{X}^0 = 0 + u_1 = u_1.$$

Similarly, $\mathbf{A}_j \cdot \mathbf{X} = u_j$ for all j. So $\mathbf{X}$ is in T. Thus we have shown: $\mathbf{X}$ is in T, i.e., $\mathbf{X}$ is a solution of (I), if and only if $\mathbf{X}$ lies in the translation of S by $\mathbf{X}^0$, and this is what Proposition 5.1 asserts.

Exercise 3. Let T be the subset of $\mathbb{R}^4$ defined by the equation

$$2x_1 - 3x_2 + x_3 + 5x_4 = 10.$$

Find a subspace S of $\mathbb{R}^4$ and a vector $\mathbf{X}^0$ in $\mathbb{R}^4$ such that T is the translate of S by $\mathbf{X}_0$.

§3. Exercises

Exercise 4. Let (x_1, y_1), (x_2, y_2), (x_3, y_3) be three non-collinear points in the plane with x_1, x_2, x_3 all different. Show that there exists a parabola P with equation

$$y = ax^2 + bx + c,$$

where a, b, c are constants and $a \neq 0$, such that each of the three given points lies on P.

Exercise 5. Find the coefficients a, b, c of a parabola $y = ax^2 + bx + c$ which passes through the points

$$(1, 6), (2, 4), (3, 0).$$

Exercise 6. (a) Show that a sphere in $\mathbb{R}^3$ has an equation:

$$x^2 + y^2 + z^2 + ax + by + cz + d = 0,$$

where a, b, c, d are constants.

(b) Given four points (x_j, y_j, z_j), $j = 1, 2, 3, 4$ in $\mathbb{R}^3$ such that they do not all lie in a plane, show that there is a sphere passing through all four points.

Exercise 7. Find an equation for the sphere which passes through the points $(1,0,1)$, $(0,2,3)$, $(3,0,4)$, $(1,1,1)$.

Exercise 8. Find a cubic curve with the equation $y = ax^3 + bx^2 + cx + d$ passing through the four points: $(1,0)$, $(2,2)$, $(3,12)$, $(4,36)$.

§4. Partial Fractions Decomposition

EXAMPLE 5. We wish to express the function

$$f(x) = \frac{1}{(x-1)(x-2)(x-3)}$$

in the form

$$f(x) = \frac{a}{x-1} + \frac{b}{x-2} + \frac{c}{x-3}, \tag{10}$$

where a, b, c are constants to be found. Multiplying both sides by $(x-1)$ $(x-2)$ $(x-3)$, we see that (10) is equivalent to

$$1 = a(x-2)(x-3) + b(x-1)(x-3) + c(x-1)(x-2)$$

which can be written as

$$1 = (a+b+c)x^2 + (-5a-4b-3c)x + 6a + 3b + 2c.$$

This is equivalent to the system

$$\begin{cases} a + b + c = 0, \\ -5a - 4b - 3c = 0, \\ 6a + 3b + 2c = 1. \end{cases} \tag{11}$$

We easily solve this and find:

$$a = \tfrac{1}{2}, \qquad b = -1, \qquad c = \tfrac{1}{2}.$$

Hence,

$$\frac{1}{(x-1)(x-2)(x-3)} = \frac{1/2}{x-1} + \frac{-1}{x-2} + \frac{1/2}{x-3}.$$

Exercise 9. Express the function

$$f(x) = \frac{1}{(x-1)(x+1)x}$$

in the form

$$f(x) = \frac{a}{x-1} + \frac{b}{x+1} + \frac{c}{x}.$$

Exercise 10. Find a, b, c, d such that

$$\frac{1}{(x-1)(x-2)(x-3)(x-4)} = \frac{a}{x-1} + \frac{b}{x-2} + \frac{c}{x-3} + \frac{d}{x-4}.$$

Exercise 11. Let $a_1, \ldots, a_n$ be n distinct numbers and set

$$f(x) = \frac{1}{(x - a_1)(x - a_2) \cdots (x - a_n)}.$$

An identity

$$f(x) = \frac{c_1}{x - a_1} + \frac{c_2}{x - a_2} + \cdots + \frac{c_n}{x - a_n} \tag{12}$$

is called a *partial fractions decomposition* of $f(x)$.

(a) Show that (12) is equivalent to an inhomogeneous system of n linear equations in the unknowns $c_1, \ldots, c_n$.
(b) Show that the corresponding homogeneous system has only the trivial solution.
(c) Use Theorem 5.3 to show that there exists constants $c_1, \ldots, c_n$ which satisfy (12).

Exercise 12. Does the system

$$\begin{cases} x_1 + 2x_2 + 3x_3 + 4x_4 = 1, \\ 2x_1 + 3x_2 + 4x_3 + 1x_4 = 0, \\ 3x_1 + 4x_2 + 1x_3 + 2x_4 = 0, \\ 4x_1 + 1x_2 + 2x_3 + 3x_4 = 0 \end{cases} \tag{13}$$

have a solution? If it does, find all solutions.

Afterword

Our path through linear algebra has emphasized spaces of vectors in dimension 2, 3, and 4 as a means of introducing concepts which go forward to $\mathbb{R}^n$ for arbitrary n. But linear algebra does not end here. Many of the ideas we have studied also carry over to other collections of objects which behave in many ways like vectors in $\mathbb{R}^n$ but which have several different properties. We list some variations which the reader may encounter in other courses in mathematics as well as in other disciplines, such as chemistry, physics, geology, economics, data analysis, and statistics.

Function Spaces. In elementary calculus, one encounters various collections of real-valued functions from the real numbers to the real numbers such as polynomials, trigonometric functions, and exponential functions. A real-valued function f is defined as soon as we know its values $f(x)$ for all x, just as we know a vector as soon as we know each of its components. We add functions f and g by defining $f + g$ to be the function with value $f(x) + g(x)$ at the real value x, and we define scalar multiplication by setting the value of rf at x to be r times $f(x)$. With these definitions of addition and scalar multiplication, the set of functions satisfies all of the properties which we found in the case of each $\mathbb{R}^n$, and this justifies our calling this set the *space* of real-valued functions. Important subspaces of this space are the polynomials, the continuous functions, and the differentiable functions.

Operations used in calculus, such as differentiation, turn out to be linear transformations on certain spaces of functions. The problem of finding eigenvalues for such transformations arises, for example, in the study of vibrating strings and in the theory of spectral lines of atoms.

Systems of Differential Equations are the generalization of the systems we studied in Chapter 2.8. The solutions of a homogeneous system of differential equations form a subspace of the space of all differentiable functions. Eigenvectors play an important role in the analysis of systems which arise in engineering and in biology.

Infinite Dimensional Spaces. It is also possible to consider vectors with infinitely many components, given as sequences $(x_1, x_2, x_3, \ldots)$ of real numbers. Again, we add sequences and multiply them by scalars componentwise as in $\mathbb{R}^n$, and once more the set of all sequences possesses the basic properties of space of vectors in $\mathbb{R}^n$. It is no longer possible to find a finite basis for this space of sequences. It has many important subspaces that occur in advanced mathematics, and also in quantum physics, such as the space of convergent sequences or the space of sequences with convergent sums.

Complex Spaces. It is also possible to consider spaces in which the scalars are not real numbers but, rather, complex numbers. Many of the same ideas apply, but many results that depend on factoring polynomials will change since there are polynomials like $\lambda^2 + 1 = 0$ which have complex roots but no real roots. Thus, a transformation which has no real eigenvectors might nonetheless have complex eigenvectors.

More General Algebraic Systems. Spaces of vectors in $\mathbb{R}^n$ are fundamental in modern algebra, which develops abstract structures such as groups, rings, and fields starting with axiom systems and proceeding formally. Many of the key examples of these abstract concepts are found in the concrete objects which we have investigated in this book, such as rings of matrices and groups of isometries.

We wish the reader success in future encounters with the concepts of linear algebra.

Index

addition
  of matrices 44
  of transformations 44, 135
  of vectors 111, 208, 230
addition of matrices 136
addition of vectors
  add 4
additive identity 7
additive inverse 5, 7
algebra of vectors 3
area of a parallelogram 21, 65, 121
associative law 7
  for matrices 43
  for scalars 7
  for vectors 7
associative law for the dot product 12

basis of a subspace 240
basis vectors 5, 112, 208

Cauchy–Schwarz inequality 213
centroid 210
characteristic equation 77, 177
column of a matrix 139
commutative law
  for dot product 12
commutative law for addition 7
commuting transformation 41
complex spaces 254
conic sections 84
coordinate 3, 111
coordinate axis 5, 112, 208
cross product 120
cube, four-dimensional 215, 220

degenerate parallelogram 37
dependent, linearly 8, 114, 238
derivative 98
determinant 60, 163, 227
diagonal matrix 47, 143
diagonal of a matrix 165
difference of vectors 7, 113, 210
differential system 97
dimension of a subspace 240
directed line segment 3
distance 10, 117
distributive law 7
  for scalars 7
  for vectors 7
distributive law for the dot product 12
dot product 11, 115, 212, 230

eigenvalue 74, 175, 226
eigenvector 74, 175

elementary matrix 47, 141
ellipse 94
ellipsoid 203
elliptical cylinder 204
equality of transformations 28
equivalence of systems 231
existence of solutions 247
exponential of a matrix 99

function space 253

homogeneous system 56, 160, 230
hyperbola 94
hyperbolic cylinder 204
hyperboloid 203
hyperplane 210, 237

identity
  matrix 31, 131
  transformation 30
image
  of a vector 23, 126, 216
image of a set 35, 134, 216
infinite dimensional spaces 254
inhomogeneous system 244
initial condition 97
inverse of a transformation 50, 145
isometry 68, 183

law of cosines 17, 116, 213
length 211
length of a vector 10, 115
line along a vector 4, 209
linear
  dependence 238
  independence 239
  transformation 30, 130, 224
linear combination 114
linear dependence 8, 114
linear independence 114
linearly independent 114

matrix 30, 130, 224
midpoint 10, 210
multiplication, scalar 4, 111, 208, 230

negatively oriented 60, 170
negative of a vector 5
nonhomogeneous system 160
null space 243

orientation of a pair of vectors 60
orthogonal matrix 189
orthogonal vectors 17, 214
orthonormal set 214

parallelogram 7, 37
parameter 9
parametric representation 9
partial fractions decomposition 251
permutation 46
permutation matrix 46, 141
perpendicular 17
polar angle 10
polar form of a vector 11
positive definite 95
positively oriented 60, 170
power of a matrix 86
preserving angle 71
preserving orientation 62, 171
product
  of transformation 39
  of transformations 137
product of matrices 42, 138
projection
  to a hyperplane 216
  to a line 18, 116, 126, 216
  to a plane 117, 127
projection to a coordinate axis 14
Pythagorean Theorem 10, 115, 122

quadratic form 88
quadric surface 202

range of a transformation 126, 216, 243
reflection 23, 126
reversing orientation 64

rotation 25, 128
  about an axis 184
  in a plane 218
row of a matrix 139

scalar 4, 111
scalar multiplication 4, 111, 208, 230
second-order differential equation 108
shear matrix 141
simplex 212
solution of an inhomogenous system 244
solutions of differential systems 104
stretching 25, 128
subspace 236
sum
  of matrices 44
  of transformations 44, 135
sum of matrices 136
symmetric matrix 78, 190
systems
  of linear equations 56, 159, 228
systems of differential equations 96, 254

transformation 23, 126, 216
translate of a subspace 249
transpose of a matrix 166
trivial solution 56, 160

uniqueness
  of solutions 110, 160, 247
uniqueness of inverses 52, 148
unit
  sphere 125, 212
  vector 125, 212
unit circle 10
  vector 10

vector 1, 3, 111, 207
vector in *n*-space 230
vector-valued function 98
volume of a parallelepiped 124, 173

zero transformation 31, 128
zero vector 4

Ross: Elementary Analysis: The Theory of Calculus.
1980. viii, 264 pages. 34 illus.

Sigler: Algebra.
1976. xii, 419 pages. 27 illus.

Simmonds: A Brief on Tensor Analysis.
1982. xi, 92 pages. 28 illus.

Singer/Thorpe: Lecture Notes on Elementary Topology and Geometry.
1976. viii, 232 pages. 109 illus.

Smith: Linear Algebra.
1978. vii, 280 pages. 21 illus.

Thorpe: Elementary Topics in Differential Geometry.
1979. xvii, 253 pages. 126 illus.

Troutman: Variational Calculus with Elementary Convexity.
1983. xiv, 364 pages. 73 illus.

Whyburn/Duda: Dynamic Topology.
1979. xiv, 338 pages. 20 illus.

Wilson: Much Ado About Calculus: A Modern Treatment with Applications Prepared for Use with the Computer.
1979, xvii, 788 pages. 145 illus.